Lecture Notes in Networks and Systems

Volume 408

The series "Lecture Notes in Networks and Systems" publishes the latest developments in Networks and Systems—quickly, informally and with high quality. Original research reported in proceedings and post-proceedings represents the core of LNNS.

Volumes published in LNNS embrace all aspects and subfields of, as well as new challenges in, Networks and Systems.

The series contains proceedings and edited volumes in systems and networks, spanning the areas of Cyber-Physical Systems, Autonomous Systems, Sensor Networks, Control Systems, Energy Systems, Automotive Systems, Biological Systems, Vehicular Networking and Connected Vehicles, Aerospace Systems, Automation, Manufacturing, Smart Grids, Nonlinear Systems, Power Systems, Robotics, Social Systems, Economic Systems and other. Of particular value to both the contributors and the readership are the short publication timeframe and the world-wide distribution and exposure which enable both a wide and rapid dissemination of research output.

The series covers the theory, applications, and perspectives on the state of the art and future developments relevant to systems and networks, decision making, control, complex processes and related areas, as embedded in the fields of interdisciplinary and applied sciences, engineering, computer science, physics, economics, social, and life sciences, as well as the paradigms and methodologies behind them.

Indexed by SCOPUS, INSPEC, WTI Frankfurt eG, zbMATH, SCImago.

All books published in the series are submitted for consideration in Web of Science.

For proposals from Asia please contact Aninda Bose (aninda.bose@springer.com).

More information about this series at https://link.springer.com/bookseries/15179

Vladimir Kurchenko · Alexei Lodygin ·
Rui Manuel Machado da Costa ·
Irina Samoylenko
Editors

Intelligent Biotechnologies of Natural and Synthetic Biologically Active Substances

XIII Narochanskie Readings

Springer

Editors
Vladimir Kurchenko
Research Laboratory of Applied Issues
Belarusian State University
Minsk, Belarus

Alexei Lodygin
Department of Applied Biotechnology
North-Caucasus Federal University
Stavropol, Russia

Rui Manuel Machado da Costa
R. Misericórdia Lagar dos Cortiços
Polytechnic Institute of Coimbra
Coimbra, Portugal

Irina Samoylenko
Department of Informational Systems
Stavropol State Agrarian University
Stavropol, Russia

ISSN 2367-3370 ISSN 2367-3389 (electronic)
Lecture Notes in Networks and Systems
ISBN 978-3-030-96640-9 ISBN 978-3-030-96641-6 (eBook)
https://doi.org/10.1007/978-3-030-96641-6

This Springer imprint is published by the registered company Springer Nature Switzerland AG
The registered company address is: Gewerbestrasse 11, 6330 Cham, Switzerland

Preface

The book is of interest to specialists who are searching for new sources of biologically active substances (BAS) for their practical use in the framework of innovative pharmaceuticals and functional foods development.

The volume contains original articles devoted to the actual problems of biologically active substances of animal, plant and fungal origin obtaining, and application. The article devoted to the study of seven types of wood-destroying fungi extracts shows the presence of BAS that can be implemented in pharmaceuticals. One of the articles presents the results of the search for promising sources of BAS from flowers of five Achillea species of the Russia and Kazakhstan native flora. Promising species for introduction and further use in pharmaceuticals have been identified. A number of the presented articles contain extensive materials on the sources of BAS production from various plant raw materials and their practical uses.

A significant part of the articles is related to the production of biologically active substances nanocomplexes and their application in medicine, food industry, and agriculture. As discussed in one of the studies, it was proposed to use cyclodextrin in the preparation of clathrates with peptides of whey proteins hydrolysates to reduce their bitter taste. The acute and chronic toxicity of curcuminoid complexes with cyclodextrins, which are highly soluble in water, has been studied. The possibility of chitosan application in the development of biocompatible wound-healing porous materials is shown. Methods of monitoring of biomaterials used in wound coatings survival are proposed. It is proposed to use chitosan-stabilized copper nanoparticles as an alternative to antibiotics in broiler feeds.

Much attention is paid to promising areas of healthy foods biotechnology development and food safety. The article on the use of chitosan to increase the shelf life of fermented dairy products is of interest. A number of articles discuss the possibility of BAS application in the functional foods development. Several articles offer innovative technologies for processing plant and animal food raw materials. The technology of whey proteins enzymatic hydrolysis for the production of hypoallergenic peptides is proposed. In a number of works, it is offered to use new original research methods for food quality control.

The book also contains articles on innovative biotechnologies for the production and application of biopolymers and their derivatives in various industries. One of the articles substantiates the production of hydrolysis products of bleached hardwood kraft pulp by carbohydrate-active enzymes. Approaches to establishing the growing location of mossy pine forests based on the spectral characteristics of environmentally dependent wood components are considered. It is proposed to use TiO_2 chitosan as a universal composite for wastewater treatment from organic solvents.

Contents

Main Stages of the Whey Semi-finished Product Technology Development 1
Mamay Dmitriy, Babenyshev Sergey, Bratsikhin Andrey, Shipulin Valentin, and Mamay Angelina

Microfiltration Processing of Raw Materials for the Fermented Milk Product Making 10
Mamay Dmitriy, Babenyshev Sergey, Bratsikhin Andrey, Mamay Angelina, and Ivanets Vadim

Development of New Types of Minced Meat Products for Preventive and Personalized Nutrition 18
Aleksandr Borisenko, Alexey Savchenko, Aleksey Borisenko, Evgeniya Belousova, and Victoria Razinkova

Prospects for the Use of Chitosan-Stabilized Copper Nanoparticles as an Alternative to Antibiotics in Broiler Feed 28
Kristina Apryatina, Larisa Smirnova, Sergey Zaitsev, Vadim Frolov, and Georgy Laptev

Proteolysis of Bovine Whey, Milk and Colostrum with Serine Endopeptidases 35
Tatsiana Halavach

TiO_2-Chitosan as Universal Composite for Waste Water Treatment from Organic Pollutants 46
Evgeniia Salomatina, Dmitriy Zuev, Sergey Zaitsev, and Larisa Smirnova

Isolation and Characterization of a New Fungi Isolates of the Genus *Trichoderma* Pers. – of Potential Producers of Biofungicides 56
Zulfira Siraeva, David Shamuradov, Alsu Mustafina, and Olga Kravtsova

Study of the Composition and *in Vivo* Effect of Biologically Active Concentrates from Secondary Raw Materials of Millet 65
Alexandr Evteev, Olga Larionova, and Anna Bannikova

Isolation of Cellulase-Containing Exosomal Vesicles from *Picea Abies* Xylem 75
Dmitry Chukhchin, Nikita Shutskiy, and Ksenia Vashukova

Composition of Biologically Active Substances of Flower Extracts from a Variety of *Achillea* Species and Their Antioxidant Properties 81
V. Kurchenko, N. Sushinskaya, K. Maiorava, E. Tarun, and A. Kupriyanov

Approaches to the Determination of the Growing Location of Mossy Pine Forests Based on the Spectral Characteristics of Ecologically Dependent Wood Components 92
Anna Khokh

A Novel Laccase from Basidiomycete *Fomes fomentarius VKPM F-1531* 99
Sofia V. Kiseleva, Ekaterina A. Cherepanova, Elnara R. Nasibullina, Lilija Kh. Khalimova, and Igor V. Maksimov

Biocompatible Materials Based on Modified Starch and Chitosan with High Mechanical Properties 106
Ekaterina Kachalova, Kristina Apryatina, and Larisa Smirnova

Products of Hydrolysis of Bleached Hardwood Kraft Pulp by Carbohydrate-Active Enzymes 114
Aleksandr Shevchenko, Andrey Aksenov, Irina Tyshkunova, Danil Falev, and Evgeniy Toptunov

Biologically Active Substances of Extracts from Various Wood-Decay Fungi and Their Antioxidant Activity 124
V. Kurchenko, N. Sushinskaya, K. Maiorava, E. Tarun, and I. Kiseleva

Novel Method of Preparation Biocompatible Porous Materials Based on Modified Chitosan with High Mechanical Properties 136
Ivan Lednev, Larisa Smirnova, Vitaliy Korokin, and Diana Fukina

Effect of Nanosized Forms of Iron-Containing Compounds on the Growth Characteristics of Raphanus Sativus Seeds 144
Blinov Andrey Vladimirovich, Gvozdenko Alexey Alexeevich, Slyadneva Kristina Sergeevna, Kravtsov Alexander Alexandrovich, and Rajabov Uktamjon Ikromjon Ugli

Study of the Possibility of Application of Acoustic Spectroscopy in Dairy Products 151
Kaishev Vladimir Grigorievich, Khramtsov Andrey Georgievich, Blinov Andrey Vladimirovich, Gvozdenko Alexey Alexeevich, and Maglakelidze David Guramievich

Technological and Nutritional Potential of Lentil in the Turkey Cutlets Production 159
Michael Schetinin, Marina Vaytanis, Olga Musina, and Zoja Khodyreva

Application of Chitosan for Fermented Dairy Products with *Lactobacillus delbrueckii* subsp. *Bulgaricus* Manufacturing 167
Vladimir Kurchenko, Aleksei Lodygin, Tatsiana Halavach, Ivan Evdokimov, and Maria Shramko

The Study of the Quantitative Content of Flavonoids and Biological Activity of the Herba Thlaspi Arvense L. 176
Kira Pupykina, Rashit Farkhutdinov, Albina Fedorova, Ekaterina Koroleva, and Pavel Akhmadiev

The Obtaining an Extract from the Herba of Monarda and Determining the Indicators of Its Quality 184
Kira Pupykina, Ekaterina Krasyuk, and Galiya Shaydullina

Using the Flax Seeds and the Flax Oil in the Production of Chopped Semi-finished Chicken Meat Products in Order to Enrich Them with Polyunsaturated Fatty Acids 191
Isabella Stefanova and Veronika Borisova

Investigation of the Possibility of Using Chondroitin Sulfate in Milk Raw Materials by the Organoleptic Method 200
Alla Novokshanova and Dmitriy Nikityuk

Properties of Cyclodextrins Nanocomplexes with Peptides and Fat-Soluble Vitamins 207
Tatsiana Halavach, Natalia Sushiskaya, LudmilaAlieva, Svetlana Lodygina, and Aleksey Lodygin

Investigation of the Acute, Subacute and Chronic Toxicity of Curcuminoids with Native and Hydroxypropylated β-cyclodextrin Complex Nanostructures 221
Maxim Kapustin, Hanna Chubarova, Vasili Tsygankow, Svetlana Lodigina, and Ekaterina Holodova

Protein-Polysaccharide Complex for the Production of an Emulsion Product 236
Elena Melnikova, Ekaterina Stanislavskaia, Ekaterina Shabalova, and Natalia Antonova

Laser Doppler Flowmetry and Tissue Oxygenation Monitoring in Assessing of the Survival Rate of an Elongated Skin Flap Under the Biomaterials Application 244
Zulfira Siraeva, Renata Mullina, Olga Kravtsova, and Ruslan Mullin

Development of Raw Semi-dry Sausages Enriched with Colloidal Chelate Complexes of Essential Nutrients 253
Aleksey Golik, Natalya Oboturova, Andrey Blinov, Tamara Bacholdina, and Uktamjon Rajabov

Analysis of the Biochemical Composition and Antioxidant Capacity of *Vaccinium Corymbosum* L. Leaves 260
Alla Deeva, Polina Shabunya, Sviatlana Fatykhava, Alena Spirydovich, and Vladimir Reshetnikov

Cryostability of Bifidobacteria in Milk and Vegetable Mixtures 270
Marina Kozhuhova, Elena Terkun, Tatiana Barkhatova, Larisa Rylskaya, and Roman Drozdov

Application of Cluster Analysis of the Genus *Pinus* Various Species Essential Oils Terpene Compounds for Their Taxonomic Identification 277
Hanna Chubarova, Maxim Kapustin, Alena Spiridovich, Svetlana Lodigina, and Alexandr Sobol

Author Index 291

Main Stages of the Whey Semi-finished Product Technology Development

Mamay Dmitriy[1(✉)], Babenyshev Sergey[1], Bratsikhin Andrey[2], Shipulin Valentin[1], and Mamay Angelina[1]

[1] North Caucasus Federal University, Stavropol, Russia
dmamai@ncfu.ru

[2] Izhevsk State Agricultural Academy, Izhevsk, Russia

Abstract. The basis for the dairy processing industry diversification is the expansion of the product range, including the functional product technologies development based on the modern methods and techniques. The aim of the made study is to develop a technology for the production of freeze-dried semi-finished products based on the milk whey. Objects of research: whey produced at the cottage cheese making, its permeates and sublimates. Subjects of research: mechanisms of the pre-clarified whey membrane separation process. A concentrate of licorice root juice was used as a source of natural polysaccharides. Generally accepted, standard methods and certified laboratory equipment were used at the performing experimental studies. As a result of the experimental studies, it was found that: preliminary clarification of whey by mixing of it with licorice root juice concentrate increased the permeability (by 10–12%) and selectivity (by 6–8%) of membranes with subsequent ultrafiltration separation of purified whey; the main operating parameters of its ultrafiltration and subsequent nanofiltration concentration were: $R_{UF} = 0.45–0.5$ MPa, $V_{UF} = 0.15–0.2$ m/s; $R_{NF} = 1.8–2.7$ MPa; $V_{NF} = 0.2–0.25$ m/s; duration of freeze drying of purified and concentrated whey was $\tau = 11.5–12.5$ h. Based on the analysis of the experimental data obtained it was concluded that the practical significance of the research was that the freeze-dried semi-finished product of pre-purified whey could be used in food production as an alternative to analogues of those ingredients. Further development of the research topic and the industrial implementation of the obtained results could be used in line of new food products technology design with freeze-dried whey permeate as an ingredient of the formulation.

Keywords: Whey · Ultrafiltration · Nanofiltration · Permeate · Freeze-dried · Semi-finished product

1 Introduction

There has been a clear trend in the structure of people's diets establishing for the last 3–4 decades that is oriented on increasing of people's demand for low-calorie biologically valuable products, especially dairy products [1–7]. In addition to skimmed milk, one of the raw material reserves of their production is whey, as the most common by-product in

V. Kurchenko et al. (Eds.): ICAETT 2021, LNNS 408, pp. 1–9, 2022.
https://doi.org/10.1007/978-3-030-96641-6_1

the industrial production of ripe cheese and cottage cheese products [8–11]. Wide using of the whey is specified by the fact that up to 50–52% of milk solids, including up to 68–70% lactose, concentrated into it [12–14]. However, those kinds of raw materials, as a rule, have unstable physicochemical properties. That is why their widespread using in the food industry is still associated with significant costs for the development of appropriate technologies for pre-clarification, subsequent fractionation included baromembrane processes as well. It is known that polysaccharides can be used for the preliminary purification of whey from some part of proteins by its precipitation [15, 16], including plant origin polysaccharides, which can be part of various dairy beverages. However, the insufficient of the available data describes effect of polysaccharides influence on the organoleptic parameters of deeply purified and then freeze-dried whey promotes to make own experimental studies.

2 Materials and Methods

The aim of the research is to develop the main stages of the technology for freeze-dried semi-finished products making from whey. To achieve the goal, the following main scientific and practical tasks had been solved:

- to develop a method of whey pre-purification from part of the protein components by their precipitation with polysaccharides contained in the concentrate of licorice root juice (CJ);
- to determine the ranges of working pressure (P) and circulation velocity (V) of pre-purified whey during its ultrafiltration separation and subsequent nanofiltration concentration on polymer membranes;
- to establish the physicochemical properties and organoleptic evaluation of the freeze-dried semi-finished product made from purified and concentrated whey.

Objects of research: whey produced at the cottage cheese making (WM) [17], its permeates and sublimates. Subjects of research: mechanisms of the pre-clarified whey membrane separation process.

As a source of natural polysaccharides, a CJ, industrial producted (Flora of the Caucasus, JSC, Russia) was used. Generally accepted, standard methods and certified laboratory equipment were used at the performing experimental studies. The degree of purification of raw materials from protein compounds at all stages of its processing was determined on the UDK 139 analyzer (VELP) according to GOST 25179-90, estimated by the optical density (D) calculated according to the generally accepted formula:

$$D = \lg(S_0/S) \quad (1)$$

where S_0 - the luminous flux, which is incident on the object, S - the luminous flux passed through the object.

The KrosFlo ® Research IIi TFF laboratory unit was used to experimentally determine the permeability Q and selectivity φ of membranes at the whey separation. The basis for the theoretical interpretation of the obtained experimental data was the sieve model of membrane filtration. The main factors affecting the permeability (Q) and protein selectivity (φ) of membranes were taken as [15, 16, 18, 19]: the operating pressure (P) and the circulation velocity of the separated system (V) in the apparatus channel.

Organoleptic parameters were determined in accordance with the nationwide standard [20]. To determine the mass fraction of the milk protein, fat, solids in the feedstock and in the final product, as well as other data necessary for experimental studies, standard certified laboratory equipment was used, including: IRF 454 B2M refractometer; Brand Titrette electronic titrator; MilkoScan FT 120 milk analyzer; MultiLine 3420 multiparameter device.

3 Results

As a result of the experimental data analysis, it was found that when mixing WM and CJ, the mixture is divided into two phases (Fig. 1): protein-carbohydrate complex and clarified whey (PW).

Fig. 1. Samples PW: № 1 - up to 15% of the volume CJ; № 2 - up to 20% of the volume CJ, № 3 – up to 25% of the volume CJ; № 4 – up to 30% of the volume CJ, № 5 – up to 35% of the volume CJ

It was established that the best clarification is obtained at a ratio of about 75:25 volume units of WM and CJ (Invention RU No. 2758352). The total mass fraction of dry substances in the PW increased up to 6.7–8.8%, and the components of the carbohydrate complex were achieved up to 70–73% of the system (Table 1). At the same time, addition of CJ improved its organoleptic characteristics (Table 2).

Analysis of the literature data showed that such kind of preliminary purification of whey should contribute to the intensification of its subsequent ultrafiltration process. Biomax polymer membranes with delay thresholds of 90–310 kDa and operating pressure up to 0.5 MPa were used for experimental verification of formulated hypothesis. It is known that the values of Q and φ of membranes are predominantly influenced

Table 1. The content of the main components in PW (p = 0.95)

Parameters (in samples №)	1	2	3	4	5
The ratio of WM and CJ, volumetric	85:15	80:20	75:25	70:30	65:35
Dry solids weight ratio ($C_{D.S.}$), %, minimum	6.7	7.0	7.6	8.2	8.8
Lactose, %	5.1	5.1	5.1	5.1	5.1
Total nitrogen, %	0.44	0.5	0.54	0.6	0.64
Inulin and pectin, %	0.5	0.7	1.2	1.7	2.1
Mineral, %	0.64	0.7	0.76	0.84	0.92
Sediment layer, mm	16	17.5	19.1	17.8	16.4

Table 2. Organoleptic characteristics of PW

Factor	WM	PW
Taste and odour	Typical for that type of raw materials	Sweet and sour, pronounced, with a faint fruit and vegetable aroma
Appearance	Yellowish-green liquid	The liquid is yellowish, with a brown tinge, slightly opalescent

by the P value [15, 16]. Data analysis (Fig. 2) obtained by experimental determination and described by dependences of the type Q = f(P) and φ = f(P) during PW ultrafiltration showed that the duration of the process (τ), its temperature (t), Q of both membranes increased up to P = 0.43–0.48 MPa at constant values of V and $C_{D.S.}$. While φ reached values of 95–96%, which was a quite satisfactory level according to the technological requirements to the purity of the finished semi-finished product (in samples of UF permeates (UFP-300 and UFP-300), the protein content (Table 3) did not exceed 0.01–0.02%).

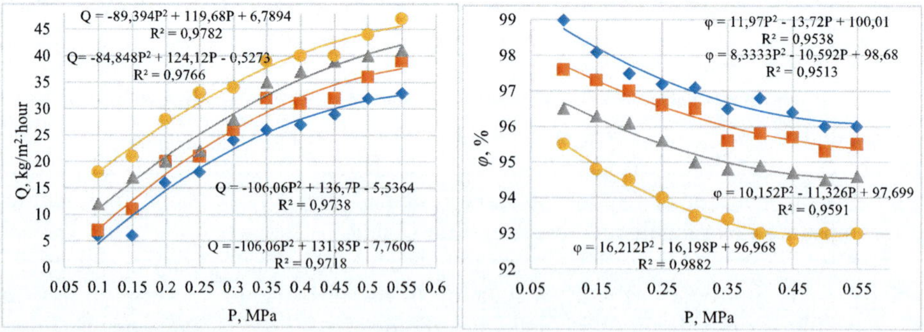

Fig. 2. Dependence Q and φ (●- Biomax 300, ▲- Biomax 100, ■- UAM-50П, ♦- UPM-10) from the value of P at ultrafiltration PW (t = 11–13 °C, V = 0.15–0.2 m/s, $C_{D.S.}$ = 6.4 ÷ 6.7%)

Table 3. Physical and chemical characteristics of the PW

Factor	UFP 100	UFP 300
Dry solids weight ratio, %	4.8–5.3	5.0–5.5
Including:		
Total nitrogen, %	0.01–0.02	0.01–0.03
Lactose, %	4.1–4.4	4.2–4.4
Mineral, %	0.6–0.8	0.6–0.8
Fats, %	0.05–0.1	0.05–0.1

It should be noted that the parameter V at the ultrafiltration of the PW was limited by the design features of the research equipment and, in accordance with the recommendations of the manufacturer, varied in the range of 0.1–0.3 m/s. The values of τ, t, and $C_{D.S.}$ were chosen based on the recommendations mentioned in the scientific publications [16, 21].

As a result of the experimental data analysis during nanofiltration concentration of UFP 100 and UFP 300 on Alfa Laval NF polymer membranes with a 200 Da (Fig. 3), it was found that the values of the increment rate of the function Q = f(P), expressed as tgφi = dQ/dP for UFP 100 were higher than for UFP 300.

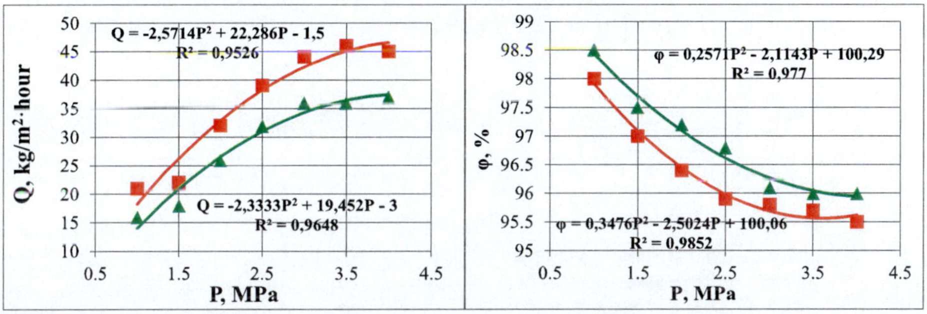

Fig. 3. Dependence of Q and φ of Alfa Laval NF 200 Da membrane from P during nanofiltration (▲) UFP 100 и (■) UFP 300 (t = 12–14 °C, V = 0.2–0.3 m/s, $C_{D.S.}$ = 6.4–6.8%, τ = 60–70 min.)

Significant increase of Q for the both of membranes was achieved up to P = 2.5–3.5 MPa. Then the values of the parameter Q remained almost constant. It is why the P = 3.0 MPa should be considered as a limit for the used conditions of the process.

In addition to the P value, V value of the separated system moving along the membrane surface were also influenced on the Q and φ parameters of the membranes. The results of the analysis of the graphical dependencies Q = f(V) and φ = f(V) obtained by processing experimental data (Fig. 4) showed that a significant increase in the Q index and a decrease in the φ of membranes with a membrane delay threshold of 200 Da occurred when V-parameter were increasing up to V = 0.2–0.3 m/s for both of the separation systems UFP-100 and UFP-300, and then practically had not changed, therefore, the value of the parameter V = 0.25 m/ s could be considered as a lower limit.

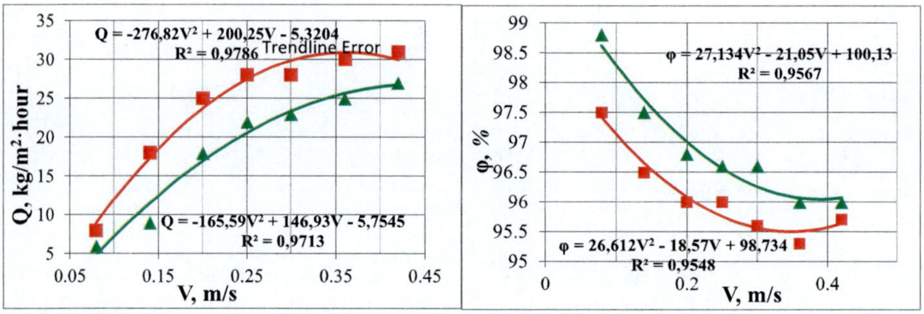

Fig. 4. Dependence of Q and φ of Alfa Laval NF 200 Da membrane from V (▲) UFP 100 and (■) UFP 300 (t = 12–14 °C, τ = 60–70 min, P = 1.7–2.8 MPa, $C_{D.S.}$ = 6.4–6.8%)

The whey permeate obtained by the PW ultrafiltration and concentrated by its nanofiltration to $C_{D.S.}$ = 22–24% (UFPN) was placed to the trays of a freeze dryer, distributed in a uniform layer of 5–6 mm and frozen at t = −(20–25) °C. Drying (at a residual pressure of (8–10) Pa) was carried out at t = −(35–45) °C. The duration of the sublimation and drying stages were determined by experimental data analyzing (Fig. 5): sublimation time = 10–10.5 h (t_s), drying time = 1.5–2 h (t_d).

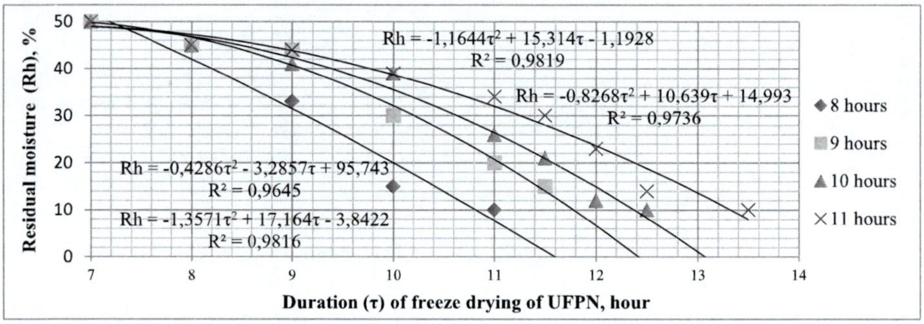

Fig. 5. Dependence of the residual moisture value (Rh) on the duration of the process (τ) of freeze-drying of UFPN

The analysis of experimental data (Table 4) showed that the dry sample of UFPN exceeded the similar products manufactured by the food industry in terms of basic physical and chemical parameters.

Table 4. Characteristics of the freeze-dried UFPN in comparison with analogues

Factor	UFPN	LACTALIS INDUSTRIE	MOL company
Weight ratio, %:			
Lactose	86 ± 0.2	80 ± 0.2	85 ± 0.2
Moisture	3.8 ± 0.5	4.0 ± 0.5	4.5 ± 0.5
Protein (N × 6,38)	0.5 ± 0.001	4.5 ± 0.001	2.0 ± 0.001
Mineral	6.5 ± 0.2	9.5 ± 0.2	7.0 ± 0.2
Inulin and pectin, %	0.5 ± 0.05	-	-
Volume density (bulk), kg/m^3	498.0 ± 2.0	611.0 ± 2.0	619.0 ± 2.0
Density (with seal), kg/m^3	737.0 ± 2.0	731.0 ± 2.0	725.0 ± 2.0
Average particle size, microns	48.0 ± 2.0	79.0 ± 2.0	81.7 ± 2.0

4 Discussion and Conclusion

As a result of the performed stages of experimental studies, it was found that the preliminary clarification of whey by mixing of it with CJ contributed to an increase in the Q (by 10–12%) and φ (by 6–8%) of the membranes with subsequent ultrafiltration separation of the PW in comparison with the whey made in the traditional way. The main operating parameters of the corresponding baromembrane processes must be provided at the followed ranges: $R_{UF} = 0.45–0.5$ MPa, $V_{UF} = 0.15–0.2$ m/s; $R_{NF} = 1.8–2.7$ MPa; $V_{NF} = 0.2–0.25$ m/s. The total duration of freeze drying is T = 11.5–12.5 h. The experimental data obtained had not contradicted the results available in scientific publications. Thus, the main stages of the whey semi-finished product making technology are: preliminary purification of the feedstock from part of the proteins by mixing of it with CJ and precipitation of the mixture, subsequent sequential ultrafiltration separation and nanofiltration concentration, and then freeze-drying. The dry samples of semi-finished whey obtained at the described technology corresponded to the main qualitative characteristics, and in some respects exceeded the close industrial analogues. Based on the analysis of the experimental data obtained it was concluded that the practical significance of the research was that the freeze-dried semi-finished product of pre-purified whey could be used in food production as an alternative to analogues of those ingredients like LACTALIS INDUSTRIE and MOL Company analogues of that products. Further development of the research topic and the industrial implementation of the obtained results could be used in line of new food products technology design based on the UFPN sublimate.

Acknowledgments. The research was performed under the contract № 075–15-2020–487.

References

1. Fatma, A., Kangming, T., Zheng-Xiang, W.: Modern techniques efficacy on tofu processing: a review. Trends Food Sci. Technol. **116**, 766–785 (2021)

2. Yasemin, I., Ayseli, N., Aytekin, M., Turan, A.: Food policy, nutrition and nutraceuticals in the prevention and management of COVID-19: advice for healthcare professionals. Trends Food Sci. Technol. **105**, 186–199 (2020)
3. Bakalis, S., Vasilis, P., Valdramidis, J., Van Impe, F.M.: Perspectives from CO+RE: How COVID-19 changed our food systems and food security paradigms. Curr. Res. Food Sci. **3**, 166–172 (2020)
4. Saguy, I.S., Roos, Y.H., Cohen, E.: Food engineering and food science and technology: forward-looking journey to future new horizons. Innov. Food Sci. Emerg. Technol. **47**, 326–334 (2018)
5. Burgain, J., Gaiani, C., Scher, J.: Encapsulation of Lactobacillus rhamnosus GG in microparticles: influence of casein to whey protein ratio on bacterial survival during digestion. Innov. Food Sci. Emerg. Technol. **19**, 233–242 (2013)
6. Karim, A., Aider, M.: Production of prebiotic lactulose through isomerisation of lactose as a part of integrated approach through whey and whey permeate complete valorisation: a review. Int. Dairy J. **126**, 105249 (2021)
7. Gulec, H.A., Cinar, K., Bagci, P.O.: Production of concentrated whey beverage by osmotic membrane distillation: comparative evaluation of feed effect on process efficiency and product quality. Int. Dairy J. **121**, 105115 (2021)
8. Wu, G., et al.: Whey protein-blackcurrant concentrate particles obtained by spray-drying and freeze-drying for delivering structural and health benefits of cookies. Innovative Food Sci. Emer. Technol. **68**, 102606 (2021)
9. Timkin, V.A., Gorbunova, Y.A.: Sequential micro- and ultrafiltration during the production of Cottage cheese (in Russian). Membr. Membr. Technol. **7**(4), 284–292 (2017)
10. Arranz, E., Corrochano, A.R., Giblin, L.: Antioxidant activity and characterization of whey protein-based beverages: effect of shelf life and gastrointestinal transit on bioactivity. Innovative Food Sci. Emer. Technol. **57**, 102209 (2019)
11. Alan, F., Pombo, W.: Cream cheese: Historical, manufacturing, and physico-chemical aspects. Int. Dairy J. **117**, 104948 (2021)
12. Melnikova, E.I., Losev, A.N., Stanislavskaya, E.B.: Microparticulation of caseic whey to use in fermented milk production. Foods Raw Mater. **5**(2), 83–93 (2017)
13. Joseph, O., Odell ,G., Wallis A.: The application of lactose in sports nutrition. Int. Dairy J. **116**, 104970 (2021)
14. Aydogdu, T., Tri Ho, Q., McCarthy, N.A.: The influence of milk minerals and lactose on heat stability and age-thickening of milk protein concentrate systems. Int. Dairy J. **118**, 105037 (2021)
15. Babenyshev, S.P., Bratsikhin, A.A., Shipulin, V.I., Mamay D.S., Permyakov, A.V., Khokha, D.S.: Determination of optimal values for the main parameters of ultrafiltration of curd whey enriched with natural polysaccharides. IOP Conf. Ser. Earth Environ. Sci. **613**(1), 012010 (2020)
16. Babenyshev, S.P., Bratsikhin, A.A., Zhidkov, V.E., Mamay, D.S., Muhamadjon, N., Khokha, D.S.: Research of the secondary milk raw materials purification from protein components by natural polysaccharides. J. Hygienic Eng. Des. **31**, 63–68 (2020)
17. Russian State Standard: Cottage cheese. Specifications (GOST 31534–2012). Standardinfom, Moscow (2014)
18. Cordova, A., Astudillo-Castro, C., Ruby-Figueroa, R., Valencia, P., Soto, C.: Recent advances and perspectives of ultrasound assisted membrane food processing. Food Res. Int. **133**, 109163 (2020)
19. Sedelkin, V.M., Surkova, A.N., Pachina, O.V., Potehina, L.N., Mashkova, D.A.: Simulation of membrane ultrafiltration of secondary raw milk. Pet. Chem. **56**(4), 367–378 (2016). https://doi.org/10.1134/S0965544116040095

20. Russian State Standard: Milk and milk products. Sensory analysis. Part 2. Recommended methods for sensory evaluation (GOST R ISO 22935-2-2011). Standardinfom, Moscow (2012)
21. Maubois, J.L.: Ultrafiltration of whey. Int. J. Dairy Technol. **33**(2), 55–58 (1980)

Microfiltration Processing of Raw Materials for the Fermented Milk Product Making

Mamay Dmitriy[1](✉), Babenyshev Sergey[1], Bratsikhin Andrey[2], Mamay Angelina[1], and Ivanets Vadim[1]

[1] North Caucasus Federal University, Stavropol, Russia
[2] Izhevsk State Agricultural Academy, Izhevsk, Russia

Abstract. The diversification of the dairy industry could be realized to increase in made products range due to the sequential separation of whole natural milk into separated fractions, each of which could be used to produce the target product, including skimmed milk raw materials with a reduced content of casein fraction. The aim of the study is to determine the rational modes of raw materials microfiltration for production of fermented milk product like katyk type, which is the most popular with people of Central Asia. Generally accepted, standard methods and certified laboratory equipment were used at the performing experimental studies. For microfiltration of skimmed milk and its subsequent use in fermented milk products making, it is advisable to use membranes with a conditional pore size in the range of 0.3–0.4 μ. Operating pressure and circulation velocity of the separated system were in the ranges P = 0.1–0.15 MPa, V = 0.2–0.25 m/s for BioMax (0.3 μ) and P = 0.075–0.125 MPa and V = 0.15–0.2 m/s for BioMax (0.4 μ) respectively. The finished product, made using technology close to traditional, had the protein content around 2.1–2.2%, and its overall organoleptic assessment was comparable to the control sample. The practical significance of the research results is to substantiate the possibility of skimmed milk permeate using for production of fermented milk products such katyk is. Further development of the research topic and the industrial implementation of the obtained results could be used in line of new low-calorie fermented milk product making which overall organoleptic evaluation would be the similar to the traditional analogue made from whole milk.

Keywords: Skimmed milk · Microfiltration · Fermented milk drinks · Protein components · Permeate · Katyk

1 Introduction

Currently, most people pay the necessary attention to protect their health, which is promoted by the resistance to the negative impact of the urbanized environment as well as due to full active longevity supporting [1–5]. Preferences of people in the daily diet has been moved towards low-calorie and especially fermented milk products containing easily digestible whey proteins instead of the casein fraction [6–9]. There are the well-known products with a reduced content of milk fat and casein, as yogurt, kefir, soft

V. Kurchenko et al. (Eds.): ICAETT 2021, LNNS 408, pp. 10–17, 2022.
https://doi.org/10.1007/978-3-030-96641-6_2

cheeses, etc. currently, they prevail in the trade assortment. That situation promotes the dairy industry diversification by expanding the range of products through the low-fat dairy raw materials using and the mass fraction of the casein fraction decreasing [10–13]. However, the insufficient of the available data describes the microfiltration method for isolating some part of the protein fraction from skimmed milk, as well as the organoleptic parameters of fermented milk products made from such raw materials, promotes to make own experimental studies.

2 Materials and Methods

The aim of the study is to determine the rational modes of raw materials microfiltration for fermented milk product making like katyk type, which is the most popular with people of Central Asia (The Republic of Tajikistan, Uzbekistan, etc.). The following main scientific and practical tasks had been solved to achieve the formulated aim:

- to study the physicochemical properties and organoleptic characteristics of the fermented milk product katyk, prepared by the traditional way from skimmed milk with a reduced content of casein fraction;
- to determine the main operating parameters ranges of the skimmed milk microfiltration for fermented milk product making.

Objects of research: samples of fermented milk product of the katyk type (FMP), produced by the traditional way [14], skimmed milk permeate (MS) [15], obtained by microfiltration (MFP). Subjects of research: mechanism of the MS microfiltration separation, organoleptic characteristics of the finished product.

Generally accepted, standard methods and certified laboratory equipment were used at the performing experimental studies. The degree of purification of raw materials from protein compounds at all stages of its processing was determined by the UDK 139 analyzer (VELP) according to GOST 25179-90, estimated by the optical density (D) calculated according to the generally accepted formula:

$$D = \lg(S_0/S). \tag{1}$$

where S_0 - the luminous flux, which is incident on the object, S - the luminous flux passed through the object.

KrosFlo ® Research IIi TFF laboratory units had been used to experimentally determine the permeability (Q) and selectivity (φ) of membranes during MS separation. The basis for the theoretical interpretation of the obtained experimental data was the sieve model of membrane filtration. The main factors affecting the parameters Q and φ of the membranes were taken as [16–19]: the operating pressure (P) and the circulation velocity of the separated system (V) in the channel of the apparatus.

Organoleptic parameters were determined in accordance with in nationwide standard [23]. To determine the mass fraction of milk protein, fat, dry solids, etc. in the feedstock, permeate and finished product, standard certified laboratory equipment was used, including: IRF 454 B2M refractometer; Brand Titrette electronic titrator; MilkoScan FT 120 milk analyzer; MultiLine 3420 multiparameter device; TV-80-1 dry-air thermostat, etc.

3 Results

The results of the analysis of data from literature sources [20, 21], as well as our own research [16, 17] showed that using BioMax brand membranes with a conditional pore diameter (CPD) from 0.1 μ to 0.5 μ for microfiltration separation of a part of the casein fraction from MS was established as advisable one. The principal possibility of practical implementation of this technological operation is determined by significant difference in the size of casein globules and whey proteins [22].

In accordance with the recommendations of [16, 21], MS microfiltration was carried out by varying the following main process parameters: operating pressure (P) and circulation velocity (V) of the separated system in the apparatus channel: P = 0.05–0.2 MPa; V = 0.1–0.2 m/s. The mass fraction of dry solids ($C_{D.S.}$) in the retentate according to technological requirements [14] had not exceed 10.4–10.6%, the temperature of the process was - t = 13–15 °C. The duration of microfiltration before washing the membranes according to the manufacturer's regulations was $\tau = 35 \div 40$ min.

It was found that the protein content in the obtained permeate samples proportionally decreases as the CPD index of the membranes decreases (Table 1) when the other parameters of the MS microfiltration conditions was the same.

Table 1. Physical and chemical characteristics of the MS and his MFP (p = 0,95)

Factor	Product					
	MS	MFP, obtained through a membrane with CPD (μ):				
		0.1	0.2	0.3	0.4	0.5
Dry solids weight ratio $C_{D.S.}$, %, including:	8.2	6.8	7.3	7.6	7.8	8.0
Total nitrogen, %	2.6	1.2	1.7	2.1	2.2	2.4
Lactose, %	5.1	5.1	5.1	5.1	5.1	5.1
Acidity, °T	18	10	11	13	13	15

At the next stage of experimental studies, a FMP had been made from every MFP samples in accordance with the requirements of the traditional method [14] and organoleptic evaluation of them was carried out (Fig. 1, Table 2).

It was found that the samples of FMP № 1 and № 2 had a homogeneous consistency with an undisturbed clot structure and an insignificant amount of separated whey, and had a uniform milky-white color and sour-milk taste. Samples FMP № 3 and № 4 were characterized by the absence of a clot structure and white color. Sample № 5 was a product with an inhomogeneous consistency due to a disturbed clot and a significant part (more than one third) of the separated whey. The analysis of the results of organoleptic evaluation and physicochemical properties of the samples of FMP (Table 3) showed that it was advisable to use MFP obtained on membranes with a CPD index of 0.3–0.4 μ for its production.

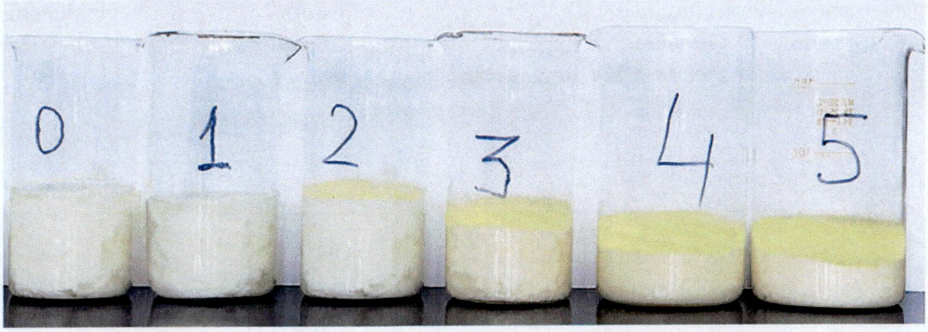

Fig. 1. FMP samples: 0 – FMP control sample (O), 1 – MFP 0.5 μ, 2 - MFP 0.4 μ, 3 - MFP 0.3 μ, 4 - MFP 0.2 μ, 5 - MFP 0.1 μ

Table 2. Results of organoleptic evaluation of FMP, received from the MFP

Factor	FMP sample numbers					
	0	1	2	3	4	5
Color	4	4	4	4	3	3
Taste	5	5	5	4	4	3
Odour	5	5	5	5	4	4
Consistency	5	5	4	4	3	3
Total score	4,75	4,75	4,5	4, 25	3,5	3,25

Table 3. Physical and chemical characteristics of the FMP from the MFP ($p = 0.95$)

Factor	FMP from MFP obtained through a membrane with CPD (μ)				
	0.1	0.2	0.3	0.4	0.5
Dry solids weight ratio $C_{D.S.}$, %, including:	5.8	6.3	6.3	7.0	7.2
Total nitrogen, %	1.2	1.7	2.1	2.2	2.4
Lactose, %	4.1	4.1	4.2	4.2	4.3
Acidity, °T	85–90	87–92	90–92	93–95	97–100

It is known [16, 21] that the microfiltration process initiated by the working pressure P in the channel of the baromembrane apparatus. But the Q of the membranes is also significantly depended on the velocity V of the separated system tangential flow, which moves along the working surface of the membrane. If the change in the concentration profile of the dispersed phase particles and the destructive effect of that flow on the layers of protein deposits would be precisely determined by its velocity in the primembrane zone, the parameter V could become comparable with P in its significance for the kinetics of the process. In accordance with the recommendations of [16, 21], the dependencies $Q = f(P)$ and $\varphi = f(P)$ were first experimentally established during MS microfiltration, as well as dependences $Q = f(V)$ and $\varphi = f(V)$ (Fig. 2 and 3).

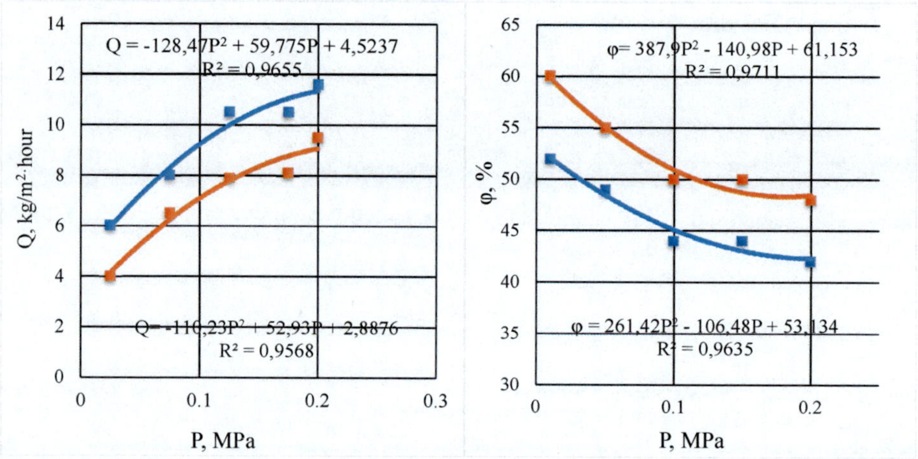

Fig. 2. Dependences of Q and φ of membranes (■- 0.3 μ, ■- 0.4 μ) on P during MS microfiltration ($t = 8–14$ °C, $V = 0.1–0.14$ m/s, $C_{D.S.} = 8.0–8.2\%$, $\tau = 35–40$ min.)

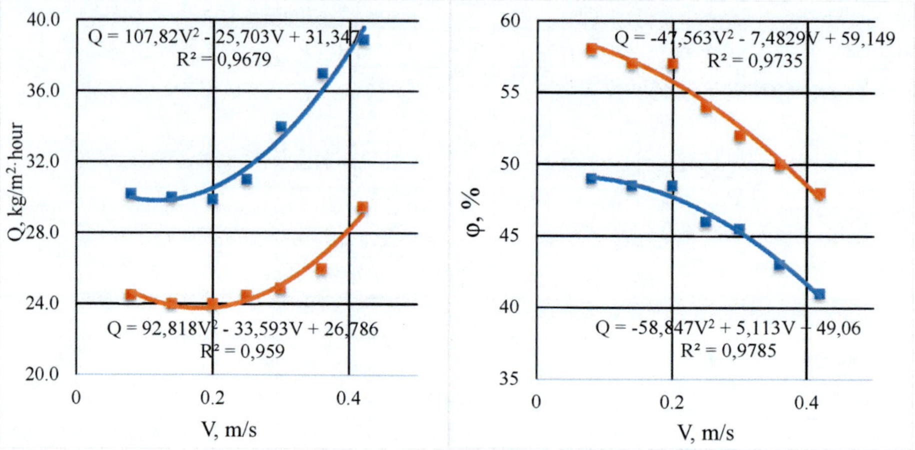

Fig. 3. Dependences of Q and φ of membranes (■- 0.3 μ, ■- 0.4 μ) on V during MS microfiltration ($t = 8–14$ °C, $P = 0.1–0.15$ MPa, $C_{D.S.} = 8.0–8.2\%$, $\tau = 35–40$ min.)

The analysis of the dependencies $Q = f(P)$ and $\varphi = f(P)$ showed that the force of the change in the parameters of Q membranes with CPD = 0.3 μ and with CPD = 0.4 quite identical and increased significa was quite identical and increased significantly with an increase in P up to 0.075–0.2 MPa. A further increase in P had not influenced on increase in Q, but the index of the membranes increased for BioMax (CPD = 0.3 μ) up to 72%, and for BioMax (CPD = 0.4 μ) - up to 60%. This is explained by the fact that at the mentioned value P the partial deformation of the membrane material structure may already occur, causing both a decrease in the value of the actual CPD index and an increase in the degree of pore blocking by casein molecules, which causes the observed changes in the Q and φ parameters. That interpretation is supported by the results of the samples analysis of the initial MS and its MFP for the total amount of protein according to the Kjeldahl method (Table 4). It was found that about 9–10% of protein particles "settle" on a membrane with CPD = 0.3 μ, and about 12–13% - on a membrane with CPD = 0.4 μ.

Table 4. The results of the analysis of samples MS of its retentate and permeate for the total amount of protein by the Kjeldahl method (p = 0.95)

Sample	Total nitrogen, %	
	Membrane with CPD = 0.3 μ	Membrane with CPD = 0.4 μ
Retentate	4.6	4.5
Permeate	2.1	2.2
MS	2.6	

In addition, it was found that the average value of the Q index of a membrane with CPD = 0.4 μ with MS microfiltration is 19–22% higher than it was for a membrane with CPD = 0.3 μ. It could be explained by the difference in CPD of the membranes 0.4 μ and 0.3 μ about 1.2–1.3 times. But at the same time, the porosity of membranes with CPD = 0.3 μ was higher, which made the values of the total filtration areas comparable. Hence, there was no significant discrepancy in the permeability of the both membranes. Another confirmation of given interpretation was in the values of φ membranes with CPD = 0.3 μ and with CPD = 0.4 μ at the initial stage of the microfiltration process that were differ about 12–16%. It should be noted that after about 3–5 min of the process this difference becomes no more than 6–8%. It confirmed that membranes with CPD = 0.4 μ with larger pores at the beginning of the operation were subjected to more significant blocking of the working surface by casein molecules compared to membranes with CPD = 0.3 μ [16]. Analysis of the functions Q = f(V) and φ = f(V) for membranes with CPD = 0.3 μ and CPD = 0.4 μ promoted conclusion that both membranes could be used for microfiltration, as significant increase in membranes indicators (dQ/dV) was observed in the same range V = 0.3–0.4 m/s.

4 Discussion and Conclusion

It is advisable to use the membranes with CPD in the range of 0.3–0.4 μ for microfiltration of MS and its subsequent using in FMP production with the next parameters of process: for BioMax 0.3 μ - P = 0.1–0.15 MPa, V = 0.2–0.25 m/s; for BioMax 0.4 μ - P = 0.075–0.125 MPa, V = 0.15–0.2 m/s. The protein content in the finished product was at the level 2.1–2.2%, and its overall organoleptic assessment was comparable to the control sample.

The practical significance of the made research results is in a practical confirmation of the possibility to use the skimmed milk permeate in fermented milk products making like katyk type. Further development of the research topic and the industrial implementation of the obtained results could be used in line of new low-calorie fermented milk product with the same organoleptic characteristics as the traditional products made from the whole milk.

Acknowledgments. The research was performed under the contract № 075–15-2020–487.

References

1. Vizireanu, M., Hruschka, D.: Lay perceptions of healthy eating styles and their health impacts. J. Nutr. Educ. Behav. **50**, 365-371. (2018)
2. Healthy Eating as We Age. J. Nutrit. Educ. Behav. **51**, 912 (2018)
3. Parsons, A.A., Monteban, M., Pat Bebo, E.L., Zubieta, A.C., Ginnetti, S.: Indicators of readiness and capacity for implementation of healthy eating strategies in child care settings serving low-income children. J. Nutr. Educ. Behav. **51**, 465–477 (2018)
4. Ruder, E.H., Lohse, B., Mitchell, D.C., Cunningham-Sabo, L.: Parent food and eating behavior assessments predict targeted healthy eating index components. J. Nutr. Educ. Behav. **51**, 711–718 (2019)
5. Morell, P., Fiszman, S.: Revisiting the role of protein-induced satiation and satiety. Food Hydrocoll. **68**, 199–210 (2017)
6. Meng, L., Li, S., Feng, Z.: The nutrient requirements of lactobacillus acidophilus LA-5 and their application to fermented milk. J. Dairy Sci. **104**(1), 138–150 (2020)
7. Yanan, X., Yu, J., Shuang, Q.: Purification and characterization of angiotensin-I-converting enzyme inhibitory peptides isolated from whey proteins of milk fermented with Lactobacillus plantarum QS670. J. Dairy Sci. **103**(6), 4919–4928 (2020)
8. Skryplonek, K., Dmytrow, I., Mituniewicz-Malek, A.: Probiotic fermented beverages based on acid whey. J. Dairy Sci. **102**(9), 7773–7780 (2019)
9. Kim, D.H., Chun, S., Lee, K.: Anti-inflammatory activities of Maillard reaction products from whey protein isolate fermented by Lactobacillus gasseri 4M13 in lipopolysaccharide-stimulated RAW264.7 cells. J Dairy Sci. **102**(9), 7707–7716 (2019)
10. Heck, A., Schäfer, J., Hinrichs, J.: Fat-free fermented concentrated milk protein-based microgel dispersions manufactured at technical scale: production parameters as drivers of textural properties. Int. Dairy J. **127**, 105195 (2021)
11. Kondybayev, A., Loiseau, G., Konuspayeva, G.: Fermented mare milk product (Qymyz, Koumiss). Int. Dairy J. **119**, 105065 (2021)
12. Park, S., Yoo, M., Lim, S.: Production of benzoic acid as a natural compound in fermented skim milk using commercial cheese starter. J. Dairy Sci. **100**(6), 4269–4275 (2017)
13. Wang, D., Fritsch, J., Moraru, C.I.: Shelf life and quality of skim milk processed by cold microfiltration with a 1.4-μm pore size membrane, with or without heat treatment. J. Dairy Sci. **102**(10), 8798–8806 (2019)
14. Uzbekistan State Standard: Katyk from cow's milk. Specification (O'z DSt 1223:2009). Uzstandart, Uzbekistan
15. Russian State Standard: Skim milk—raw materiel. Spedticatlone (GOST 31658—2012). Standardinfom, Moscow, Russia (2011)
16. Babenyshev, S.P., Evdokimov, I.A., Bratsikhin, A.A., Zhidkov, V.E., Anisimov, G.S., Mamay, D.S.: Experimental determination of parameters for milk whey microfiltration process. J. Hyg. Eng. Des. **28**, 85–95 (2019)

17. Babenyshev, S.P., Bratsikhin, A.A., Zhidkov, V.E., Mamay, D.S., Muhamadjon, N., Khokha, D.S.: Research of the secondary milk raw materials purification from protein components by natural polysaccharides. J. Hyg. Eng. Des. **31**, 63–68 (2020)
18. Cordova, A., Astudillo-Castro, C., Ruby-Figueroa, R., Valencia, P., Soto, C.: Recent advances and perspectives of ultrasound assisted membrane food processing. Food Res. Int. **133**, 109163 (2020)
19. Sedelkin, V.M., Surkova, A.N., Pachina, O.V., Potehina, L.N., Mashkova, D.A.: Simulation of membrane ultrafiltration of secondary raw milk. Pet. Chem. **56**(4), 367–378 (2016). https://doi.org/10.1134/S0965544116040095
20. Brandon, C.: Determination of the efficiency of removal of whey protein from sweet whey with ceramic microfiltration membranes. J. Dairy Sci. **104**(7), 7534–7543 (2021)
21. Maubois, J.L.: Ultrafiltration of whey. Int. J. Dairy Technol. **33**(2), 55–58 (1980)
22. Patel, K.: Primary information, health benefits, side effects, usage, and other important details. Turkish Green Crescent Soc. **2015**, 6847–6858 (2015)
23. Russian State Standard: Milk and milk products. Sensory analysis. Part 2. Recommended methods for sensory evaluation (GOST R ISO 22935–2–2011). Standardinfom, Moscow, Russia

Development of New Types of Minced Meat Products for Preventive and Personalized Nutrition

Aleksandr Borisenko(✉), Alexey Savchenko, Aleksey Borisenko, Evgeniya Belousova, and Victoria Razinkova

North Caucasus Federal University, Stavropol, Russia
{alaborisenko,aborisenko,esmolko}@ncfu.ru

Abstract. Particular relevance nowadays is of the development of personalized food products with functional ingredients that provide the human body with the main sources of essential nutrients - complete proteins, unsaturated fatty acids, vitamins, micro- and macro-elements, and dietary fiber. The purpose of scientific research was to create minced meat products for preventive and personalized nutrition with a balanced nutrient composition, high consumer characteristics, nutritional, biological value and yield. As a result of the research, a purposeful combination of the following types of animal and plant raw materials for recipes of minced meat products has been substantiated: duck meat in a 4: 1 ratio to mechanically deboned chicken meat, various vegetables (carrots, white cabbage, sweet peppers), semolina, egg powder, soybean oil. The expediency of the complex application of the Mobi-Lux universal food fortifier and electrochemically activated water (catholyte-alkaline fraction with pH = 11,2) in the recipes of the developed minced meat products has been proved. Finished products have a high yield, are a source of complete protein (the minimum amino acid rate is 102,5%), unsaturated fatty acids (the ratio of saturated to unsaturated fatty acids is 0.4), calcium, iron, iodine, vitamin C, vitamin A, vitamin B_{12} and dietary fiber.

Keywords: Personalized nutrition · Functional food products · Nutritional balance · Food fortification · Dietary fiber · Iodized milk protein · Electrochemically activated water

1 Introduction

Alimentary products for preventive and personalized nutrition in the world market already occupy a fairly large share of sales, which is growing significantly every year [3, 14, 18]. Due to the high growth in the number of food-dependent diseases around the world, in recent years, great attention has been paid to foods and healthy diets. In this regard, the main trend in the creation of new and improvement of existing food products is their personalization and transfer to the segment of functional products [14, 17].

Personalized nutrition is a new direction that covers technologies for analyzing and modeling the nutritional composition of food products, assessing the state of the human

V. Kurchenko et al. (Eds.): ICAETT 2021, LNNS 408, pp. 18–27, 2022.
https://doi.org/10.1007/978-3-030-96641-6_3

body using means of collecting data on its nutritional status, including with the possibility of using genomic and post-genomic methods, developing personalized food products, providing services for the selection of a personal diet, as well as logistics of delivery services [11, 17]. According to analysts' forecasts, each food product in the future will become personalized not only in chemical composition, but in organoleptic characteristics. [10, 18].

The main platform for the realization of personalized nutrition technologies is FoodNet – a market for new developments, breakthrough technologies, consumer products and services (from the «microbe» to the plate) [16]. The subject of the market working is high-precision production and delivery of food and biotechnological products to the consumer, development, production, packaging, logistics to the final form consumer and waste disposal. The market includes manufacturing technologies, equipment, formulations, software, as well as digital platform solutions.

Analyzing different points of view on the problem under study at the present stage of development, five main features (principles) of the FoodNet market development can be identified [1, 16]:

- market development is directly linked to the task of meeting the basic needs of the population - in high-quality healthy food;
- the market is focused on the specific needs of people, as end consumers, it involves the improvement of technologies and processes for the production of food products, provided that the usual style and methods of their consumption are preserved;
- the market is characterized by the absence of generally accepted standards, while out of more than thirty market segments, only a few are at the stage of forming their own standards;
- it is already obvious that the market will be a trading network in which intermediaries are replaced by management software and online services;
- in the near future, the market will become visible and significant on a global scale. In the medium term, the market volume by 2035 is projected at USD 3,5 trillion.

Thus, taking into account current trends and prospects, there is an urgent need to create new food products for personalized nutrition and to analyze the possibilities of using modern computer design tools in order to transfer existing already produced food products to the healthy nutrition segment.

2 Materials and Methods

The aim of the research was to create minced meat products for personalized nutrition with preventive properties, stable and high consumer characteristics, nutritional and biological value, organoleptic indicators and yield due to the balance of the nutrient composition, specially selected components of animal and plant origin, which increase the manufacturability and productivity of production.

The main raw material for the developed food products was the meat of chickens and ducks. To obtain a balanced nutrient composition, various vegetables (carrots, white cabbage, sweet peppers), semolina, egg powder, and a composition of specially selected functional ingredients were introduced into the foodstuffs formulation.

Modeling of recipes, calculation of the nutritional composition and nutritional value of minced meat products were carried out using the «Etalon» software package (certificate of registration №2005610751 of Rospatent of the Russian Federation).

To determine the degree of swelling, a sample of a food fortifier in the amount of 2 g was placed in a 50–100 ml centrifuge tube, 15 ml of distilled water was added at room temperature, mixed thoroughly and kept for 1 h. Then centrifugation was carried out for 5 min at 1000 rpm. The centrifugate was poured off and the moisture content in the sediment was determined by drying in a drying oven.

Water absorption capacity (WAC) and Fat absorption capacity (FAC) capacity of the food fortifier was determined using a stainless steel mesh glassful (height 80 mm, mesh hole diameter 1,5 mm, number of holes per 1 cm^2 – 10–20). The bottom and walls of the glassful were covered with filter paper to avoid loss of small particles. The glass was moistened with water (vegetable oil was used instead of water in determining the fat absorption capacity (FAC), then the water (vegetable oil) water dripped off within 20 min and the glass was weighed. It was filled with 2 g of a food fortifier, after which the glass with a sample portion was immersed in water (vegetable oil) at room temperature for 20 min. After flow down for 20 min, the outer walls and the bottom of the beaker were wiped with filter paper, weighed and calculated WAC (FAC) in % was calculated as the ratio of the weight of the product with the beaker after soaking to the weight of the product with the beaker before soaking.

To determine the moisture content in minced meat, a sample in an amount of 3 g, weighed with an accuracy of 0,01 g, was distributed in an even layer on the bottom of the weighing bottle and oven drying in drying chamber at 130 °C for 80 min, after which the weighing bottle were cooled in a desiccator and weighed. The moisture content was calculated in% as the ratio of the difference between the weight of the weighing bottle with minced meat before drying and after drying to the weight of the weighing bottle with minced meat before drying.

When determining the Water binding capacities (WBC) of meat-growing minced meat, before the study, the ashless filter was placed on a glass plate 10 × 10 mm in size. Minced meat (0,3 g) was weighed on a torsion balance on a piece of polyethylene 40 mm in size and transferred to an ashless filter so that the minced meat was under a circle of polyethylene. The minced meat was covered on top with the same glass plate and a weight of 1 kg was placed on it and pressing continued for 10 min. After that, the filter with minced meat was released from the load, and then the contour of the spot around the pressed minced meat was outlined with a chemical pencil. The outer contour was drawn when the filter paper was dried in air. Using graph paper, the areas of the spot formed by the minced meat and released moisture absorbed by the filter paper were determined. The size of the wet spot was calculated from the difference between its area and the area of the spot formed by the minced meat. WBC to the mass of minced meat was calculated by the formula:

$$X_1 = \frac{a - 8,4 \cdot b}{m} \cdot 100, \quad (1)$$

где X_1 this is the WBC to the mass of minced meat, %; a this is the total mass of moisture in the minced meat, mg; b this is the area of the wet spot, mg; 8, 4 this is a reference value for the mass of moisture in 1 cm^2 of wet spot; m this is the mass of minced meat, mg. WBC to the total moisture was calculated using the formula:

$$X_2 = \frac{a - 8,4 \cdot b}{a} \cdot 100, \quad (2)$$

где X_2 this is WBC to total moisture, %.

The yield of finished minced meat products was determined as the ratio of the mass of the finished product after heat treatment and cooling to 40 °C to the mass of the finished product before heat treatment.

To obtain catholyte-alkaline with pH = 11,2 of Electrolyzed Reduced Alkaline Water (ERAW) installation based on a flow-through electrochemical modular element PEM-3, developed by the Vitold Bahir Institute of Electrochemical Systems and Technologies, Moscow, Russia were used [4].

The flow-through electrochemical modular element PEM-3 (developed on the basis of the PEM-1 cell, UK patent 2 253 860) is a diaphragm electrochemical reactor operating at elevated pressure. The anode of the element is made of titanium with a special coating. The titanium cathode has an increased catalytic activity due to special surface treatment. The ceramic diaphragm is located between the anode and the cathode of the element and serves to separate the electrolysis products.

The PEM-3 is designed for electrochemical unipolar treatment of fresh or slightly mineralized water with the aim of directional changes in its chemical composition due to the synthesis of stable and metastable products of anode or cathode electrochemical reactions, directional changes in the composition of dissolved gases, catalytic activity in various physico-chemical reactions.

It can be used as independent electrochemical devices (PEM-3 reactor), and in the form of high performance units called RPE-03 reactors. Therefore, the research results obtained using the PEM-3 module are scaled to industrial-scale installations.

3 Results

The meat of farm animals and poultry is one of the main sources of complete protein, as well as micro- and macronutrients, especially easily digestible iron, vitamins A and B, which contribute to the normal functioning of the endocrine, cardiovascular, nervous and other systems of the human body [19]. At the same time, duck meat is one of the most balanced in terms of nutrient composition, it contains twice the amount of vitamin A compared to other types of meat, and is also more fatty in relation to chicken meat [5]. Duck fat contains a large amount of ω-3 unsaturated fatty acids, which have a beneficial effect on the cardiovascular system and improve brain function. In addition to fatty acids, the chemical composition of duck meat contains a large amount of various vitamins and minerals: vitamins A, PP, E, all vitamins of group B. Especially duck meat is considered useful for people suffering from anemia and some nervous disorders [19].

The prophylactic focus of the developed minced meat products, ensuring the possibility of their inclusion in the diets of personalized nutrition, was carried out by combining natural meat and vegetable raw materials, as well as the use of a complex food fortifier «Mobi-Lux Universal» (Scientific and Production Company «Mobitek-M», Russia).

The composition of this food supplement includes natural substances of plant and animal origin – blood plasma proteins, hemoglobin, calcium mineral fortifier, dietary fiber, iodized milk whey protein «Bioiod».

The fortifier contains purified hemoglobin of the blood of farm animals, including iron in the heme form Fe2+, iodized proteins of milk serum, which is a mixture of complete proteins containing 2,5% of covalently bound iodine atoms, the guaranteed amount of which in products does not decrease under the action of high temperatures. Due to the physiologically accessible form for the body, organic iodine normalizes the balance of iodine in the body and improves the mental, mental and physical condition of a person, and has no side effects associated with the consumption of inorganic iodine.

«Mobi-Lux universal» also contains a mineral (calcium) fortifier, which includes not only about 36% calcium, but also a whole range of irreplaceable minerals: magnesium, potassium, copper, zinc and others, in the form of natural compounds. This has a positive effect on protein, lipid and mineral metabolism, significantly improves the formation of the human skeleton during the growth period, and ensures the normal level of functioning of the body's adaptive-protective systems. Only 1,5% of it by weight provides the body's daily requirement for calcium and does not impair the taste of fortified foods. The availability of calcium is 2,6 times higher than from salt mixtures and ranges from 80 to 90%. It has also been found that natural sources of calcium reduce blood cholesterol levels and enhance the absorption of inorganic iron from food.

«Mobi-Lux universal» contains dietary fiber, the source of which is insoluble wheat or apple fibers, as well as soluble fiber – polyfructose inulin, obtained from chicory. Fibers normalize the activity of the gastrointestinal tract and restore beneficial microflora, improve the absorption of nutrients, including calcium, and the elimination of toxic compounds. Improves metabolism and contributes to the prevention of many alimentary-dependent diseases [12, 15].

Nutritional value of the complex food fortifier per 100 g is: proteins - 50 g, carbohydrates - 15 g, fats – 0,5 g, dietary fiber - 20 g (insoluble - 15 g, soluble - 5 g), calcium - 20 000 mg, iron – 12 mg, iodine - 40000 μg, the energy value is 265 kcal or 1108 kJ.

Research carried out of swelling (120%) and WAC (185,2%) of the food fortifier «Mobi-Lux Universal» (Table 1) made it possible to recommend it for use in formulations of minced meat products. Separately, it is necessary to note the high value of the FAC of the investigated product (243,8%), which is a favorable factor for its use in the formulations of duck meat products, including the possibility of using vegetable oils for enrichment of finished minced meat products with unsaturated fatty acids.

Table 1. Functional and technological properties of the complex food fortifier «Mobi-Lux Universal»

The name of indicators	Meaning indicator	Standard deviation
Swelling degree, %	120,0	±5,0
WAC, %	185,2	±4,1
FAC, %	243,8	±13,5

The studies have shown that the use of the food fortifier «Mobi-Lux universal» provides not only an increase in the nutritional value of finished products, but also an increase in their yield, improves the appearance and organoleptic characteristics of products, increases the preservation time and presentation of products during storage, allows use finished products in order to prevent the deficiency of complete protein, calcium, iron, iodine and dietary fiber in the diet of various categories of the population.

In the production of prophylactic minced meat products, it is important not only to reduce the total fat content, but also to replace saturated fatty acids (SFA) of animal origin with vegetable oils, which contain an increased amount of monounsaturated (MUFA) and polyunsaturated fatty acids (PUFA) of the ω-6 family and ω-3. In this regard, soybean oil was introduced into the formulation of products, which contains (mg/100 g) organic choline – 0,2; fatty acids (g/100 g): palmitic acid – 10,3; stearic – 3,5; linoleic acid – 50,9; linolenic – 10,3 and oleic (ω-9) – 19,8, which can significantly improve the functioning of the liver and heart. The introduction of soybean oil, which is also a source of vitamin E (17,1 mg/100 g), into the formulation of new types of minced meat products makes it possible to increase the mass fraction of vegetable fat in them and improve the balance of unsaturated fatty acids (UFA). Soybean oil has a positive effect on brain function, normalizes blood cholesterol levels, and improves sexual function in men. In addition, this oil is recommended for the prevention of diseases of the digestive system, immune diseases and metabolic disorders.

Table 2 shows the indicators of nutritional balance, nutritional and energy value of the developed minced meat products.

Table 2. Indicators of nutritional balance, nutritional and energy value of minced meat products

The name of indicators	Meaning
Protein, g/100 g product	12,17
Minimum amino-acid score, %	102,5
Utility coefficient, share of units	0,856
Fat, g/100 g product	12,05
SFA, g/100 g lipids	24,8
PUFA, g/100 g lipids	25,4
MUFA, g/100 g lipids	36,9
SFA/UFA	0,4
Dietary fiber, g/100 g of product	3,1
Energy value, kJ/kcal	775,8/185,3

The energy value of the developed minced meat products is 185,3 kcal, of which 26,3% is protein, which allows them to be classified as food products with a high protein content. High values of the minimum amino acid rate and the utilitarian coefficient show that the amino acid composition of the protein is full value.

The ratio between saturated and unsaturated fatty acids in the product corresponds to the recommended value [9].

In accordance with the Technical Regulations of the Customs Union (TR of the CU 022/2011 [8]) the developed product is a source of dietary fiber, the amount of which in the obtained food products from duck meat is 3,1 g/100 g of the product.

The developed product is also a source of vitamins and micro-, macroelements (Table 3), namely calcium, iron, iodine, vitamin C, vitamin A, and is also a product with a high content of vitamin B_{12}.

Table 3. The content of micro-, macroelements and vitamins in the product

Name of nutrients	Content in 100 g of finished minced meat products, taking into account losses during heat treatment	% of the recommended daily requirement according to TR CU 022/2011
Calcium, mg	151,714	15,17
Phosphorus, mg	117,566	14,69
Potassium, mg	204,049	5,83
Magnesium, mg	26,981	6,74
Iron, mg	1,697	12,12/16,97*
Zinc, mg	1,344	8,96
Iodine, mcg	24,604	16,40
Vitamin C, mg	10,918	18,19
Riboflavin (B_2), mg	0,108	6,75
Vitamin B_6, mg	0,134	6,69
Vitamin B_{12}, mcg	0,592	59,2
Vitamin A, mcg	182,325	22,79
Thiamine (B_1), mg	0,074	5,32
Niacin, mg	2,303	12,79
Pantothenic acid (B_5), mg	0,392	6,53
Vitamin E, mg	0,708	7,08
Vitamin D, mcg	0,16	3,2

*the value of the recommended daily requirement for men in accordance with MR 2.3.1.0253–21 «Norms of physiological needs for energy and nutrients for various groups of the population of the Russian Federation».

To assess the feasibility of using electrochemically activated water [2, 4, 13] in the formulation of duck meat products, a study of its effect on the water-binding capacity (WBC) of minced meat was carried out. The experiments were carried out by the pressing method (Table 4).

The analysis of the obtained results showed that when using ERAW (at a pH of 11,2) in the minced meat recipe, there was a significant increase in the value of the WBC indicator (by 6% to the mass of minced meat and by 8% to the total moisture).

Table 4. Results of the study of WBC of meat and vegetable mince

Type of water used in the minced meat recipe	Mass fraction overall moisture,%	Moisture content in sample, mg	Square wet spot, cm^2	WBC to the mass of minced meat, %	WBC to total moisture, %
Drinking water (pH 7,0)	70,0	210,0	3,5	60,2	86,0
ERAW (pH 11,2)	70,2	210,6	1,5	66,0	94,0

On the basis of the conducted experiments, it was concluded that the use of electrochemically activated water in the formulation of minced meat products makes it possible to increase the WBC of minced meat by 8%, which may be associated with the previously established by us specific distribution of charges and electrostatic potential on the surface of protein molecules under the influence of catholyte [6, 7]. The established increase in the value of the WBC indicator can have a significant impact on the output of finished products.

The results of determining the losses during heat treatment of minced meat products (average data with three replicates) made using ERAW (experiment) and drinking (control) water are presented in Table 5.

Table 5. Results of determining losses during heat treatment of minced meat products

The name of the indicator	Experience		Control	
	g	%	g	%
Mass of minced meat product prepared for heat treatment	82,92	100,0	82,72	100,0
Weight of the finished product after heat treatment	77,36	93,3	73,92	89,4
Heat treatment losses	5,56	6,7	8,80	10,6
Weight of the finished product after cooling	76,25	92,0	72,85	88,1
Heat treatment losses, taking into account cooling losses	6,67	8,0	9,87	11,9

According to the results of the studies, it can be concluded that the use of ERAW in the formulation of developed food products from duck meat allows an increase in the yield of finished products (by reducing losses during heat treatment) by 4% compared to the use of drinking water.

Based on the research carried out, a formulation of minced meat products has been developed containing poultry meat, mechanically deboned chicken meat, carrots, white cabbage, sweet peppers, semolina, egg powder, wheat bran, bread crumbs, table salt, ground black pepper, ERAW. Duck meat was used as poultry meat in a ratio of 4:1 to mechanically deboned chicken meat, while as an additional source of animal protein and components exhibiting prophylactic properties, the «Mobi-Lux universal» food fortifier was included in the minced meat formulation and soybean oil in the following ratio of components, %: duck meat 31,5–32,0, mechanically deboned chicken meat 7,875–8,0, carrot 11,8–12,0, white cabbage 8,8–9,0, sweet pepper 8,6–8,8, semolina 5,0–5,2, «Mobi-Lux universal» 5,8–6,0, egg powder 2,0–2,2, soybean oil 1,8–2,0, wheat bran 2,3–2,5, bread crumbs 2,3–2,5, table salt 1,2, ground black pepper 0,1, catholyte to 100.

A technology for the production of minced duck meat products has been developed. In accordance with it, the meat is grinded on a top with a lattice diameter of 2–3 mm. All plant raw materials after preliminary processing are crushed on a top with a lattice diameter of 2–3 mm. Dry products are hydrated with catholyte. Mixing the ingredients according to the recipe is carried out in a minced meat mixer for 5–6 min. After molding, breading is carried out with a mixture of rusks with wheat bran. Formed minced meat products are produced frozen in a ready-to-eat form (after warming up). Freezing is carried out at a temperature of minus 18 °C to a temperature inside the product of minus 10 °C.

4 Discussion and Conclusion

As a result of the research, a formulation and highly effective technology of nutrient-balanced products with functional and prophylactic properties have been developed to strengthen the health of consumers and prevent the most common nutritional diseases. Purposeful combination of meat, vegetable and functional ingredients made it possible to obtain minced meat products for personalized nutrition of the following groups of consumers: those in need of prevention and treatment of osteoporosis, atherosclerosis, hypertension, diabetes mellitus, iodine deficiency, cardiovascular diseases; with a low level of hemoglobin in the blood; with a high volume of mental work and mental stress.

The developed minced meat products are distinguished by an increased yield, high organoleptic characteristics while reducing the cost of finished products in comparison with traditional recipes. Minced meat products can be produced both in the meat industry in frozen form and in the public catering network in the form of finished culinary products, which makes it possible to most fully cover various market segments in the field of personalized food.

References

1. Agalarova, E.G., Gunko, J.A., Kusakina, O.N., Cherednichenko, O.A., Dotdyeva, Z.S.: The national technological initiative FoodNet market: regional perspectives. In: Bogoviz, A.V. (eds.) The Challenge of Sustainability in Agricultural Systems. Lecture Notes in Networks and Systems, pp. 527–534, 206. Springer, Cham (2021)
2. Aider, M., Kastyuchik, A., Gnatko, E., Benaly, M., Plutakhin, G.: Electro-activated aqueous solutions: theory and application in the food industry and biotechnology. Innov. Food Sci. Emerg. Technol. **15**, 38–49 (2012)

3. Arroyo, P.E., Linan, J., Vera Martinez, J.: Who really values healthy food? British Food J. **123**(2), 720–738 (2021)
4. Bahir, V.M.: Electrochemical Activation: Inventions, Engineering, Technology. VIVA-STAR Publ, Moscow (2014)
5. Basic, M., et al.: Duck meat in human nutrition. Tehnologija mesa **56**, 50–57 (2015)
6. Borisenko, A.A., Bratsikhin, A.A., Khramtsov, A.G., Borisenko, A.A., Saricheva, L.A.: Impact of dispersion medium on functional properties of the proteins. Res. J. Pharm., Biol. Chem. Sci. **9**(1), 296–300 (2018)
7. Borisenko, A.A., Bratsikhin, A.A., Saricheva, L.A., Borisenko, A.A., Kostenko, K.V.: Forecasting of functional properties of the soya protein based on the molecular dynamics method. J. Hygienic Eng. Des. **20**, 82–85 (2017)
8. Eurasian Economic Commission: Technical Regulations of the Customs Union «Food products in terms of their labeling» (TR CU 022/2011), Moscow, Russia (2011)
9. Fehily, A., Pickering, J., Yarnell, J., Elwood, P.: Dietary indices of atherogenicity and thrombogenicity and ischaemic heart disease risk: The Caerphilly Prospective Study. Br. J. Nutr. **71**, 249–257 (1994)
10. German, B., Zivkovic, A., Dallas, D., Smilowitz, J.: Nutrigenomics and personalized diets: what will they mean for food? Annu. Rev. Food Sci. Technol. **2**, 97–123 (2011)
11. Gibney, M., Walsh, M., Goosens, J.: Personalized nutrition: paving the way to better population health. In: Eggersdorfer, M., Kraemer, M., Vordaro, J.B., et al. (eds.) Good Nutrition: Perspectives for the 21st Century, pp. 235–248. Karger Publishers (2016)
12. He, Y., et al.: Effects of dietary fiber on human health. Food Sci. Human Wellness **11**, 1–10 (2022)
13. Henry, M., Chambron, J.: Physico-Chemical, Biological and Therapeutic Characteristics of Electrolyzed Reduced Alkaline Water (ERAW) settings. Water **5**(4), 2094–2115 (2013)
14. Hidayat, A.W., Guruh, S.Z.: Model of the consumer switching behavior related to healthy food products. Sustainability **13**, 3555 (2021)
15. Jahan, K., Qadri, O.S., Younis, K.: Dietary fiber as a functional food. In: Ahmad, S., Al-Shabib, N.A. (eds.) Functional Food Products and Sustainable Health, pp. 155–167. Springer, Singapore (2020). https://doi.org/10.1007/978-981-15-4716-4_10
16. Kusakina, O., Kazarova, A., Kosinova, E., Agalarova, E., Aydinova, A.: Current state and development trends of the Foodnet market in Russia. IOP Conf. Ser. Earth Environ. Sci. **745**, 012005 (2021)
17. Ordovas, J.M., Ferguson, L.R., Tai, E.S., Mathers, J.C.: Personalised nutrition and health. *BMJ*, bmj.k2173 (2018)
18. Sujata, M., Kopal, S.: Functional foods as personalised nutrition: definitions and genomic insights. In: Functional Food and Human Health, pp. 513–535. Springer, Singapore (2018)
19. Wozniak, E., Banaszewska, D., Biesiada-rzazga, B.: Health-promoting properties of poultry meat in the context of diseases of civilization. Folia Pomeranae Universitatis Technologiae Stetinensis Agricultura, Alimentaria, Piscaria et Zootechnica **328**, 235–246 (2016)

Prospects for the Use of Chitosan-Stabilized Copper Nanoparticles as an Alternative to Antibiotics in Broiler Feed

Kristina Apryatina[1] (✉), Larisa Smirnova[1], Sergey Zaitsev[1], Vadim Frolov[2], and Georgy Laptev[3]

[1] National Research Lobachevsky State University of Nizhny Novgorod, Nizhny Novgorod, Russia
apryatina_kv@mail.ru
[2] LLC "Agrochitin", Nizhny Novgorod, Russia
[3] LLC "BIOTROF", St. Petersburg, Russia

Abstract. In this work stable complexes of chitosan-copper nanoparticles were obtained. The effect of the obtained complex of chitosan-copper nanoparticles and the feed antibiotic "Maxus" on the microflora of the gastrointestinal tract of 35-day-old Ross 308 meat broilers was compared using the NGS-sequencing method. It has been shown that the introduction of the complex into the broilers' main diet allows to completely exclude the antibiotic from the feed composition and to maintain the microflora balance at a high level.

Keywords: Chitosan · Copper nanoparticles · Broiler feed · Antibiotic

1 Introduction

Antibiotics are used in the livestock, meat and dairy industries not only to cure animal diseases but also to prevent them and to stimulate their growth. Broiler chickens are among the leading consumers of antibiotics in particular because of naturally developing microbial infections and possible microbial contamination of produced feed [1].

Overuse of antibiotics in farm animals has serious public health consequences. It contributes to the emergence of antibiotic-resistant bacteria and resistance genes not only in animals but also in humans [2–4]. This usually occurs through the consumption of food, but can also occur through direct contact with animals or through environmental objects. Ultimately, this can lead to infectious diseases in humans caused by antibiotic-resistant bacteria, which can be difficult or even impossible to cure.

Despite all the dangers of antibiotics, there is a need for their use and, therefore, effective alternatives that exclude the use of vaccines and drugs are needed.

The most expected and promising option at present is the use of nanoparticles of various metals with pronounced antimicrobial properties in doses not exceeding the authorized ones. Nanoparticles can be used as an alternative to antibiotics in poultry. To date, various types of nanoparticles have been investigated for use in poultry by feeding (nano-feeding), watering or other delivery routes to improve bird health.

V. Kurchenko et al. (Eds.): ICAETT 2021, LNNS 408, pp. 28–34, 2022.
https://doi.org/10.1007/978-3-030-96641-6_4

Copper is a vital trace element involved in various physiological and biochemical processes of living organisms. Copper (Cu) is an important mineral for chickens because it is required for proper physiological functions - among them hemoglobin synthesis, oxygen transport and participation in enzyme-coenzyme catalytic reactions. Copper is involved in the production of antibodies, white blood cells, and the synthesis of antioxidant enzymes [5–7], is an integral part of enzymes involved in iron metabolism, red blood cell synthesis, and immune function. It also plays a key role in the synthesis of connective tissues, such as collagen and elastin, and improves nervous system development by synthesizing dopamine [8]. In addition, copper affects the productive performance of animals because it acts as a growth enhancer [6].

The digestibility of copper salts is very low and approximately 80% of copper is excreted in feces, causing environmental pollution [9]. However, excess copper is also harmful and can negatively affect bird growth [10].

It was hypothesized that copper nanoparticles, due to their high physiological activity, can become growth stimulants and a promising alternative to antibacterial agents, and they can also be used in much lower doses due to their lower toxicity than the salt forms of copper in animal feed [11]. In addition to high bioavailability, such effects of copper nanoparticles as growth stimulating, antibacterial, antiviral, antifungal and immunomodulatory effects of copper nanoparticles have been proven [12, 13]. One of the most important advantages of copper nanoparticles is that they do not lead to bacterial resistance, which is the most important advantage over antibiotics in the treatment of various bacterial diseases in animals, for example, bacterial mastitis in cattle [14].

The problem of stabilization and penetration of copper nanoparticles from the intestinal cavity into the blood, their targeted delivery to various organs and prolonged release in the body remains urgent [15]. One of the promising ways to solve this problem is the use of natural biocompatible polymers as matrix carriers of nanoparticles to create a biologically active drug capable of controlled targeted delivery of nanoparticles in the body. Among them, chitosan can be distinguished due to its nontoxicity and polyfunctional biological properties [16, 17]. It is known that it can perform a transport and protective function for the delivery of nanoparticles and dietary supplements to the body during oral administration.

Thus, it can be assumed that the combination of the properties of copper and chitosan nanoparticles will make it possible to create highly effective antimicrobial drugs as an alternative to antibiotics and possessing adaptogenic, immunomodulatory and antioxidant properties of prolonged action for oral administration. The main goal of this work was to determine the possibility of replacing feed antibiotics in the diet of broiler chickens with chitosan complexes with copper nanoparticles.

2 Materials and Methods

We used chitosan ("Bioprogress", p. Biokombinat, Moscow region,) with a molecular weight 2.0×10^5 and the degree of deacetylation ~0.8, copper chloride (II) $CuCl_2 \times 2H_2O$ (Closed Joint Stock Company "Khimreaktiv"), glacial acetic acid (chemically pure grade 99.5%), pure grade ascorbic acid (Closed Joint Stock Company "Khimreaktiv").

Copper nanoparticles were obtained directly in all solutions of chitosan, which acts as a stabilizer. In all experiments, we used solutions of chitosan with a mass content of 3% in

1.2% acetic acid. The concentration of $CuCl_2$ (as a precursor of copper nanoparticles) in the chitosan solution is 0.04 mol × L^{-1}. Ascorbic acid (0.04 mol × L^{-1}) was introduced into the chitosan solution with continuous stirring, the solution was purged with argon, kept for 10 min, then a $CuCl_2$ solution was added dropwise, and the system was stirred for 0.5 h, then the solution was purged again with argon to remove them. oxygen. Copper nanoparticles were obtained under the action of microwave radiation at an installation power of 900 W for 10 min. The kinetics of the reduction of copper ions and the formation of nanoparticles were studied spectrophotometrically, observing the appearance and growth of plasmon resonance in the $\lambda \sim 590$ nm region, which is characteristic of copper nanoparticles. At the end of the process, the solution had a red-brick color.

The effect of the chitosan-copper nanoparticle complex on the gastrointestinal tract microflora of broiler chickens of the Ross 308 breed at 35 days of age was studied. The analysis of the bacterial community was carried out using the molecular genetic method NGS sequencing. These studies make it possible, in a short time, to determine the composition of the microflora of the digestive tract of birds as fully and accurately as possible, at the early stages to identify pathogens - causative agents of diseases, to evaluate the effect of feed additives (probiotics, acidifiers, feed enzymes, antibiotics) that affect the intestinal microflora.

NGS sequencing technique.

Amplification for subsequent NGS sequencing was performed using a DNA amplifier Verity («Life Technologies, Inc.», USA) using eubacterial primers (IDT), 343F (5'-CTCCTACGGRRSGCAGCAG-3') and 806R (5'-GGACTACNVGGGTWTCTAAT-3'), flanking the V1V3 region of the 16S rRNA gene. Metagenomic sequencing was performed on a MiSeq genomic sequencer («Illumina, Inc.», USA) with a MiSeq Reagent Kit v3 («Illumina, Inc.», USA). The maximum length of the obtained sequences was 2 × 300 nt. Chimeric sequences were excluded from the analysis using the «USEARCH 7.0» program (http://drive5.com/usearch/). The processing of the obtained 2 × 300 nt reads was carried out using a bioinformatics platform «CLC Bio GW 7.0» («Qiagen», Netherlands) and included overlapping, quality filtering (QV > 15), primer trimming. Determination of the taxonomic affiliation of microorganisms to genus was carried out using the RDP Classifier program. The error of the MiSeq instrument used for NGS sequencing was 5%.

Experimental groups:

Group 1 - control group, main diet + feed antibiotic "Maxus" - 100 g/t;
Group 2 - control group, main diet without of feed antibiotic;
Group 3 - experimental group, main diet + chitosan complex - 100 g/t;
Group 4 - experimental group, main diet + chitosan complex with copper nanoparticles - 100 g/t;
Group 5 - experimental group, main diet + chitosan complex (via the drinking water) - 1 ml/l of water;
Group 6 - experimental group, main diet + chitosan complex with copper nanoparticles (via the drinking water).

3 Results and Discussion

The present work introduces a one-pot microwave-assisted method of synthesis of copper nanoparticles in aqueous medium using ascorbic acid as reducing agent and chitosan as the stabilizer. The synthesis of copper nanoparticles has remained a problem due to their oxidation under ambient conditions, as a result, in order to prevent their oxidation, the solution was purged with argon.

The kinetics of the formation of copper nanoparticles in chitosan solutions was monitored spectrophotometrically by the appearance and increase in the intensity of absorption bands corresponding to the plasmon resonance of copper nanoparticles - λ ~ 590 nm. In Fig. 1 shows a typical absorption spectrum of copper nanoparticles.

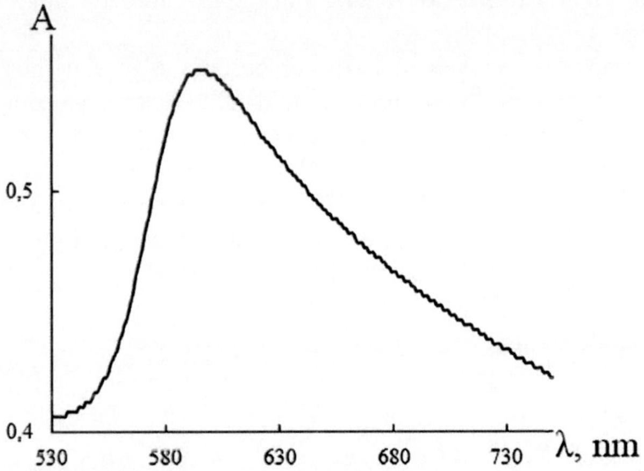

Fig. 1. Plasmon absorption band of copper nanoparticles obtained by microwave-assisted synthesis in an acetic acid solution of chitosan.

A comparison was made of the effect of the obtained complex of chitosan-copper nanoparticles and the feed antibiotic "Maxus" on the microflora of the gastrointestinal tract of broiler chickens using the NGS-sequencing method.

Normoflora.

Investigated the proportion of cellulolytic bacteria in the microflora sample. Cellulolytic bacteria are bacteria that break down the fiber in diets. Among the main representatives of these bacteria, the phylum Bacteroidetes, Thermoanaerobacteriales and the families Peptostreptococcacea, Clostridiaceae, Eubacteriaceae, Lachnospiraceae, Ruminococcaceae were identified. The proportion of cellulolytics was high in all samples and exceeded 80% when chitosan-copper nanoparticles were added to the drinking water of broilers, and also reached 80% in samples from the control group. The smallest amount of cellulolytic microorganisms was found in samples where broilers were fed the chitosan complex without copper nanoparticles. The group with the addition of copper nanoparticles to drinking water significantly exceeded the group in the main diet of which the feed antibiotic "Maxus" was introduced in the number of detected cellulolytic microorganisms (by 5% of all detected bacteria).

Thus, it can be concluded that copper nanoparticles have a positive effect on the vital activity of microorganisms of the cellulolytic group.

Bacteria are pathogen antagonists and immunomodulators.

Lactobacillus.

In all studied samples, lactobacilli were represented by the genera Lactobacillus sp., Lactococcus sp., Leuconostoc sp. and others related to Lactobacilliales.

The proportion of lactobacilli in all experimental groups of broilers was small. It was shown that the number of beneficial lactobacilli in the group with the feed antibiotic "Maxus" and the group of broilers, which were added a complex of chitosan-copper nanoparticles to drinking water, was similar (1.32% and 1.53%, respectively).

VFA-synthesizing bacteria (lactate-utilizing).

VFA-synthesizing bacteria (selenomonads) ferment lactic acid formed by bacteriodes and lactic acid bacteria to volatile fatty acids used by poultry in metabolic processes.

The largest amount of VFA-synthesizing microorganisms was found in the sample from the experimental group No. 4 - broilers, in the diet of which a complex of chitosan with copper nanoparticles (9.56%) was added. In the group of broilers in the diet of which the antibiotic was introduced, the amount of VFA-synthesizing bacteria was 3.6%.

Consequently, the complex of chitosan-copper nanoparticles promotes the development of lactate-utilizing microorganisms, which contributes to the vital activity of broilers.

Pathogens.

As can be seen from Fig. 2, pathogenic species were detected in significant quantities, depending on the samples studied.

The following types of bacteria were identified that can be participants in pathogenic processes:

- bacteria of the genus Helicobacter (up to 4%) - causative agents of disorders of the gastrointestinal tract of birds and causative agents of gastroenteritis.
- bacteria of the genus Mycoplasma (up to 0.15%) - pathogens of chicken mycoplasmosis, affecting the gastrointestinal tract and respiratory system;
- bacteria of the Clostridiaceae family (up to 0.55%) - causative agents of anaerobic diseases and frequent participants in pathogenic processes involving toxins produced by microorganisms;
- bacteria of the Enterococcus family (up to 0.1%) - pantropic pathogens of inflammatory processes in the body of a bird, affecting the gastrointestinal tract and the musculoskeletal system.
- other types.

It can be seen from the studies that the number of pathogenic microorganisms in samples from most groups of birds did not exceed 2% (Fig. 2). An exception is the group with the feed antibiotic "Maxus" in the main diet, where the largest number of pathogenic microorganisms was found - 4.3%. As seen from Fig. 2, the overwhelming number of pathogens in this group are microorganisms of the genus Helicobacter, in particular, Helicobacter pullorum (4% of all microorganisms). The lowest relative content of pathogens was observed in samples from birds from the group whose diet included a complex of chitosan-copper nanoparticles - 0.32% of all detected bacteria. In the group with the addition of copper nanoparticles to the main broiler diet, the number of pathogenic microorganisms detected is 1.88%, which is more than two times less than in the group where the feed antibiotic "Maxus" was used.

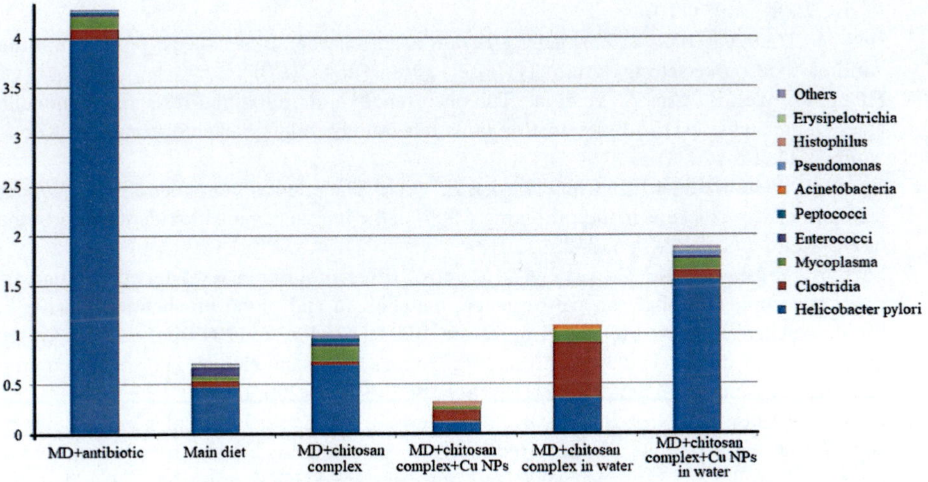

Fig. 2. The relative abundance of pathogens in the samples, %.

4 Conclusion

We found that the balance of microflora in birds of all samples is maintained at the proper level, however, a certain pathogenic load is present in the intestines of the studied birds. From the point of view of the microflora balance, the intestines of chickens are in the best condition in groups with the introduction of copper nanoparticles into the composition of feed.

In general, it can be noted that the chitosan-copper nanoparticle complex is a valuable feed additive for broiler chickens. Such a complex can be introduced both in dry form in compound feed, and in dissolved form in drinking water. The introduction of the complex into the broiler diet allows to completely eliminate the antibiotic from the feed, which means to reduce the burden on public health and reduce the appearance of antibiotic-resistant bacteria and resistance genes.

Acknowledgments. This work has been performed with the support of Research and Educational Center of Nizhny Novgorod region in the framework of the agreement No 16–11-2021/48 and financial support via State assignment in the Research scientific laboratory of "Chemistry of natural products and their synthetic analogues" of Scientific Educational Centre "Technoplatform 2035".

References

1. Mehdi, Y., Létourneau-Montminy, M.-P., Gaucher, M.-L.: Use of antibiotics in broiler production: global impacts and alternatives. Anim. Nutr. **4**(2), 170–178 (2018)
2. Burow, E., Grobbel, M., Tenhagen, B.-A., et al.: Antibiotic resistance in escherichia coli from broiler chickens after amoxicillin treatment in an experimental environment. Microb. Drug Resist. **26**(9), 1098–1107
3. Hedman, H.D., Vasco, K.A., Zhang, L.: A review of antimicrobial resistance in poultry farming within low-resource settings. Animals (Basel) **10**(8), 1264 (2020)
4. Shang, K., Wei, B., Cha, S.-Y., et al.: The occurrence of antimicrobial-resistant salmonella enterica in hatcheries and dissemination in an integrated broiler chicken operation in Korea. Animals **11**, 154 (2021)
5. Ognik, K., Sembratowicz, I., Cholewińska, E., et al.: The effect of administration of copper nanoparticles to chickens in their drinking water on the immune and antioxidant status of the blood. Anim. Sci. J. **89**(3), 579–588 (2018)
6. Scott, A., Vadalasetty, K.P., Lukasiewicz, M., et al.: Effect of different level of copper nanoparticles and copper sulphate on performance, metabolism and blood biochemical profile in broiler chicken. J. Anim. Physiol. Anim. Nutr. **102**(1), e364–e373 (2017)
7. Sharma, M., Joshi, C., Pathak, N., Kaur, H.: Copper status and enzyme, hormone, vitamin and immune function in heifers. Res. Vet. Sci. **79**, 113–123 (2005)
8. Mroczek-sosnowska, N., Batorska, M., Lukasiewicz, M., et al.: Effect of nanoparticles of copper and copper sulfate administered in OVO on hematological and biochemical blood markers of broiler chickens. Ann. Warsaw Univ. Life Sci. SGGW. Anim. Sci. **52**, 141–149
9. Leeson, S.: Copper metabolism and dietary needs. World's Poultry Sci. J. **65**, 353–366 (2009)
10. Sharif, M., Rahman, M., Ahmed, B., Abbas, R.Z., Hassan, F.: Copper nanoparticles as growth promoter, antioxidant and anti-bacterial agents in poultry nutrition: prospects and future implications. Biol. Trace Elem. Res. **199**(10), 3825–3836 (2020). https://doi.org/10.1007/s12011-020-02485-1
11. Boyles, M.S.P., Ranninger, C., Reischl, R., et al.: Copper oxide nanoparticle toxicity profiling using untargeted metabolomics. Partic. Fibre Toxicol. **13**, 49 (2016)
12. El-Kassas, S., El-Naggar, K., Abdo, S.E., et al.: Dietary supplementation with copper oxide nanoparticles ameliorates chronic heat stress in broiler chickens. Anim. Product. Sci. **60**(2), 254–268 (2019)
13. Sizova, E., Miroshnikov, S., Lebedev, S., Usha, B., Shabunind, S.: Use of nanoscale metals in poultry diet as a mineral feed additive. Anim. Nutr. **6**(2), 185–191 (2020)
14. Kalińska, A., Jaworski, S., Wierzbicki, M., Gołębiewski, M.: Silver and copper nanoparticles—an alternative in future mastitis treatment and prevention? Int. J. Mol. Sci. **20**(7), 1672 (2019)
15. Scott, A., Vadalasetty, K.P., Chwalibog, A., Sawosz, E.: Copper nanoparticles as an alternative feed additive in poultry diet: a review. Nanotechnol. Rev. **7**(1), 69–93 (2018)
16. Mohammed, A.S.A., Naveed, M., Jost, N.: Polysaccharides; classification, chemical properties, and future perspective applications in fields of pharmacology and biological medicine (a review of current applications and upcoming potentialities). J. Polym. Environ. **29**(8), 2359–2371 (2021). https://doi.org/10.1007/s10924-021-02052-2
17. Lyu, F., Xu, X., Zhang, L.: Natural polysaccharides with different conformations: extraction, structure and anti-tumor activity. J. Mater. Chem. B **8**, 9652–9667 (2020)

Proteolysis of Bovine Whey, Milk and Colostrum with Serine Endopeptidases

Tatsiana Halavach(✉)

Belarusian State University, Minsk, Belarus
halavachtn@gmail.com

Abstract. The objective of this study is to characterize physical and chemical properties and antioxidant activity of milk peptide fractions, resulting from partial hydrolysis with serine endopeptidases (alcalase and trypsin). Protein and peptide profile of the cleaved milk proteins was determined according to electrophoretis analysis in polyacrylamide gel and chromato-mass-spectrometry. Fluorometric method was used to determine antioxidant activity of hydrolysates and native dairy substrates. Comparative analysis of whey, milk and colostrum hydrolysates obtained with alcalase and trypsin was carried out. Based on experimental study, alcalase was chosen to produce colostrum hydrolysates with average and extensive degree of hydrolysis, and high antioxidant effect. New data on the features of enzymatic cleavage of whey and casein protein fractions from whey, milk and colostrum, and the level of their antioxidant activity were obtained. Milk protein partial hydrolysates with low antigenicity and high antioxidant potential are prospective components of specialized foodstuffs (infant, dietary and sport formulas).

Keywords: Whey · Milk · Colostrum · Alcalase · Trypsin · Proteolysis · Antioxidants

1 Introduction

Proteolysis, or the cleavage of peptide bonds with specific enzymes (proteases), permits the development of protein hydrolysates with improved nutritional and bioactive characteristics [13, 27, 32]. Serine proteases, characterized by the presence of a serine residue in the catalytic center, belong to the most common class of peptidases with industrial importance [1, 25, 32]. Alcalase and trypsin, which are serine proteases of bacterial and animal origin, respectively, possess activity under neutral and alkaline conditions and exhibit wide substrate specificity. Trypsin predominantly cleaves peptide bonds in the carboxyl site of amino acids Arg and Lys, while alcalase has broad site-specificity (preferably a large uncharged residues) [4, 27, 28].

Milk proteins are potential source of bioactive peptides possessing antihypertensive, antioxidant, antithrombotic, opioid, and immunomodulating action. However, whey and casein fractions possess high allergenic potential, which is important in the development of safe milk products for infant food [2, 6, 31]. Biologically active properties of

V. Kurchenko et al. (Eds.): ICAETT 2021, LNNS 408, pp. 35–45, 2022.
https://doi.org/10.1007/978-3-030-96641-6_5

cleaved protein are determined by hydrolysis degree and enzyme used [3, 5]. It should be noted that careful choice of protease and filtration techniques can significantly reduce protein allergenicity. The molecular mass cut-off value of the filters must be in range 3–5 kDa to remove allergenicity of whey protein hydrolysates [30]. The previous study showed an increase in the antioxidant activity and the reduced antigenicity of whey and colostrum proteins with an enhancement in the degree of their proteolysis. The colostrum hydrolysates obtained with alcalase had a greater antioxidant potential than neutrase [7, 8]. In accordance with previous experimental [8] and literature data [16], temperature 50–60 °C and pH 8 are optimal for obtaining tryptic and alcalase hydrolysates of whey proteins with active acidity preferred for food components (pH 6.5–7.0 at the end of enzymatic process). The optimal technological process usually include the minimum number of stage, however ultrafiltration is obviously a necessary step to preparing safe hypoallergenic dairy hydrolysates [5, 6].

The aim of this work is a comparative study of partial hydrolysates of milk proteins obtained using serine proteases (alcalase and trypsin), characterization of their protein and peptide composition and antioxidant activity.

2 Materials and Methods

Concentrate of whey proteins (specifications BY 100377914.550–2008), dry defatted milk (STB 1858–2009), dry defatted colostrum (VNIMI, Moscow, Russia), alcalase (EC 3.4.21.62, Alcalase® 2.4L, activity 2.64 U/g; Sigma, USA), and trypsin (EC 3.4.21.4, $\geq$6.0 BAEE U/g; Sigma, USA) were applied in proteolysis reaction. Hydrolysis of milk proteins was carried out at enzyme/substrate ratio 0.25–8.0%, pH 7.4/8.0, temperature 50 °C during 2–3 h. Filters Spin X–UF Concentrator 20 (cut-off value 5/10 kDa; Corning, UK) were used in samples fractionation.

Protein and peptide profile of obtained hydrolysates was defined by native and denaturing (with sodium dodecyl sulfate, SDS) electrophoresis in polyacrylamide gel [18]. Processing of gels was conducted using gel-documentation system Image Master VDS-SL and ImageMaster ID Software 4.20 (Amersham Bioscience, USA). Protein amount in experimental samples was determined according to calibration β-lactoglobulin (β-lg), α-lactalbumin (α-la), and bovine serum albumin (BSA) plots. The depth of proteolysis (%) was determined as the ratio of hydrolysed protein content in test sample to total protein amount in control sample. Total (TN) and α-amino nitrogen (AN) content in experimental samples were defined according to ISO 8968–1: 2014 [14] and ISO 11402: 2004 [20], respectively, while solids ratio – in accordance with ISO 6731: 2010 [15]. Chromato–MS system Agilent 1290 (Agilent, USA) with mass spectrometric detector Q–TOF 6550 (ESI^+ mode) was used to obtain mass spectra, spectral recording range was 100–3200 m/z. Antioxidant activity (AOA) was evaluated with fluorimetric method according to the technique developed by E.I. Tarun (2014) [26], and $ABTS^{\bullet+}$ method as described by B. Hernández-Ledesma et al. (2007) [10]. The results of independent experiments were presented as the mean value ± confidence interval ($p \leq 0.05$, $n = 3$).

3 Results

This work presents variants for obtaining hydrolysates of whey proteins with average degree of hydrolysis using serine proteases. Enzymatic cleavage of 2% whey solution with trypsin/alcalase was performed at protease amount 0.02%, pH 8, and temperature 50 C for 2 h. Obtained samples were analyzed by electrophoresis.

The protein and peptide profile of tryptic hydrolysates was studied depending on the reaction duration (15–120 min). According to native electrophoresis after 60 min of proteolysis, β-lg and α-la were not detected in the samples, whereas bovine serum albumin (BSA) was retained throughout the enzymatic process (Table 1). The SDS-electrophoregrams of whey samples hydrolyzed for 120 min reveal trace amounts of native whey proteins, except for BSA, whose content corresponds to that of the original level (Fig. 1A, lane 6).

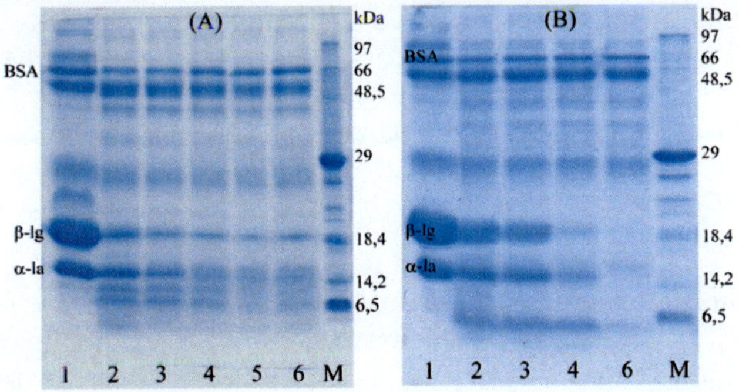

1 – hydrolysis 0 min, 2 –15 min, 3 – 30 min, 4 – 60 min, 5 – 90 min, 6 – 120 min, M – marker

Fig. 1. SDS-electrophoregrams of whey protein hydrolysates with trypsin (**A**) and alcalase (**B**)

Moreover, under these conditions trypsin most effectively cleaves β-lg, than α-la. Analysis of the peptide fraction of whey hydrolysate with trypsin showed the formation of intermediate peptides with 6.5 ≤ molecular range (mr) < 14.2 kDa, whose quantity decreased throughout the enzymatic process (Fig. 1A, lanes 2–6).

In the case of alcalase, proteolysis was performed under similar conditions. It was shown that within 90 min almost all β-lg and α-la were hydrolyzed (Table 1), and after 120 min of the enzymatic reaction only trace amounts of the predominant whey proteins and native BSA were detected (Fig. 1B, lane 6). Analysis of the peptide composition of whey hydrolysate with alcalase showed the formation of a discrete peptide fraction with mr < 6.5 kDa, the amount of which decreases significantly from 90 to 120 min of proteolysis (Fig. 1, lanes 2–6). Under the experimental terms, alcalase quite effectively cleaves β-lg and α-la, except for BSA.

According to electrophoretic analysis, the serine endopeptidases studied efficiently hydrolyze β-lg, which is the main milk allergen, and α-la. Different hydrolysis intermediates were detected in the whey samples obtained using alcalase and trypsin (Fig. 1A, lane 6; Fig. 1B, lane 6). However, the hydrolysis degree (AN/TN) with trypsin (18.2 ± 0.7) and alcalase (17.8 ± 1.1)% was comparable.

Table 1. Characteristics of whey protein hydrolysates obtained using trypsin and alcalase

Hydrolysis duration, min	Amount of cleaved whey proteins[a], %			
	Tryptic hydrolysis		Alcalase hydrolysis	
	β-lg	α-la	β-lg	α-la
15	93.6 ± 3,8	72.6 ± 5.8	67.6 ± 4.0	47.9 ± 5.4
30	100	94.6 ± 3.8	91.8 ± 3.8	63.2 ± 5.7
60	100	100	100	93.7 ± 4.3
90–120	100	100	100	100
AN/TN, %	18.2 ± 0.7		17.8 ± 1.1	
Fraction with mr ≤ 10 kDa, %	92.4 ± 1.7		93.0 ± 0.7	
Residual AG[b], 10^{-3} RU	3.7 ± 0.2		4.9 ± 0.1	
$IC_{50(W)}/IC_{50(WH)}$ [b, c]	3.6 ± 0.1		5.0 ± 0.4	
Cleavage sites (P_1)	Arg and Lys [27, 28]		Phe, Trp, Tyr, Glu, Met, Leu, Ala, Ser, and Lys [4, 27, 28]	

P_1 – carboxyl site of the cleaved peptide bond, [a]data on the amount of cleaved proteins were obtained according to native electrophoresis, [b]data for filtrate samples [8], [c]indicator reflects how many times the AOA of whey hydrolysates (WH) exceeds the AOA of the initial whey (W).

In the case of BSA hydrolysis, the use of these serine proteases is inefficient. Consequently, there was a necessity for an additional stage of BSA removal from hydrolysates using filters with a permeability ≤10 kDa. Ultrafiltrates of whey hydrolysates with trypsin and alcalase were obtained and analyzed. According to competitive ELISA, a decrease in residual antigenicity of hydrolysate samples as well as an increase in their antioxidant potential in the $ABTS^{\bullet +}$–Trolox test system were found (Table 1). In general, ultrafiltrates of whey hydrolysates (mr ≤ 10 kDa) represent a protein component with low antigenic potential and high AOA.

The features of milk proteins (casein and whey fractions) hydrolysis with alcalase and trypsin at enzyme/substrate ratio of 0.25–8.0%, pH 8, temperature 50 °C for 2 h were determined at the next stage. It was found that alcalase and trypsin effectively cleaved the casein fraction. Thus, native casein was not detected at an enzyme/substrate ratio of 0.25% in trypsin hydrolysate and 1% in alcalase sample (Fig. 2). The whey fraction was actively treated with alcalase and to a lesser extent with trypsin. Partial proteolysis products were detected in all trypsin hydrolysate samples (Fig. 2A), whereas protein compounds with mr < 6.5 kDa were not detected on the electrophoregram at alcalase concentration ≥4% (Fig. 2B).

The proteolysis depth of milk proteins depending on the enzymatic process duration with alcalase was studied. Hydrolysis was carried out at defatted milk concentration 5%, enzyme/substrate ratio 1%, pH 8, and temperature 50 °C for 2 h. During the first 60 min of the enzymatic reaction, the casein fraction cleaved into intermediate peptides, whereas the whey fraction was almost completely hydrolyzed after 2 h (Fig. 2C). In addition, cleavage of the peptide fraction with increasing proteolysis duration up to 120 min, as well as an increase in the amount of α-amino nitrogen from 31.5 to 56 mg% were showed for alcalase hydrolysates.

The high site-specificity of trypsin (Arg and Lys) which accounts for the formation of intermediate high-molecular-mass hydrolysis products with mr > 6.5 kDa (Table 1, Fig. 1A) should be noted. At the same time, the broad site-specificity of alcalase determines the cleavage of substrates into multiple peptides with the formation of discrete peptide fractions, which undergo further cleavage with increasing duration of hydrolysis (Table 1; Fig. 1B, 2B) and endopeptidase concentration (Fig. 2C). Due to the high proteolytic activity of the commercial enzyme preparation alcalase, its broad substrate and site-specificity, stability during storage, and affordability, this endopeptidase was chosen to obtain partial hydrolysates of whey proteins with average as well as extensive hydrolysis degree.

A way of producing hydrolysate of defatted colostrum with alcalase was proposed. Enzymatic protein cleavage was carried out at protein substrate concentration 3% and enzyme/substrate ratio 1%, temperature 50 °C, pH 8 for 3 h. Filters with separation capacity 10 kDa were used for fractionation of hydrolysates.

The electrophoregram shown in Fig. 2D reflects the typical composition of first milk. For samples of native colostrum, a relatively high content of immunoglobulins (Igs), the amount of casein comparable with milk, the presence of α-la, β-lg and other proteins of the whey fraction (BSA and lactoferrin) were shown.

According to SDS-electrophoregram (Fig. 1A), almost complete proteolysis of β-lg, α-la, and minor proteins into intermediate peptides was established in the whey hydrolysate. Numerous products of partial immunoglobulin proteolysis were detected in colostrum hydrolysate, and cleavage of casein fraction, β-lg, α-la, and minor whey proteins was established (Fig. 2D). According to total protein determination, unfiltered colostrum hydrolysate contains (29.1 ± 1.2)% of the fraction with mr ≤ 10 kDa. Thus, due to the content of partial proteolysis products of the Igs fraction with mr > 10 kDa, cleaved colostrum contains a smaller amount of low-molecular-mass component than whey hydrolysate.

As part of the optimization of obtaining enzymatic whey and colostrum hydrolysates with extensive cleavage degree, a higher enzyme/substrate ratio 5% as well as filters with a separating capacity 5 kDa were used to increase the content of the low-molecular-mass peptide fraction. Thus, according to an earlier experiment related to the defatted milk hydrolysate preparation, intermediate proteolysis products were not detected at enzyme/substrate ratio ≥4% (Fig. 2C).

During the process of preparing extensive colostrum hydrolysates enzymatic cleavage was carried out at enzyme/substrate ratio 5%, temperature 50 °C, pH 7.4 for 3 h, the obtained samples were subjected to filtration with permeability 5 kDa. The content of low-molecular-mass fraction (mr ≤ 5 kDa) in unfiltered extensive hydrolysates of whey and colostrum was found to reach (39.6 ± 0.5) and (30.3 ± 0.2)%, respectively. The obtained data indicate an increase in the cleaved component proportion in whey hydrolysate by 1.3 times in compare to hydrolysed colostrum.

According to chromato-mass spectrometry, differences in the peptide fraction were found in the range 100–1500 Da. High signal level was established for single-charged ions with m/z values 680–900 Da, which is proportional to whey peptides of 6–8 amino

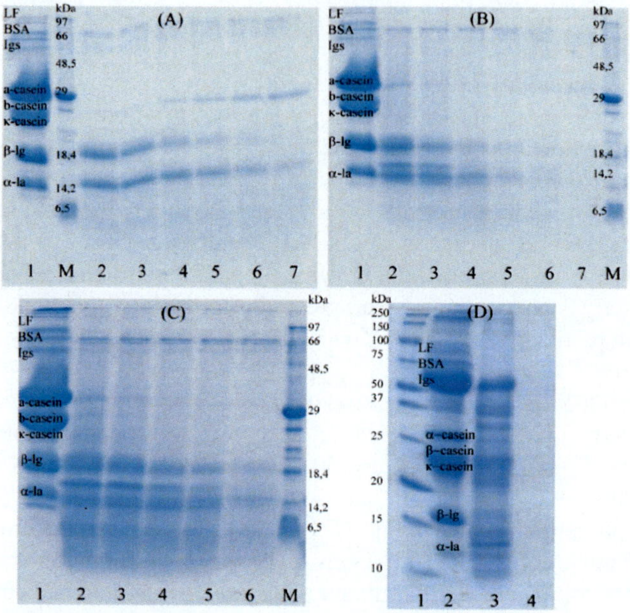

A and **B**: M – marker, 1 – milk (control), 2 – enzyme amount 0.25 %, 3 – 0.5 %, 4 – 1 %, 5 – 2 %, 6 – 4 %, 7 – 8 %; **C**: 1 – hydrolysis 0 min, 2 – 15 min, 3 – 30 min, 4 – 60 min, 5– 90 min, 6 – 120 min, M – marker; **D**: 1 – marker, 2 – colostrum, 3 – colostrum hydrolysate, 4 – filtrate

Fig. 2. Electrophoregrams of defatted milk hydrolysates with trypsin (**A**) and alcalase (**B** and **C**), hydrolysate of defatted colostrum with alcalase (**D**)

acid residues length. The maximum signal per the mass spectra was established at m/z values 560–650 (5–6 amino acid residues). Thus, peptides with a lower molecular mass are characteristic of the colostrum sample. The obtained data are due to differences in the composition of the protein component of whey and colostrum, as well as to the peculiarities of substrate and site-specific of alcalase.

Comparative analysis of the antioxidant activity of whey and colostrum hydrolysates (ultrafiltrates 5 and 10 kDa) by ORAC and $ABTS^{\bullet+}$ methods is shown in Fig. 3. A decrease in IC_{50} (sample concentration corresponding to 50% inhibition of reactive oxygen species) indicates improving AOA of the tested compound. Calculations were made according to protein and solids content in the samples.

The differences in AOA level of whey and colostrum are caused by the peculiarities of protein-peptide composition (the ratio of casein and whey proteins) and the content of non-protein component. Thus, the amount of protein in whey reaches 80%, while defatted colostrum – 70%. In addition, colostrum is enriched with non-protein component (vitamins, minerals) with antioxidant potential.

Significant AOA increase in comparison with native substrates was found for samples of hydrolyzed whey and colostrum. The maximum antiradical effect (per protein content) was shown in experiments with filtrates of cleaved colostrum. In general, samples of hydrolysed whey and colostrum with different degree of hydrolysis were obtained, possessing a confirmed antioxidant potential.

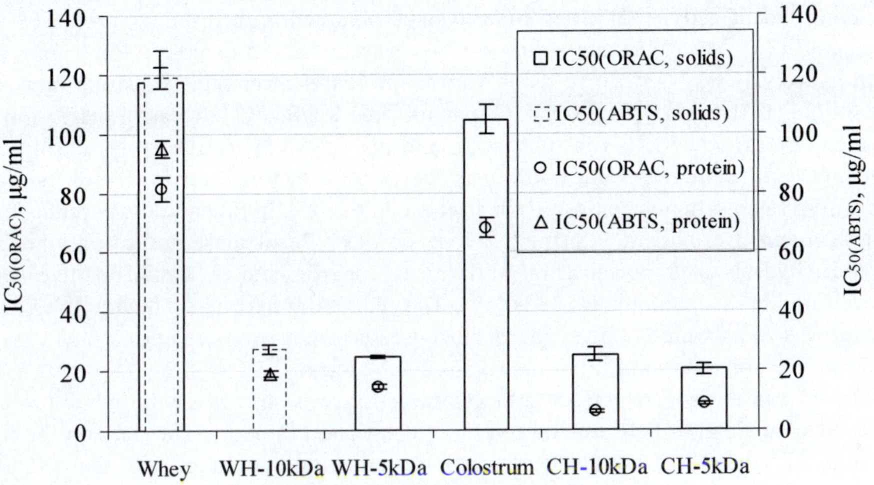

Fig. 3. IC_{50} determination by ORAC and $ABTS^{\bullet+}$ methods (WH – whey hydrolysate, WH–5/10 kDa – 5/10 kDa WH filtrates; CH – colostrum hydrolysate, CH–5/10 kDa – 5/10 kDa CH filtrates)

4 Discussion

According to the literature [4, 16, 19] and own research [7–9], the choice of the suitable enzyme provides a specific peptide profile and the depth of hydrolysis of protein substrates. An increase in hydrolysis degree of milk proteins leads to the raise in their antioxidant activity and a decrease in residual antigenicity [17, 22, 29, 33, 34].

Extensive hydrolysate of whey proteins was characterized by Doucet et al. (2003) [4]. The 20% whey solution (pH 8.0) was hydrolyzed with alcalase at temperature 45 °C and enzyme/protein ratio 1:10 for 5 h. High proteolysis of whey isolate with alcalase caused gelation mainly through hydrophobic interactions. Aggregates were formed by low-molecular-mass peptides (<2 kDa) the share of which reached 80%. 130 peptides were identified in accordance with reversed-phase high-performance liquid chromatography (HPLC) and mass spectrometry [4].

M.V.T. Mota et al. (2004) [16] analyzed the hydrolysis products of whey proteins with alcalase, trypsin and pepsin obtained at different pH and temperature values (37 °C and pH 7–8–9, pH 8 and 30–37–50 °C). Proteolysis of β-lg and α-la to intermediate peptides was retrieved in all experimental samples cleaved using alcalase. 12 major peptide fractions were revealed on HPLC profiles of alcalase hydrolysate, while 3 and 9 fractions for pepsin and trypsin, respectively [16].

According to studies [19, 34], the 5% whey solution was cleaved at 50 °C, pH 5.5/7.0/8.0 for papain/neutrase/alcalase during 4–5 h, the enzyme ratio was 3% (relative to protein amount). The amino acid content of papain-treated hydrolysates was higher than that of alcalase, whereas free cysteine and proline were not detected after alcalase hydrolysis [19]. It was shown that cleavage with alcalase could reduce the antigenicity of β-lg and α-la effectively. Temperature and pH had the greatest effect on anti-α-la and anti-β-lg IgG binding inhibition, respectively [34].

Antioxidant activity of whey hydrolysates obtained with different proteases was evaluated [33]. Whey protein solution (3%) was preheated at 90 °C for 5 min, and then hydrolysis was performed using various proteases at enzyme/substrate ratio 2%, pH 8.0/50 °C (trypsin), pH 7.0/55 °C (papain), pH 8.5/65 °C (alcalase), pH 7.0/50 °C (protamex), pH 7.0/50 °C (flavourzyme), and pH 7.0/55 °C (protease N) during 4 h. The pretreatment increased the hydrolysis degree of whey protein with all enzymes, and alcalase-treated samples possessed the highest AOA [33]. In other study, peptides from whey proteins were liberated using neutrase, corolase PP, alcalase or flavourzyme. The obtained hydrolysates showed a high antiradical properties and a potential positive action on cultured human endothelial cells [17]. The maximum hydrolysis degree (63%) and the highest AOA were obtained for alcalase-treated whey as compared to flavourzyme and enzyme combination [22].

In the present study, greater antioxidant action was shown for the whey proteins cleaved with alcalase than for the tryptic hydrolysate (Table 1). The data on the high antiradical potential of the alcalase-treated samples are consistent with the results of several literature sources [17, 22, 29, 33]. The intensive hydrolysis of the casein and whey fractions using alcalase with the formation of numerous low-molecular-mass peptides presented in electrophoregrams (Fig. 1 and 2) should be noted, which explains the low antigenic potential of the proteolysis products (Table 1). The broad substrate and site-specificity of alcalase in the cleavage of milk proteins were uncovered in a number of scientific studies [4, 16, 19, 34].

There is a known method of whey protein hydrolysate production with average degree of hydrolysis [11], which includes preparation of 5% solution of whey protein (pH 7.9–8.3), hydrolysis with pancreatine at enzyme/substrate ratio 1.5–2.5%, temperature 48–54 °C for 3.0–3.5 h, and ultrafiltration with membrane permeability 20 kDa. The method is aimed at obtaining whey protein hydrolysate with high biological value, improved organoleptic properties and reduced allergenic potential.

A new method of obtaining partial hydrolysate of whey and colostrum proteins which includes using alcalase at enzyme/substrate 1%, hydrolysis at pH 7.4/8.0, temperature 50 °C for 2–3 h, and subsequent ultrafiltration with separation of the non-hydrolysed fraction with mr > 10 kDa was proposed. The resulting hydrolysate has reduced antigenicity and high antioxidant potential.

The method [23] also involves the use of alcalase, but in a larger amount (10% of protein mass). However, the application of 20 kDa filters [11, 23] does not allow the separation of residual non-cleaved whey proteins (β-lg and α-la with mr 18.4 and 14.2 kDa, respectively). It should be note the method [21], which includes hydrolysis with protamex or alcalase (4% by whey solids) at pH 7.5–8.0, temperature 40–55 °C for 4 h, and subsequent use of the membranes with discharge capacity 10 kDa.

Methods for obtaining hydrolysates of whey proteins with a high degree of hydrolysis were proposed [12, 24]. According to the method [12], proteolysis is carried out using the enzyme preparation flavourzyme at enzyme/substrate ratio 5% for 20 h. Then the resulting crude hydrolysate is ultrafiltrated on membranes with molecular masses of 5 kDa and 2 kDa. The final hydrolysate is lactose-free with more than 75% protein concentrated on low-molecular-mass fraction (0.5–2.0 kDa).

Other method of extensive whey hydrolysate production with low residual antigenicity [24] includes introduction of enzyme composition of pancreatin 1.5–2.0%, protamex 0.5–1.0% and flavourzyme 2.0–2.5%, performing hydrolysis at 45–48 °C during 18–24 h till amine nitrogen content 500–600 mg%. The resulting hydrolysate is low-lactose, and more than 80% of the protein fraction is presented by low-molecular-mass peptides and free amino acids (<2.0 kDa).

Preparation of extensive whey and colostrum hydrolysates proposed in present paper provides the use of highly active protease alcalase at enzyme/substrate ratio 5%, hydrolysis at pH 7.4/8.0, temperature 50 °C for 2–3 h, and subsequent ultrafiltration for separation of the fraction with mr > 5 kDa. The method makes it possible to obtain a hypoallergenic protein component with high antioxidant activity.

5 Conclusion

In this work, the use of serine proteases (alcalase and trypsin) for obtaining partial hydrolysates of whey and milk with average degree of hydrolysis was proposed, as well as the necessity of ultrafiltration with separation of fractions over 10 kDa was proved. At the same time, hydrolysis of colostrum is advisable to be performed with alcalase possessing broad site-specificity, which is directed to increase the amount of the cleaved whey fraction. Obtaining extensive hydrolysates of whey and colostrum involves the use of alcalase under optimized conditions, followed by ultrafiltration with separation of the fraction with a molecular mass over 5 kDa. Alcalase hydrolysis produces a wide range of low-molecular-mass peptides, while ultrafiltration removes the non-cleaved protein fraction. Experimental samples of whey and colostrum hydrolysates have reduced allergenic potential and confirmed antioxidant activity. In general, the work presents a comparative study of partial hydrolysates of dairy proteins (whey, milk and colostrum) obtained using alcalase and trypsin, characterization of their protein and peptide profile and antiradical properties.

The practical significance of the work consists in obtaining partial hydrolysates of milk proteins with different degree of hydrolysis, which can be used as a component with high nutritional and biological value to create new specialized products. The prospect of further research is connected with the characterization of bioactivities spectrum of partial whey and colostrum hydrolysates.

References

1. Chandrasekaran, M. (ed.): Enzymes in Food and Beverage Processing (S. I, II). Taylor & Francis Group, London, UK (2016)
2. Chourasia, R., Phukon, L.C., Singh, S.P., Rai, A.K., Sahoo, D.: Role of enzymatic bioprocesses for the production of functional food and nutraceuticals. Biomass, Biofuels, Biochemicals, pp. 309–334. Elsevier B.V, Amsterdam, Netherlands (2020)
3. Dos Santos Aguilar, J.G., Sato, H.H.: Microbial proteases: production and application in obtaining protein hydrolysates. Food Res. Int. **103**, 253–262 (2018)
4. Doucet, D., Otter, D.E., Gauthier, S.F., Foegeding, E.A.: Enzyme-induced gelation of extensively hydrolyzed whey proteins by alcalase: peptide identification and determination of enzyme specificity. J. Agric. Food Chem. **51**(21), 6300–6308 (2003)

5. Du Toit, G., et al.: Identifying and managing cow's milk protein allergy. Arch. Dis. Childhood – Educ. Pract. **95**(5), 134–144 (2010)
6. El-Agamy, E.I.: The challenge of cow milk protein allergy. Small Rumin. Res. **68**(1–2), 64–72 (2007)
7. Halavach, T.M., Dudchik, N.V., Tarun, E.I., Zhygankov, V.G., Kurchenko, V.P., Romanovich, R.V., et al.: Biologically active properties of hydrolysed and fermented milk proteins. J. Microbiol. Biotechnol. Food Sci. **9**(4), 714–720 (2020)
8. Halavach, T.M., Kurchenko, V.P.: Milk protein hydrolysis with enzyme preparation and proteolytic systems of lactic acid bacteria. Proc. BSU **7**(1–2), 106–126 (2012). (in Russian)
9. Halavach, T.N., Kurchenko, V.P., Zhygankov, V.G., Evdokimov, I.A.: Determination of physicochemical, immunochemical and antioxidant properties, toxicological and hygienic assessment of whey protein concentrate and its hydrolysate. Foods Raw Mater. **3**(2), 105–114 (2015)
10. Hernández-Ledesma, B., Quirós, A., Amigo, L., Recio, I.: Identification of bioactive peptides after digestion of human milk and infant formula with pepsin and pancreatin. Int. Dairy J. **17**(1), 42–49 (2007)
11. Kruglik, V.I., Zorin, S.N., Gmoshinskij, I.V., Ponomarev, D.V., Nikitina, N.E., Abramova, A.A., et al.: Patent of Russia 2375910 (2009)
12. Kruglik, V.I., Zorin, S.N., Gmoshinskiy, I.V., Nikitina, N.Ye., Volkova, I.N., Revyakina, N.V., et al.: Patent of Russia 2428047 (2011)
13. Lee, A., Pontin, M.C.F., Kosmerl, E., Jimenez-Flores, R., Moretti, D.B., Ziouzenkova, O.: Assessment of adipogenic, antioxidant, and anti-inflammatory properties of whole and whey bovine colostrum. J. Dairy Sci. **102**, 8614–8621 (2019)
14. Milk and milk products: Determination of nitrogen content. Part 1: Kjeldahl principle and crude protein calculation. ISO 8968–1:2014. International Organization for Standardization, Geneva, Switzerland (2014)
15. Milk, cream and evaporated milk: ISO 6731:2010. Determination of total solids content (Reference method). International Organization for Standardization, Geneva, Switzerland (2010)
16. Mota, M.V.T., Ferreira, I.M.P.L.V.O., Oliveira, M.B.P., Rocha, C., Teixeira, J.A., et al.: Enzymatic hydrolysis of whey protein concentrates: peptide HPLC profiles. J. Liq. Chromatogr. Relat. Technol. **27**(16), 2625–2639 (2004)
17. O'Keeffe, M.B., FitzGerald, R.J.: Antioxidant effects of enzymatic hydrolysates of whey protein concentrate on cultured human endothelial cells. Int. Dairy J. **36**(2), 128–135 (2014)
18. Osterman, L.A.: Electrophoresis. Isoelectric focusing ultracentrifugation. Methods of Protein and Nucleic Acid Research, vol. 1. Springer, Heidelberg (1984). ISSN: 978-3-642-87487-1
19. Ou, K., Liu, Y., Zhang, L., Yang, X., Huang, Z., Nout, M.J.R., et al.: Effect of neutrase, alcalase, and papain hydrolysis of whey protein concentrates on iron uptake by Caco-2 cells. J. Agric. Food Chem. **58**(8), 4894–4900 (2010)
20. Phenolic, amino and condensation resins – Determination of free-formaldehyde content. ISO 11402:2004. International Organization for Standardization, Geneva, Switzerland (2004)
21. Prosekov, A.J., Ul'rikh, E.V., Poturaeva, N.L., Koroleva, O.V., Budrik, V.G., Botina S.G., et al.: Patent of Russia 2528068 (2014)
22. Souza, R.S.C. de, Tonon, R.V., Stephan, M.P., Silva, C.M., Penteado, A.L., Cabral, L.M.C., et al.: Evaluation of the antioxidant potential of whey protein concentrated by ultrafiltration and hydrolyzed by different commercial proteases. Braz. J. Food Technol. **22**, e2018021 (2019)
23. Sviridenko, J.J., Abramov, D.V., Myagkonosov, D.S., Ovchinnikova, E.G., Tutelyan, V.A., Mazo, et al.: Patent of Russia 2663583 (2018)
24. Sviridenko, J.J., Abramov, D.V., Myagkonosov, D.S., Tutelyan, V.A., Mazo, V.K., Zorin, S.N.: Patent of Russia 2529707 (2014)

25. Tacias-Pascacio, V.G., Morellon-Sterling, R., Siar, E.-H., Tavano, O., Berenguer-Murcia, Á., Fernandez-Lafuente, R.: Use of Alcalase in the production of bioactive peptides: a review. Int. J. Biol. Macromol. **165**, 2143–2196 (2020)
26. Tarun, E.I.: Comparison of antioxidant activities of gallic, coffee and chlorogenic acids. Proc. BSU **9**, 186–191 (2014). (in Russian)
27. Tavano, O.L.: Protein hydrolysis using proteases: an important tool for food biotechnology. J. Mol. Catal. B Enzym. **90**, 1–11 (2013)
28. Torkova, A., Ryazantzeva, K., Agarkova, E.Yu., Tsentalovich, M., Kruchinin, A., Fedorova, T.V.: Cheese whey catalytic conversion for obtaining a bioactive hydrolysate with reduced antigenicity. Curr. Res. Nutr. Food Sci. **4**(SI.2), 182–196 (2016)
29. Torkova, A.A., Ryazantseva, K.A., Agarkova, E.Y., Kruchinin, A.G., Tsentalovich, M.Y., Fedorova, T.V.: Rational design of enzyme compositions for the production of functional hydrolysates of cow milk whey proteins. Appl. Biochem. Microbiol. **53**(6), 669–679 (2017). https://doi.org/10.1134/S0003683817060138
30. Van Beresteijn, E.C.H., Peeters, R.A., Kaper, J., Meijer, R.J.G.M., Robben, A.J.P.M., Schmidt, D.G.: Molecular mass distribution immunological properties nutritive value of whey protein hydrolysates. J. Food Prot. **57**(7), 619–625 (1994)
31. Vandenplas, Y., Munasir, Z., Hegar, B., Kumarawati, D., Suryawan, A., Kadim, M., et al.: A perspective on partially hydrolyzed protein infant formula in nonexclusively breastfed infants. Korean J. Pediatr. **62**(5), 149–154 (2019)
32. Whitehurst, R.J., van Oort, M. (eds.): Enzymes Food Technology, vols. 1, 13, 2nd edn. Blackwell Publishing Ltd, Iowa, USA
33. Zhang, Q.-X., Wu, H., Ling, Y.-F., Lu, R.-R.: Isolation and identification of antioxidant peptides derived from whey protein enzymatic hydrolysate by consecutive chromatography and Q-TOF MS. J. Dairy Res. **80**(3), 367–373 (2013)
34. Zheng, H., Shen, X., Bu, G., Luo, Y.: Effects of pH, temperature and enzyme-to-substrate ratio on the antigenicity of whey protein hydrolysates prepared by Alcalase. Int. Dairy J. **18**(10 11), 1028–1033 (2008)

TiO_2-Chitosan as Universal Composite for Waste Water Treatment from Organic Pollutants

Evgeniia Salomatina(✉), Dmitriy Zuev, Sergey Zaitsev, and Larisa Smirnova

National Research Lobachevsky State University of Nizhny Novgorod, Nizhny Novgorod, Russia

Abstract. Modification of TiO_2 surface with silver nanoparticles is an urgent task in the light of the development of new types of photocatalysts for cleaning the environment from organic pollutants. For achieving this goal, it is required to obtain systems in which the deposition of silver nanoparticles on the titanium dioxide surface has been carried out by a nontrivial, simple, non-labor intensive method. The essence of the method is that at first stage formation of Ag nanoparticles from $AgNO_3$ under UV-irradiation in a solution of a stabilizing polymer – chitosan occurs, than TiO_2 particles disperse in the resulting colloidal solution, and subsequent enzymatic destruction of polysacharide. As a result, Ag nanoparticles, the size of which is 22.0 ± 0.25 nm, completely settle on the TiO_2 surface. It was found that in the reactions of decomposition of methylene blue, methylene orange and congo red in an aqueous solution under UV irradiation, modified TiO_2 is 2–2.5 times more photocatalytically active than the initial titanium dioxide.

Keywords: Titanium dioxide · Silver nanoparticles · Chitosan · Fermentative destruction · Photocatalytic properties · Methylene orange · Methylene blue · Congo red · UV-irradiation

1 Introduction

The aggravation of ecological problems dictates the need to develop new technologies to protect the environment from the anthropogenic, negative impact on it of emissions of existing industrial plants, utilities and vehicles. In this connection, the use of photocatalytic processes for decontamination of pollutants has gained particular socio-economic importance in recent years [1–6]. The advantages of photocatalysis are well known [7–11]:

1) the possibility of oxidation of almost any organic carcinogenic pollutants in water and air (pesticides, pharmaceuticals, cyan-toxins, chlorophenols, azo-dyes) as well as some inorganic ones, such as CO, H_2S, HCN, NH_3, NO_x, under mild conditions - at room temperature and atmospheric pressure - to safe products [12–17];
2) economic advantages compared to other methods, related to the possibility of decomposition of organic pollutants of waste water of enterprises, even in the case of low concentrations of these substances [1–5];

V. Kurchenko et al. (Eds.): ICAETT 2021, LNNS 408, pp. 46–55, 2022.
https://doi.org/10.1007/978-3-030-96641-6_6

3) implementation of the photocatalytic oxidation method does not require additional reagents, since the oxidizer is air oxygen [1–5, 18].

Nanocomposites containing TiO_2 of anatase polymorphic modification occupy a special place among photocatalytic materials [19–22]. This material is a typical semiconductor and is capable of reversible UV-induced reaction $Ti^{4+} + e^- \rightleftarrows Ti^{3+}$ accompanied by Ti-O bond breaking, electron transfer from valence to conduction band and generation of electron-hole pairs [1–5, 23, 24]. Free electrons are good reducing agents while "holes" are strong oxidizing agents which lead to formation of hydroxyl radicals (•OH), superoxide-anion radicals (O_2^-•), hydroperoxide radicals (OH_2•) and singlet oxygen in the presence of air oxygen or vaporous water. The latter are involved in transformations of various organic substances, decomposition of dirt, soot, toxic drugs, cigarette smoke, are capable of causing death of many bacteria and microorganisms in water, air or accumulated on surfaces [25].

The introduction of small amounts of active additives into the structure of titanium dioxide can lead to an increase in the quantum yield of the reaction.

$Ti^{4+} + e^- \rightleftarrows Ti^{3+}$ and shifting its absorption spectrum to the visible wavelength range. This contributes to an increase in the concentration of electrons and "holes" in the surface layer of the oxide, which leads to an increase in its efficiency [1, 2, 26]. One option to improve the properties of TiO_2 is to alloy its surface with nanoparticles (NPs) of various metals, such as Pt, Ag, Au, Pd, Ni, Cu and Rh [27–30]. It is known that the photocatalytic activity of the catalysts containing NFs studied so far changes in the following series: $Au/TiO_2 > Ag/TiO_2 \geq Pt/TiO_2 > TiO_2$ [31]. The final result depends largely on the size of the NPs and the method of their deposition on the titanium oxide.

Currently, to modify the surface of TiO_2 metal nanoparticles are used such methods as - high temperature sputtering of TiO_2 and other substances in a vacuum [32], hydrothermal method [33], cold plasma treatment [34], sol-gel method [35], electron beam interaction [36]. From these examples, we can see that all these methods are quite labor-intensive, none of them combines the ease of production, the uniformity of nanoparticle distribution in the material, and, consequently, does not use the whole reserve of photocatalytic activity.

Thus, the development of non-labor-intensive methods of obtaining composites based on TiO_2 of anatase polymorphic modification and NPs of various metals remains relevant. In this regard, one of the promising methods seems to be the deposition of NPs obtained in chitosan solutions (CTS) on the surface of titanium dioxide during the enzymatic destruction of polysaccharide. The unique properties of CTS, viz: high stabilizing capacity due to the presence of amino- and hydroxyl groups in the link [37] and susceptibility to enzymatic degradation with cleavage of the O-glycoside bond between neighboring links in the polymer chain to form oligomeric macromolecules or to β-(1–4) D-glucosamine, soluble in water [38] - offer excellent opportunities for obtaining metal particles with sizes less than 20 nm [39] and their deposition on the TiO_2 surface at room temperature. The possibility of using the TiO_2/chitosan system for the decontamination of a wide range of environmental pollutants, both organic and inorganic, is not excluded due to the ability of the latter to sorb heavy metal ions.

The aim of this research is obtaining TiO_2/Ag composites by enzymatic action on silver nanodispersions stabilized by chitosan and studying the photocatalytic activity of the obtained systems under UV-irradiation in the decomposition reactions of methylene orange (MO), methylene blue (MB) and congo red (CR), which are natural pollutants in the wastewater of large-scale dye productions.

2 Materials and Methods

2.1 Modification of Powdered TiO_2

We used titanium dioxide TiO_2 of anatase polymorphic modification (KRONOS 1001 brand from KRONOS TITAN GmbH & Co. OHG (Germany), $\rho = 4.05$ g/cm^3, average particle size 2.1 ± 0.2 μm). Silver nitrate, analytical grade (Sibproekt; $AgNO_3$ content not less than 99.8%) was a precursor for the formation of silver NPs. Chitosan (CTS) with a molecular weight of 1×10^5 and a degree of deacetylation of 80%, obtained from crab shells (JSC "Bioprogress", Moscow, Russia, the mass fraction of minerals in CTS did not exceed 0.1%, moisture - 6%, insoluble substances - 0.1%) was used as a polymer-stabilizer in the formation of metal NPs.

We prepared aqueous acetic acid solutions of CTS (acetic acid (AA), chemically pure grade), into which the calculated amount of NPs precursors was added. Chitosan-stabilized metal nanodispersions were obtained by UV irradiation of polysaccharide solutions containing $AgNO_3$. The source of UV light was a high-pressure arc discharge tube mercury lamp "DRT-240" lamp with a power of. 1600 mW/m^2 at a distance of 27 cm. In 2 h, after the end of the NPs formation process, powdered TiO_2 was added to the dispersion.

To isolate and precipitate NPs on titanium dioxide, enzymatic destruction of the stabilizer polymer was carried out using chitosanase, an enzyme of chitosan hydrolysis, which first breaks it down to oligosaccharides and, ultimately, to glucosamine («Sigma Aldrich», activity 10 units per gram). Previously, 50 mg of chitosanase was dissolved in 6 ml of 50 wt. % glycerin solution in water.

Thus, a method has been developed in which, after the formation of NPs at various concentrations of $AgNO_3$ and TiO_2, as well as the minimum concentration of the stabilizer polymer, the metal NPs are completely deposited onto the oxide surface.

2.2 Optical Spectroscopy

Spectra in the UV and visible wavelength range were recorded on a SHIMADZU spectrophotometer, model UV-1650PC. Scanning range from 190 to 1100 nm.

2.3 Scanning Electron Microscopy

The powder sample morphology was obtained by scanning electron microscopy (SEM) on JSM-IT300LV (JEOL) with the electron probe diameter of about 5 nm, probe current below 0.5 nA, and operating voltage 20 kV. The study of the sample surface topography was performed using the low-energy secondary electrons and backscattered electrons under high vacuum mode. The investigation of elemental composition was performed by the method of energy dispersive X-ray microanalysis (EDX) with detector X-MaxN 20 (Oxford Instruments).

2.4 Study of Photocatalytic Activity

The photocatalytic properties of the obtained samples were studied in the decomposition reactions of methylene blue (MB) (clean for analysis grade), methylene orange (MO) (clean for analysis grade) and congo red (CR) (clean for analysis grade) in aqueous solutions when exposed to light. For this, solutions of MB, MO and CR in distilled water were prepared. The concentrations of the substances were 6.25, 15.6 and 0.015 mmol/l, respectively. The decomposition cells were filled with 0.02 g of photocatalyst samples and 20 ml of azo-dyes solutions.

The cells covered with quartz glass were kept for 30 min in the dark to establish the adsorption-desorption equilibrium. Than the MB, MO or CR aqueous solutions were placed under a high-pressure arc discharge tube mercury lamp "DTR-240" UV lamp (the power 1600 mW/m^2 at a distance of 27 cm, the most intense emission bands were 313 nm, 330 nm, 366 nm) and irradiated with constant stirring at atmospheric pressure and 30 °C. After every 30 min of irradiation, spectra of MB, MO or CR solutions were recorded to control their decomposition. At the end of the process, the dependences of the change in the concentration of MB, MO or CR solutions at a specific wavelength for each substance on the time of their irradiation were plotted: $C/C_0 = f(t)$ (C is the current concentration, mmol/l, C_0 is the concentration of the initial solution of the substance, mmol/l).

3 Results

As noted in the introduction, a promising method for modifying the surface of TiO_2 by silver nanoparticles in terms of the possibility of practical use is their formation in situ with subsequent deposition on titanium dioxide particles, eliminating the need for cleaning the products from foreign inclusions. The process was carried out in 2 stages. At the first stage there was the formation of silver NFs in acetic acid aqueous solutions of CTS during $AgNO_3$ reduction under UV-light of according to our previously developed method [40, 41]. The process of formation of Ag NPs in CTS solutions was monitored spectrophotometrically by the appearance and growth of the plasmon resonance band at $\lambda = 390 - 430$ nm, characteristic of silver NPs (Fig. 1).

The intensity of the absorption band maximum corresponding to plasmon resonance of silver NPs reaches a constant value after 120 min of irradiation, which indicates the completion of the NPs formation process (Fig. 5(b)). The average size of silver NPs determined by the method of X-ray scattering at small angles is -22.0 ± 0.25 nm; the nanoparticles are characterized by a relatively narrow size distribution (Fig. 5(c)). Full adsorption of CTS macromolecules on silver nanoparticles and the absence of free polysaccharide in the nanodispersion solution were proved by thin-layer chromatography, which indicates the formation of CTS-NPs complex.

At the second stage in chitosan solutions containing silver nanoparticles, powdered titanium dioxide was introduced under ultrasonic influence on the systems (using ultrasonic bath "Sapphire" UZV-2,8, operating frequency 35 kHz), which increases dispersion of TiO_2 particles and contributes to the uniformity of their distribution over the system volume.

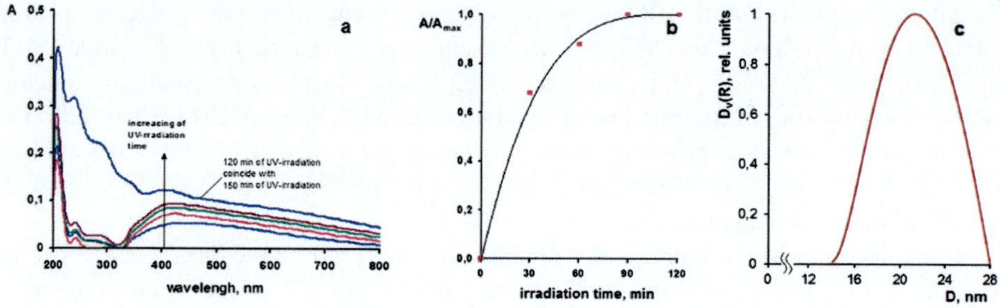

Fig. 1. (a) Absorption spectra of CTS solution containing $AgNO_3$ during UV-induced formation of Ag particles; (b) relative change in optical density of the system at 418 nm during UV-induced formation of silver particles in a solution of 3 wt% CTS in 1.5 wt% AA containing 2 wt% $AgNO_3$; (c) distribution by size of silver particles formed in CTS solution

Thus, under these conditions a "blank" was formed in which titanium dioxide and metal nanoparticles are separated by a polymer-stabilizer layer.

Deposition of Ag NPs on the surface of powdered TiO_2 occurred spontaneously after removal of chitosan with its complete enzymatic destruction at 50–55 °C for 2 h using chitosanase. For this purpose, an aqueous-glycerol solution of the enzyme was introduced into the dispersions containing metal and TiO_2 NPs. As a result, the formation of a stained precipitate was observed. The resulting systems were then heated at 70–80 °C to deactivate the enzyme. The absence of silver NPs plasmon resonance band in the supernatant spectrum indicated almost complete deposition of NPs on the oxide surface (Fig. 2).

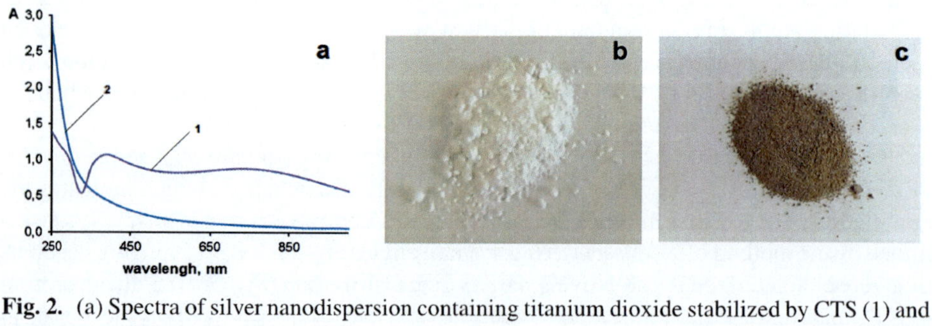

Fig. 2. (a) Spectra of silver nanodispersion containing titanium dioxide stabilized by CTS (1) and supernatant (2) after deposition of silver powder on TiO_2 surface; (b) photo of commercial TiO_2 powder; (c) photo of TiO_2/Ag

The precipitates were separated by centrifugation and repeatedly treated with distilled water to neutral water pH, dried in a vacuum oven until constant weight, and ground in an agate mortar (Fig. 2(c)). The structural characteristics of the powders were examined by scanning electron microscopy with X-ray fluorescence attachment (Fig. 3). It was found that the TiO_2 powder consists of nanoparticles with an average size of about 100–200 nm. This morphology is preserved after obtaining nanocomposites with Ag. The initial composition of TiO_2 powder is 65.4 at.% (Ka, O) and 34.6 at.% (Ka, Ti), which corresponds to the stoichiometric ratio of elements in the oxide within the method sensitivity ~0.1 at.% (σ). EDX results did not reveal any carbon impurities in the TiO_2/Ag samples. The EDX element maps and the images obtained with the backscattered electron detector of the TiO_2/Ag powder surface allow to show the atomic number contrast and confirm the presence of Ag particles on the titanium dioxide surface as "light" spots (Fig. 3).

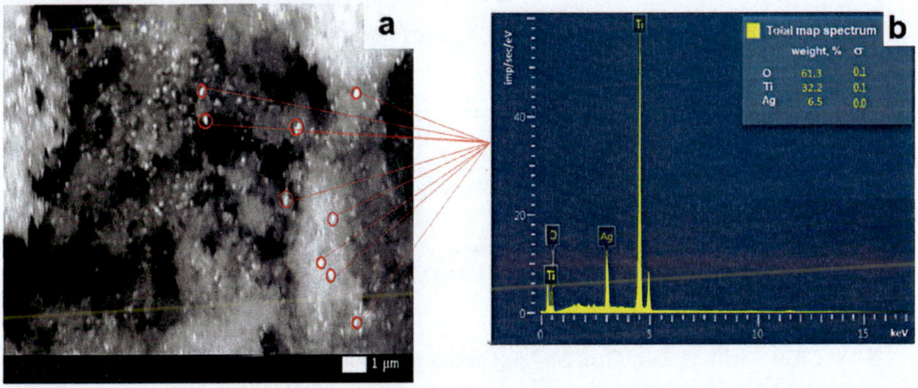

Fig. 3. SEM image (a) and X-ray fluorescence analysis (b) of the TiO_2/Ag sample surface

The photocatalytic properties of the initial TiO_2 and TiO_2/Ag were studied in the decomposition reactions of MB, MO and CR in aqueous solutions when exposed to UV light. For this purpose, azo-dye solutions in distilled water were prepared. The choice of objects is due to the fact that they are common pollutants of waste water from textile and paint and varnish production and are non-biodegradable toxic substances.

The process of decomposition of MO, MB and CR with the participation of powdered photocatalysts was monitored spectrophotometrically by the decrease in the absorption band intensity at 470 nm for MO, at 660 nm for MB and at 493 nm for CR (Fig. 4, 5 and 6).

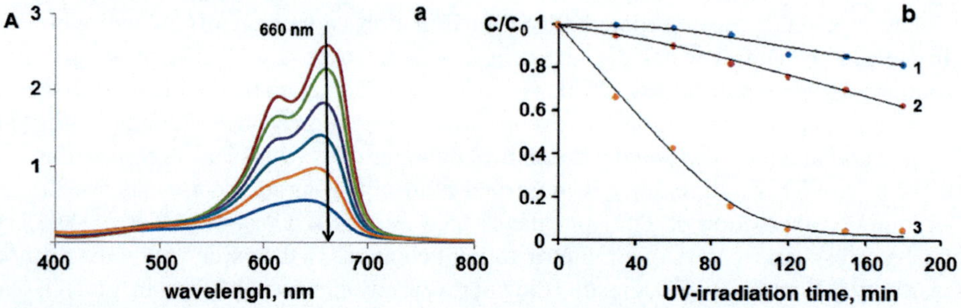

Fig. 4. a) Change of MB solution spectrum at its UV-irradiation in the presence of catalysts; b) Relative change in concentration of MB aqueous solution at UV-irradiation: curve 1 – «empty» experiment; curve 2 - in the presence of TiO_2 powder; curve 3 - in the presence of TiO_2/Ag powder

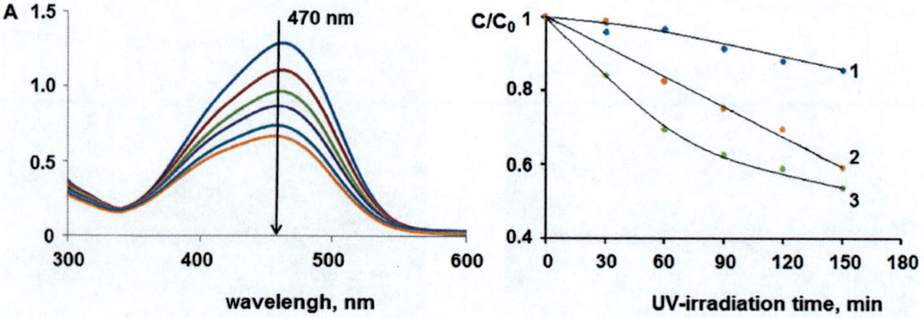

Fig. 5. a) Change in the spectrum of MO solution at its UV-irradiation in the presence of catalysts; b) relative change in the concentration of MO aqueous solution at UV-irradiation: curve 1 - «empty» experiment; curve 2 - in the presence of TiO_2 powder; curve 3 - in the presence of TiO_2/Ag powder

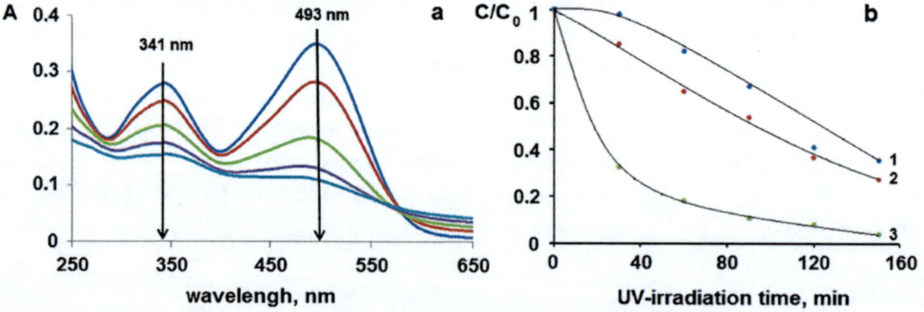

Fig. 6. a) Change in the spectrum of CR solution at its UV-irradiation in the presence of catalysts; b) relative change in the concentration of CR aqueous solution at UV-irradiation: curve 1 - «empty» experiment; curve 2 - in the presence of TiO_2 powder; curve 3 - in the presence of TiO_2/Ag powder.

4 Discussion and Conclusion

It was found that the obtained materials exhibit photocatalytic activity. MO and MC under UV-irradiation in the absence of catalysts almost do not decompose, as evidenced by almost constant value of their concentration. A different picture is observed under UV-exposure to CR solution - the decrease of its concentration in the solution under light exposure occurs even in the absence of photocatalysts. Photocatalytic activity of titanium dioxide samples doped with Ag NPs is higher compared to pure TiO_2 (anatase) in all cases. Thus, using TiO_2 (anatase) decomposes approximately 20% of the dyes in 150 min, whereas using TiO_2/Ag, the conversion depth of MO, MC, and CR is 55%, ~90%, and ~97%, respectively.

Thus, a low-temperature method of doping the surface of titanium dioxide particles with silver nanoparticles formed in situ by UV irradiation of aqueous solutions of the corresponding precursors in the presence of chitosan as a polymer stabilizer was developed. The advantages of the developed method of deposition of gold or silver nanoparticles on the surface of titanium dioxide with enzymatic degradation of the carrier polymer nanoparticles are affordability and ease of implementation, when large energy costs are not required. The simplicity of the method of obtaining modified forms of titanium dioxide and the results on the decomposition of azo dyes in aqueous solutions in their presence show the promise of using TiO_2/Ag systems as highly efficient photocatalysts for the neutralization of organic pollutants in water, significantly exceeding the properties of TiO_2 when exposed to ultraviolet light.

Acknowledgments. This research was funded by Council of grant of the President of the Russian Federation (grant number MK-2195.2021.1.3), partly by Russian Foundation for Basic Research (project № 19–03-00843) and the Ministry of Science and Higher Education of the Russian Federation within the framework of the basic part of the state task (project No. 0729-2020-0039).

References

1. Marcì, G., Palmisano, L. (eds.): Heterogeneous Photocatalysis. Relationships with Heterogeneous Catalysis and Perspectives. Elsevier, Netherlands (2019)
2. Thomson, C., Lee, A.-L., Vilela, F.: Heterogeneous photocatalysis in flow chemical reactors. Beilstein J. Org. Chem. **16**, 1495–1549 (2020)
3. Boukherroub, R., Ogale, S.B., Robertson, N. (eds.): Nanostructured Photocatalysts. From Materials to Applications in Solar Fuels and Environmental Remediation. A Volume in Micro and Nano Technologies. Elsevier, Netherlands (2020)
4. Wei, Z., Liu, J., Shangguan, W.: A review on photocatalysis in antibiotic wastewater: pollutant degradation and hydrogen production. Chin. J. Catal. **41**, 1440–1450 (2020)
5. Rueda-Marquez, J.J., Levchuk, I., Ibañez, P.F., Sillanpää, M.: A critical review on the application of photocatalysis for toxicity reduction of real wastewaters. J. Clean. Prod. **258**, 120694 (2020)
6. Zhu, S., Wang, D.: Photocatalysis: basic principles, diverse forms of implementations, and emerging scientific opportunities. Adv. Energy Mater. **7**, 1700841 (2017)
7. Reddy, P.V.G., et al.: A review on multicomponent reactions catalyzed by zero-dimensional/one-dimensional titanium dioxide (TiO_2) nanomaterials: promising green methodologies in organic chemistry. J. Environ. Manage. **279**, 111603 (2021)

8. Bernardini, C., Cappelletti, G., Dozzi, M.V., Selli, E.: Photocatalytic degradation of organic molecules in water: photoactivity and reaction paths in relation to TiO_2 particles features. J. Photochem. Photobiol. Chem. **211**, 185–192 (2010)
9. Achudhan, D., et al.: The antibacterial, antibiofilm, antifogging, and mosquitocidal activities of titanium dioxide (TiO2) nanoparticles green-synthesized using multiple plant extracts. J. Environ. Chem. Eng. **8**, 104521 (2020)
10. Indekeu, A., et al.: Synthesis of modified TiO_2-based catalysts for the photocatalytic production of solar fuels from synthesis gas. Catal. Today **10**, 227 (2020)
11. Michalkiewicz, B., Majewska, J., Kądziołka, G., Bubacz, K., Mozia, S., Morawski, A.W.: Reduction of CO_2 by adsorption and reaction on the surface of TiO_2-nitrogen modified photocatalyst. J. CO_2 Utiliz. **5**, 47–52 (2014)
12. Chen, D., et al.: Photocatalytic degradation of organic pollutants using TiO_2-based photocatalysts: a review. J. Clean. Prod. **268**, 121725 (2020)
13. Barsukov, D.V., Pershin, A.N., Subbotina, I.R.: Increase of CO photocatalytic oxidation rate over anatase TiO_2 particles by adsorbed water at moderate coverages: the role of peroxide species. J. Photochem. Photobiol. A: Chem. **324**, 175–183 (2016)
14. Alonso-Tellez, A., Robert, D., Keller, N., Keller, V.: A parametric study of the UV-a photocatalytic oxidation of H_2S over TiO_2. Appl. Catal. B **115–116**, 209–218 (2012)
15. Brüger, A., Fafilek, G., Neumann-Spallart, M.: Treatment of cyanide: photoelectrocatalytic degradation using TiO_2 thin film electrodes and the influence of volatilization. Sol. Energy **205**, 74–78 (2020)
16. Wu, H., Ma, J., Zhang, C., He, H.: Effect of TiO_2 calcination temperature on the photocatalytic oxidation of gaseous NH_3. J. Environ. Sci. **26**, 673–682 (2014)
17. Serpone, N.: Heterogeneous photocatalysis and prospects of TiO_2-based photocatalytic detoxing the atmospheric environment. Catalysts **8**, 553 (2018)
18. Tsang, C.H.A., et al.: Titanium oxide-based photocatalytic materials development and their role in air pollutants degradation: overview and forecast. Environ. Int. **125**, 200–228 (2019)
19. Fujishima, A., Zhang, X., Tryk, D.: Heterogeneous photocatalysis: from water photolysis to applications in environmental cleanups. Int. J. Hydrogen. Ener. **14**, 2664–2672 (2007)
20. Binas, V., Venieri, D., Kotzias, D., Kiriakidis, G.: Modified TiO_2 based photocatalysts for improved air and health quality. J. Materiom. **3**, 3–16 (2017)
21. Srikanth, B., Goutham, R., Badri Narayan, R., Ramprasath, A., Gopinath, K.P., Sankaranarayanan, A.R.: Recent advancements in supporting materials for immobilized photocatalytic applications in wastewater treatment. J. Environ. Manage. **200**, 60–78 (2017)
22. Kedves, E.-Z., Pap, Z., Hernadi, K., Baia, L.: Significance of the surface and bulk features of hierarchical TiO_2 in their photocatalytic properties. Ceram. Int. **47**, 7088–7100 (2020)
23. Oluwafunmilola, O., Maroto-Valer, M.: Review of material design and reactor engineering on TiO_2 photocatalysis for CO_2 reduction. J. Photochem. Photobiol., C **24**, 16–42 (2015)
24. Fujishima, A., Rao, T.N., Tryk, D.A.: Titanium dioxide photocatalysis. J. Photochem. Photobiol. C: Photochem. Rev. **1**, 1–21 (2000)
25. Ángel-Sanchez, K.D., Vázquez-Cuchillo, O., Aguilar-Elguezabal, A., López, A.C., Herrera-Gómez, A.: Photocatalytic degradation of 2,4- dichlorophenoxyacetic acid under visible light: effect of synthesis route. Mater. Chem. Phys. **139**, 423–430 (2013)
26. Xiong, L.-B., Li, J.-L., Yang, B., Yu, Y.: Ti^{3+} in the surface of titanium dioxide: generation, properties, and photocatalytic application. J. Nanomater. **2012**, 831524 (2012)
27. Sanzone, G., et al.: Ag/TiO_2 nanocomposite for visible light-driven photocatalysis. Superlattices Microstruct. **123**, 394–402 (2018)
28. Sornalingam, K., McDonagh, A., Zhou, J.L., Johira, M.A.H., Ahmed, M.B.: Photocatalysis of estrone in water and wastewater: comparison between Au-TiO_2 nanocomposite and TiO_2, and degradation by-products. Sci. Total Environ. **610**, 521–530 (2018)

29. Sakthivel, S., Shankar, M.V., Palanichamy, M., Arabindoo, B., Bahnemann, D.W., Murugesan, V.: Enhancement of photocatalytic activity by metal deposition: characterization and photonic efficiency of Pt, Au, and Pd deposited on TiO_2 catalyst. Water Res. **38**, 3001–3008 (2004)
30. Ismael, M.: A review and recent advances in solar-to-hydrogen energy conversion based on photocatalytic water splitting overdoped-TiO_2 nanoparticles. Sol. Energy **211**, 522–546 (2020)
31. Tang, K.Y., et al.: Gold-decorated TiO_2 nanofibrous hybrid for improved solar-driven photocatalytic pollutant degradation. Chemosphere **265**, 129114 (2021)
32. Atef, N., Emara, S.S., Eissa, D.S., El-Sayed, A., Abdelraouf, O.A.M., Allam, N.K.: Well-dispersed Au nanoparticles prepared via magnetron sputtering on TiO_2 nanotubes with unprecedentedly high activity for water splitting. Electrochem. Sci. Adv. **1**, e2000004 (2020)
33. Pandiyaraj, K.N., et al.: Dye wastewater degradation by the synergetic effect of an atmospheric pressure plasma treatment and the photocatalytic activity of plasma-functionalized Cu–TiO_2 nanoparticles. J. Hazard. Mater. **405**, 124264 (2020)
34. Anaya-Esparza, L.M., et al.: Synthesis and characterization of TiO_2-ZnO-MgO mixed oxide and their antibacterial activity. Mater. **12**, 698 (2019)
35. Haruta, M., Tsubota, S., Kobayashi, T., Kageyama, H., Genet, M.J., Delmon, B.: Low-temperature oxidation of CO over gold supported on TiO_2, α-Fe_2O_3, and Co_3O_4. J. Catal. **144**, 175–192 (1993)
36. Haruta, M.: Catalysis of gold nanoparticles deposited on metal oxides. CATTECH **6**, 102–115 (2002)
37. Muggli, D.S., McCue, J.T., Falconer, J.L.: Mechanism of the photocatalytic oxidation of ethanol on TiO_2. J. Catal. **173**(2), 470–483 (1998)
38. Rinaudo, M.: Chitin and chitosan: properties and applications. Prog. Polym. Sci. **31**, 603–632 (2006)
39. Mohamed, N., Madian, N.G.: Evaluation of the mechanical, physical and antimicrobial properties of chitosan thin films doped with greenly synthesized silver nanoparticles. Mater. Today Commun. **25**, 101372 (2020)
40. Gracheva, T.A., Kuz'micheva, T.A., Perevezentsev, V.N., et al.: Kinetics and mechanisms of the UV-radiation-assisted formation of gold nanoparticles in $HAuCl_4$-doped chitosan solutions. Tech. Phys. **62**, 1228–1232 (2017)
41. Apryatina, K.V., Mochalova, A.E., Gracheva, T.A., Kuz'micheva, T.A., Smirnova, L.A., Smirnova, O.N.: Iinfluence of the molecular mass of chitosan on the dimensional characteristics of silver nanoparticles. Poly. Sci. Ser. B **57**, 145–149 (2015)

Isolation and Characterization of a New Fungi Isolates of the Genus *Trichoderma* Pers. – of Potential Producers of Biofungicides

Zulfira Siraeva[1,2](✉), David Shamuradov[2], Alsu Mustafina[1,3], and Olga Kravtsova[1]

[1] Institute of Fundamental Medicine and Biology, Kazan (Volga Region) Federal University, Kazan, Russia
zsiraeva@yandex.ru, alsu_i_mustafina@mail.ru, okravz@yandex.ru
[2] Kazan State Medical University, Kazan, Russia
david_shamuradov@bk.ru
[3] Almetyevsk State Oil Insitute, Almetyevsk, Russia

Abstract. One of the perspective directions in the plant defense against mycoses is the use of preparations based on fungi, in particular from the genus *Trichoderma* Pers. However, limiting factors for their safe application consider colonizing and fungistatic activities as well as the absence of toxic effects. The aim of this work is to search and substantiate the possibility of using new non-pathogenic isolates of fungi from the genus *Trichoderma* Pers. with a high growth speed and fungistatic activity. 97 micromycete isolates, phenotypically assigned as *Trichoderma* Pers., were isolated from the various soil samples of the Russian Federation. Based on kinetic parameters and colonizing activity, three *Trichoderma* Pers. strains (*T. cerinum* strain 84, *T. tomentosum* strain 189 and *T. asperellum* strain 195) were selected to check their toxic effects in vivo model (rats). It was shown that *per os* application LD50 exceed dose $1.0 \pm 0.001 \times 10^8$ spores per kg, no cumulative effect was found after 8 weeks' exposure of suspension at dose $1.0 \pm 0.001 \times 10^7$ spores/kg and skin irritating effect at dose $0.25 \pm 0.001 \times 10^8$ spores/kg. The results show the prospects for further study of the properties of *T. cerinum* strain 84, *T. tomentosum* strain 189 and *T. asperellum* strain 195 with the aim of creating biofungicides.

Keywords: Trichoderma · Growth speed · Biofungicide · Fungistatic activity · Acute toxicity · Cumulative effect

1 Introduction

One of the leading problems in crop production is a high level of soil infestation [16], seed [18] and planting material [22] by plant pathogens. One of the main reason is application of chemical pesticides, leading to the selection of resistant strains of pathogens [2, 13, 15] and to the toxic effect on the bacterial community, as a consequence, on the biological activity of the soil [5, 15, 16]. In this regard, the attention of researchers is draw to the development and use of environmentally safe biological products [16, 22, 24].

V. Kurchenko et al. (Eds.): ICAETT 2021, LNNS 408, pp. 56–64, 2022.
https://doi.org/10.1007/978-3-030-96641-6_7

Trichoderma Pers. fungi are known for their antagonistic activity against plant pathogens [3, 22, 24] and their ability to produce of biologically active compounds [12, 24]. At the same time, colonizing and fungistatic activities are considered limiting characteristics for their use in agricultural production [4, 12, 24], as well as the absence of toxic effects on the human body [10, 11].

The aim of this work is to determine and substantiate the possibility of using new non-pathogenic isolates of fungi from the genus Trichoderma Pers. with a high growth rate and fungistatic activity. The objectives of the study include the micromycete isolation from different soil samples (Russian Federation), phenotypically related to the genus *Trichoderma* Pers; determination and selection of isolates with the highest growth rate and fungistatic activity against test cultures of phytopathogenic micromycetes with primary toxicological safety assessment in vivo model.

2 Materials and Methods

2.1 Preparation of *Trichoderma* Pers. Monosporic Culture

Trichoderma Pers. strains were isolated from soil samples collected in the Republic of Tatarstan (RT), Volgograd and Sverdlovsk regions of Russia during 2019–2021. To isolate primary cultures from pooled soil samples, we used the serial dilution method [1]. Sowing of soil suspensions at a dilution of 10^4 and 10^5 was carried out superficially on potato-glucose agar (PGA) (composition, g/l: potato – 200.0 glucose – 20.0, agar – 20.0).

Microscopic analysis of colonies on solid media and preparations with sporulating fungal mycelium was analyzed using a Primo Star microscope (Carl Zeiss, Germany) at a total magnification of ×400. Colonies with typical *Trichoderma* morphological characteristics [7, 19, 20] was used to obtain monospore cultures.

Isolation of monospore cultures was carried out on potato-glucose medium (PGM) with the addition of glycerol (0.05% vol). The inoculation suspension was prepared by dilution to a titer of 20–30 colonies per 1 ml of PGM. The cultivation was carried out for 24 h until the suspension formation. Microcolonies were removed with a sterile loop and incubated on PGA for 10 days. The obtained monospore clones were seeded on PGA and used in further work.

2.2 Morphological and Molecular Identification of Fungi Isolates

Fungal isolates were primary identified on the basis of routine cultural and morphological characteristics. For morphological characterization, the isolates were cultured on PGA. Morphological signs were studied after 7 days of cultivation at a temperature of 20 °C and alternating darkness (12 h) and illumination (12 h). The analysis of morphological characters of the isolates was carried out according to [7, 19, 20]. Molecular identification was performed by sequencing of ITS region according to the protocol described earlier [17] using BLAST service (https://blast.ncbi.nlm.nih.gov).

2.3 Kinetic Characteristics of Growth Rate in Isolates *Trichoderma* Pers

Growth parameters were determined on Petri dishes with a bottom diameter of 90 mm, containing 20 ml of nutrient medium. The cultivation was carried out on PGA medium at a temperature of 20 °C and alternating darkness (12 h) and illumination (12 h) until the sporulation area covered the entire surface of the cups. During incubation, the diameter of the colonies along the outer edge was measured every day and the growth rate (V) was calculated according the formula (1):

$$V = \frac{\mathrm{d2} - \mathrm{d1}}{\mathrm{t2} - \mathrm{t1}} \quad (1)$$

where V is growth rate, mm/h; $d_2 - d_1$ is the colony diameter, mm; $t_2 - t_1$ is the time, h.

2.4 Fungistatic Activity of Isolates *Trichoderma* Pers

Determination of the fungistatic activity of *Trichoderma* Pers. regional phytopathogenic micromycetes such as *Bipolaris sorokiniana* 75SZ, *Fusarium sporotrichioides* 62SZ, *Alternaria alternata* 22SZ. from the collection of the agrobiotechnological laboratory of the Department of Biochemistry, Biotechnology and Pharmacology of the Kazan Federal University were used as test cultures.

Micromycetes were plated by a prick on agar medium in Petri dishes in diametrically opposite direction. The control was pure cultures of the studied fungi. The cultivation was carried out at 20.0 ± 1.0 °C for 7 days. The growth and nature of the interaction of micromycetes was monitored for 14 days every 24 h.

To describe the types of relationships between the studied organisms, the Jackson and Karl scale was used [21] (Table 1).

Table 1. The scale of the types of relationships between the studied micromycetes

Type of reaction		Score
Designation	Characteristic	
A	Mixed growth of two organisms	0
B	Mutual suppression on contact; after the contact of the colonies, the growth of both stops	1
B1	Mutual suppression on contact; after a while, the antagonist continues to grow on top of the colony of the suppressed organism	2
C	Mutual suppression at a distance	3
D	Suppression of one organism on contact; the antagonist overgrows the suppressed colony	4
E	Suppression of one organism on contact; the antagonist continues to grow at a constant or lower speed over the colony of the suppressed organism	5

For each isolate antagonism index (AI) was calculated according to formula (2):

$$AI = \mathrm{B}\,(n \times 1) + \mathrm{B1}\,(n \times 2) + \mathrm{C}\,(n \times 3) + \mathrm{D}\,(n \times 4) + \mathrm{E}\,(n \times 5), \quad (2)$$

where A, B, B1, C, D, E are types of reactions; *n* is frequency of reactions; 1–5 are score.

2.5 Toxic Effect of *Trichoderma* Pers

Primary toxicological studies were carried out on white Wistar rats weighing 200 ± 20 g in accordance with the standards [6, 9]. Acute toxicity parameters were determined by [14] with a single intragastric administration of a spore suspension of *Trichoderma* at a dose of $1.0 \pm 0.001 \times 10^8$ spores/kg of animal weight (3 males, 3 females). The choice of dose was made in accordance with the expected consumption rate and in accordance with physiologically grounded standards for administration to laboratory animals. The observation period for the animals is 14 days.

Determination of acute dermal toxicity was carried out by applying a napkin soaked in the suspension (at a dose of $0.25 \pm 0.001 \times 10^8$ spores/kg) on the epilated skin area of the lateral surface of the rat's torso (3 males, 3 females) with an area of 16 cm^2. After 4 h, the suspension was washed off the skin with sterile water.

The study of cumulative properties was carried out under conditions of intragastric administration of the drug 5 times a week for 8 weeks at a dose of $1.0 \pm 0.001 \times 10^7$ spores/kg, equal to 1/10 of that tested in an acute experiment (3 males, 3 females). In all experiments, control animals were injected with sterile water.

The general condition of the animals in the control and experimental groups was assessed by motor activity, the nature of food and water consumption, the state of the mucous membranes, hair and skin, body weight, and clinical symptoms of intoxication were noted. The rats were weighed before and after the end of drug administration and the recovery period; when studying the cumulative effect – once a week.

Clinical analyzes of the peripheral blood of white rats were performed according to standard hematological methods. Before the start of the experiments, the analysis of the intestinal microflora of rats was carried out in all variants. In an acute experiment, the material for analysis was taken before and 24 h after the introduction of the suspension; when studying the cumulative effect, the material for analysis was taken monthly; after the end of the course of administration and the recovery period, a final study of the intestinal microbocenosis was carried out. The inoculum suspension was prepared in phosphate-buffered saline, pH 7.2–7.4. Inoculations of 10-fold dilutions were carried out on selective media no later than 1 h after sampling the material. Species identification of the isolated bacteria was carried out according to [8, 23, 25], fungi – according to [20].

At the end of the observation period, the animals were euthanized for the purpose of autopsy and postmortem examination, including analysis of the macroscopic picture of the organs of the thoracic and abdominal cavities. Microscopic examination of the structure of the heart, lungs, immune defense organs (spleen, thymus, lymph nodes), liver, kidneys, intestines was carried out after the preparation of histological preparations (paraffin sections 7 μm thick, stained with hematoxylin-eosin).

The dissemination effect in acute and cumulative experiments was determined by the method of imprint sections of organs on PGA. The growth of *Trichoderma* colonies on PGA was considered as a criterion for dissemination.

2.6 Statistical Analysis

The experiments were carried out in three to six replicates. Statistical processing of the experimental results and assessment of the reliability of differences were carried out according to the Student's test for a probability level of at least 95% using the Microsoft Excel 2007 and Statistica 8.0 software package.

3 Results

3.1 Kinetic Parameters of Fungi Growth

97 isolates of fungi belonging to the genus *Trichoderma* Pers. were isolated from soil samples.

The linear growth rate of the isolates varied over a wide range, from 0.04 ± 0.01 to 0.49 ± 0.02 mm/h, depending on the isolate and the cultivation time. Nine isolates of all studied had the highest growth rate. As shown in Fig. 1 data, the rate of mycelial growth of nine isolates on the first day of incubation did not exceed 0.125 ± 0.01 mm/h (Fig. 1). From 24 to 72 h of incubation, the growth rate increased significantly and ranged from 0.128 ± 0.01 to 0.49 ± 0.02 mm/h, depending on the isolate. Starting from 72 h, the growth rate became linear, after 120 h of incubation, it decreased. Isolates 189, 188, 190 70, 84 completely colonized the surface of the Petri dish in 120 h (Fig. 1A), and isolates 191, 195, 184, 92 - in 144 h (Fig. 1B).

Taking into account the results obtained, the isolated isolates can be arranged in a row according to the criterion of increasing growth rate:

$92 < 184 < 195 < 191 < 84 < 70 < 190 < 188 \leq 189$.

3.2 Fungistatic Activity of *Trichoderma* Pers

The most common types of relationship between *Trichoderma* isolates and test cultures were type B (Fig. 2A), B1 (Fig. 2B), and D (Fig. 2C) responses. An E-type interaction (Fig. 2D) was characteristic of only three isolates – 84, 189, 190. None of the isolates showed type A and C reactions.

As can be seen from Fig. 3, the highest AI against test cultures (AI values from 22 to 27) was characteristic of *Trichoderma spp.* 84, *Trichoderma spp.* 189, *Trichoderma spp.* 195, isolated from the soils of the Elansky district of the Volgograd region (84) and the Kukmorsky district of Tatarstan (189, 195).

Based on the results of PCR identification, it was established that the isolates belong to the species *T. cerinum* strain 84, *T. tomentosum* strain 189, *T. asperellum*, strain 195.

3.3 Toxicological Effectiveness of Studied Fungi in Vivo

The study of acute toxicity with intragastric administration of suspensions of *T. cerinum* 84, *T. tomentosum* 189 and *T. asperellum* 195 did not reveal signs of intoxication of animals both immediately after administration and during the subsequent observation period. Consumption of food and water, physical activity corresponded to the norm. When analyzing hematological parameters, no statistically significant differences from

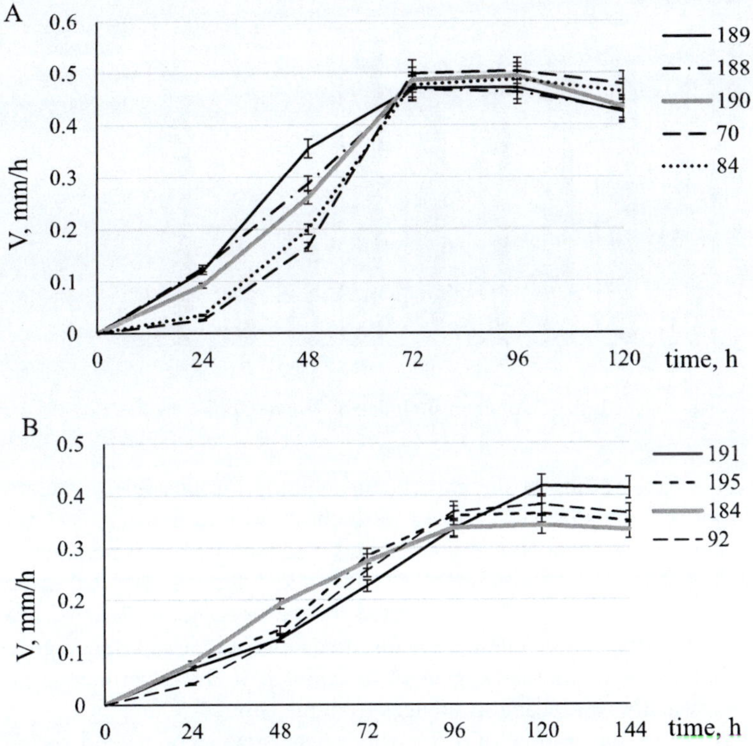

Fig. 1. Growth of *Trichoderma* Pers. isolates. A – strains 189, 188, 190 70, 84; B – strains 191, 195, 184, 92.

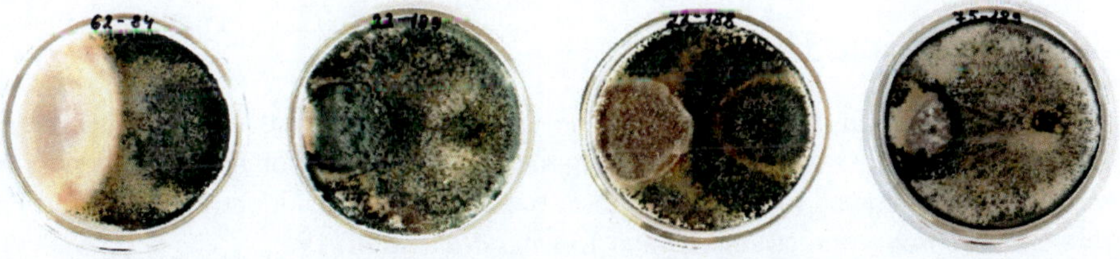

Fig. 2. Representative images of type B (A), B1 (B), D (C), E (D) relationships

the control were revealed. The study of the macro- and microscopic structure of the internal organs of animals showed the absence of morphological changes. The death of laboratory animals during the entire observation period was not revealed. Due to the absence of animal death, the LD50 exceeds the dose of $1.0 \pm 0.001 \times 10^8$ spores/kg of animal body weight.

When determining acute dermal toxicity under the conditions of cutaneous application, no irritating effect was revealed either after the application of the suspension or during the recovery period.

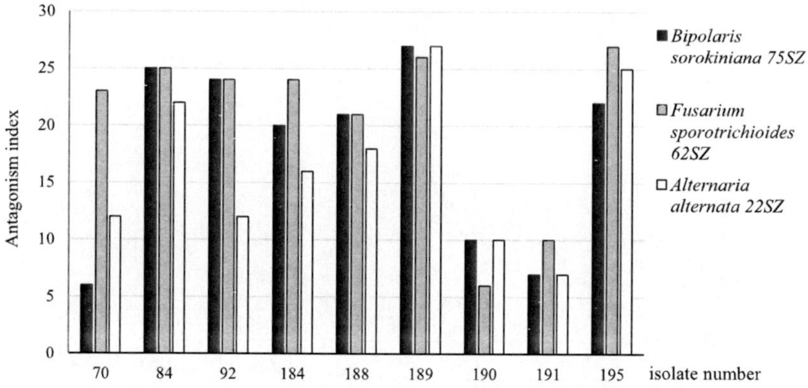

Fig. 3. Antagonism index of *Trichoderma* Pers.

Analysis of the results of the experiment to study the possible cumulative effect showed a positive dynamics of changes in the body weight of animals in the experimental and control groups. Hematological and biochemical blood parameters did not differ significantly from the control level. Anatomical and pathological disorders and deviations from the norm were not detected by microscopic analysis of histological preparations from the side of internal organs; no deaths of animals were registered. The mass coefficients of the internal organs of the animals in the experimental groups did not have statistically significant differences from the control.

Intragastric administration of suspensions was not accompanied by significant changes in aerobic and facultative anaerobic intestinal microorganisms.

During microbiological analysis of inoculations from sections of internal organs, in none of the series of experiments a pure culture of *Trichoderma* was isolated.

4 Discussion and Conclusion

97 micromycete strains were isolated, phenotypically assigned to the genus *Trichoderma* Pers. Study of the growth kinetics and fungistatic activity of isolates of the genus *Trichoderma* Pers. allowed to select the most competitive isolates identified as *T. cerinum* strain 84, *T. tomentosum* strain 189 and *T. asperellum* strain 195.

According to the results of the assessment of acute oral toxicity, the LD50 of suspensions for white rats with a single intragastric administration exceeds the dose of $1.0 \pm 0.001 \times 10^8$ spores/kg. The strains do not have an irritating effect when applied to the skin in doses less than or equal to $0.25 \pm 0.001 \times 10^8$ spores/kg. The cumulative properties of the drug when administered orally at a dose of $1.0 \pm 0.001 \times 10^7$ spores/kg for 8 weeks are not expressed.

The results obtained indicate the prospects for further study of the properties of *T. cerinum* strain 84, *T. tomentosum* strain 189 and *T. asperellum* strain 195 for the purpose of creating biofungicides.

Acknowledgments. This paper has been supported by the Kazan Federal University Strategic Academic Leadership Program (PRIORITY-2030).

References

1. Al-Dhabaan, F.A.M., Bakhali, A.H.: Analysis of the bacterial strains using Biolog platesin the contaminated soil from Riyadh community. Saudi J. Biol. Sci. (2016) (In press)
2. Bosch, F., Blake, J., Gosling, P., Helps, J., Paveley, N.D.: Identifying when it is financially beneficial to increase or decrease fungicide dose as resistance develops: an evaluation from long-term field experiments. Plant. Pathol. **4**, 631–641 (2020)
3. Chou, C., Castilla, N., Hadi, B., Tanaka, T., Chiba, S., Sato, I.: Rice blast management in Cambodian rice fields using *Trichoderma harzianum* and a resistant variety. Crop Protect. **135**, 104864 (2020)
4. Daguerre, Y., Siegel, K., Edel-Hermann, V., Steinberg, C.: Fungal proteins and genes associated with biocontrol mechanisms of soil-borne pathogens: a review. Fungal Biol. Rev. **28**(4), 97–125 (2014)
5. De Santiago, A., Recena, R., Perea-Torres, F., Moreno, M.T., Carmona, E., Delgado, A.: Relationship of soil fertility to biochemical properties under agricultural practices aimed at controlling land degradation. Land Degrad. Dev. **30**, 1121–1129 (2019)
6. Directive 2010/63/EU of the European Parliament and of the Council of 22 September 2010 on the protection of animals used for scientific purposes. Text with EEA relevance. http://eur-lex.europa.eu/legal-content/EN/TXT/?uri=celex:32010L0063
7. Gams, W., Bissett, J.: Morphology and identification of *Trichoderma*. In: Kubicek, C.P., Harman, G.E. (eds.) *Trichoderma* and Gliocladium. Basic Biology, Taxonomy and Genetics, vol. 1, pp. 3–34. Taylor and Francis Ltd., London, UK
8. Garrity, G.M., Brenner, D.J., Krieg, N.R., Staley, J.T.: Bergey's manual of systematic bacteriology. In: Brenner, D.J., Krieg, N.R., Staley, J.R. (eds.) Volume Two The Proteobacteria Part B The Gammaproteobacteria. New York: Springer (2005). https://doi.org/10.1007/0-387-28022-7
9. Guide for the Care and Use of Laboratory Animals. The National academies press, USA (2011)
10. Kredics, L., et al.: The *Trichoderma* reesei proteasome. In: Mach, R.L., Zeilinger, S. (eds.) 9th International Workshop on *Trichoderma* and Gliocladium. Vienna University of Technology, Austria (2006)
11. Munoz, F.M., Demmler, G.J., Travis, W.R., Ogden, A.K., Rossmann, S.N., Rinaldi, M.G.: *Trichoderma longibrachiatum* infection in a pediatric patient with aplastic anemia. J. Clin. Microbiol. **35**, 499–503 (1997)
12. Naher, L., Yusuf, U.K., Ismail, A., Hossein, K.: *Trichoderma spp.*: a biocontrol agent for sustainable management of plant diseases. Pakistan J. Botany **46**(4), 1489–1493(2014)
13. Nicolopoulou-Stamati, P., Maipas, S., Kotampasi, C., Stamatis, P., Hens, L.: Chemical pesticides and human health: the urgent need for a new concept in agriculture. Front. Public Health **4**, 1–8 (2016)
14. OECD/OCDE 420 adopted 17 December 2001: OECD Guidelines for the testing of chemicals acute oral toxicity-method by acute toxic class
15. Pogacean, O., Gavrilescu, M.: Plant protection products and their sustainable and environmentally friendly use. Environ. Eng. Manag. J. **8**(3), 607–627 (2009)
16. Prashar, P., Shah, S.: Impact of fertilizers and pesticides on soil microflora in agriculture. Sustain. Agric. Rev. **19**, 331–361 (2016)
17. Pryce, T.M., Palladino, S., Kay, I.D., Coombs, G.W.: Rapid identification of fungi by sequencing the ITS1 and ITS2 regions using an automated capillary electrophoresis system. Med. Mycol. **41**(5), 369–381 (2003)
18. Ribot, C., et al.: Susceptibility of rice to the blast fungus, *Magnaporthe grisea*. J. Plant Physiol. **165**, 114–124 (2008)

19. Samuels, G.J., Petrini, O., Kuhls, K., Lieckfeldt, E., Kubicek, C.P.: The *Hypocrea schweinitzii* complex and *Trichoderma* sect *Longibrachitum*. Stud. Mycol. **41**, 1–54 (1998)
20. Samuels, G.J.: *Trichoderma*: systematic, the sexual state, and ecology. Phytopathology **96**(2), 195–206 (2006)
21. Seiketov, G.: Fungi of the genus *Trichoderma* and their use in practice. Science, Alma-Ata (1982). (In Russian)
22. Siraeva, Z.: Efficiency of adaptive-integrated potato protection on high infection background. Biomics **12**(3), 367–375 (2020). (In Russian)
23. Vos, P., et al.: Bergey's Manual of Systematic Bacteriology. Springer, New York (2009).https://doi.org/10.1007/978-0-387-68489-5
24. Waghunde, R., Sabalpara, A.N., Shelake, R.M.: *Trichoderma*: a significant fungus for agriculture and environment. Afr. J. Agric. Res. **11**(22), 1952–1965 (2016)
25. Whitman, W., et al.: Bergey's Manual of Systematic Bacteriology. Springer, New York (2012). https://doi.org/10.1007/978-0-387-68233-4

Study of the Composition and *in Vivo* Effect of Biologically Active Concentrates from Secondary Raw Materials of Millet

Alexandr Evteev, Olga Larionova, and Anna Bannikova(✉)

Saratov State Agrarian University, named after N.I. Vavilov, Saratov, Russia

Abstract. This work is related with the study of the properties of concentrates of biologically active substances (BAS) - polyphenols and xylooligosaccharides (XOS) from the secondary raw materials of millet (husk), and an assessment of their effect on the organism of laboratory animals. BAS was characterized by the content of the mass fraction of protein, moisture, starch, fiber, reducing substances, the qualitative and quantitative composition of phenolic substances, the fractional composition of carbohydrates, the monosaccharide composition of polysaccharides, the qualitative and quantitative compositions of XOS concentrates. The obtained BAS concentrates contain protein from 0.9%, carbohydrates - 91.50%, including XOS with prebiotic properties - 68.50% and ash 6.30%. The polyphenol concentrate is mostly represented by ferulic acid (33.47%). In the polyphenol concentrate, the yield of ferulic acid increased by 19%, gallic acid - by 2.5%, as compared to the BAS concentrate. Autopsy of laboratory animals, which were injected with 4x-chloride intraperitoneally four times during the experiment, did not reveal any visible pathological changes. Due to the action of the tested additive, the liver, with a pronounced clinical picture of acute hepatosis, which turned into a chronic form, retained its functions, which makes it possible to recommend the obtained concentrate for gastrointestinal diseases. With the introduction of epinephrine hydrochloride, which provokes degenerative changes in the structure of the heart muscle, no atomic changes were found. Thus, the possibility of practical use of millet husk as a source of antioxidants and prebiotics has been shown.

Keywords: Polyphenols · Biologically active substances · Secondary raw materials · Xylooligosaccharides · Ferulic acid · Laboratory animals · Hepatosis

1 Introduction

The processing of agricultural products is always associated with the formation of a significant amount of secondary processing products. Thousands of tons of millet grain are processed annually in the world, and a significant part of the millet husk, as a by-product of industrial processing, is processed. Millet husk is an unused waste of processing millet grain into cereals, which, on average, makes up about 17% of the grain mass [1].

V. Kurchenko et al. (Eds.): ICAETT 2021, LNNS 408, pp. 65–74, 2022.
https://doi.org/10.1007/978-3-030-96641-6_8

It is known that a significant part of antioxidants is contained in the outer shell of the grain of millet cereals. In this regard, products containing whole grains or bran are of considerable interest [2]. In addition, millet shell contains vitamins, trace elements, mono- and oligosaccharides, polyphenolic compounds and other biologically active substances. The antioxidants of millet include hydroxyaromatic acids, represented mostly by derivatives of benzoic and cinnamic acids, such as gallic, ferulic, caffeic, lilac, n-hydroxybenzoic, vanillic, coumaric, salicylic, protocatechic and others [4, 7, 8]. These acids are contained in the shells of millet grain both in a free and bound state, while a significant part of free acids falls on the outer shell and can be easily extracted with organic solvents. Bound oxyaromatic acids mostly include acids contained in the cell membrane, and acid or alkaline hydrolysis is applicable for their release [11, 13].

Xylans, polysaccharides contained in by-products of millet husk processing, can be transformed by external factors into XOS. From the analysis of literature data, it is known that xylooligosaccharides, to a greater extent, exhibit prebiotic and antioxidant properties, and are also capable of suppressing the activity of some pathogenic and enterotid intestinal bacteria, and act as anti-inflammatory and anti-allergic agents. Of these properties, the prebiotic activity of CBS, aimed at stimulating the growth of probiotic microflora of the intestines of animals and humans, is of greater interest [6–9, 12, 14, 15].

Despite the abundance of information on the chemical composition of millet grain, secondary products of millet processing are poorly studied in the future of their use as a source of biologically active substances (BAS). The study of the chemical composition and the study of the possibility of biotransformation of millet husk is an urgent direction, since it will allow assessing the possibility of using these secondary products as a new source of biologically active substances of plant origin. Biomodification of millet polymers in this work is aimed at obtaining new food functional ingredients and biologically active substances, and the method of enzymatic hydrolysis is a method of direct action on the protein-carbohydrate matrix of raw materials by depolymerase enzymes [2].

Thus, the purpose of this study is to study the complex of properties of biologically active substances from secondary raw materials of millet and to assess their effect on the organism of laboratory animals.

2 Materials and Methods

2.1 Raw Materials and Preparations Used in Research

The following were chosen as raw materials and enzyme preparations used in the experimental study:

- millet of the "Saratovskoe yellow" variety, harvest of 2020, obtained from the Povolzhye Scientific and Production Association of the Saratov State Agrarian University;

– industrial enzyme preparations, "GlukoLux A" (Glucoamylase activity (units/ml): 30 °C - 13000 ± 1300, 60 °C - 80,000 ± 8000), "CelloLux A" (Xylanase activity (KsA) - 1700 units/ml, Cellulolytic activity (ClA) - 6000 units/ml), "AmyloLux-A" (3000 units/ml) (Amylolytic activity (units/ml) - 3200 ± 320), "Protosubtilin G3x" (A - 120 units/d) (proteolytic activity - 120 units/g) produced by PA "Sibbiopharm".

2.2 Obtaining of the Concentrates

The fraction containing the outer shells of the millet grain "husk" was selected for subsequent grinding on a rotary mill LMC-1M with replaceable sieves, sieves were used 0.2 mm, the separation of fractions was carried out using sieves with a mesh size of 1.0 mm, 0.80 mm, 0, 63 mm, 0.56 mm, 0.315 mm. The resulting "flour" from husk, with the content of the grain part of millet, was used for extraction of biologically active substances, the quality control of grinding "flour" was carried out by sieving through a sieve of 0.132 mm.

The process of obtaining BAS concentrates includes multistage extraction using enzyme preparations "AmiloLux-A (3000 units/ml)", "GlukoLux A", "CelloLux A", "AmiloLux A (3000 units/ml)" and ultrasonic treatment according to the protocol described earlier, with minor changes [2].

The XOS and PF concentrates were obtained by the method of alcohol extraction. The alcohol from the combined ethanol extracts is distilled off on a rotary evaporator in a discharged medium at 60 ± 5 °C, with drying up to 30%, and freeze-dried to a moisture content of 8 ± 1%. Polyphenol (PF) concentrate is a yellow - brown or light brown fine powder with a faint vanilla - grain odor. Xylooligosaccharides precipitated by ethanol solution, similarly, were lyophilized to a moisture content of 8 ± 1% and look like a light-brown powder with a weak grain-sweet odor.

2.3 Methods

The mass fraction of crude protein was determined by the Kjeldahl method in accordance with GOST 10846-91 "Grain and products of its processing. Protein determination method";

The mass fraction of fat was determined by the Soxhlet extraction method in accordance with GOST 29033-91 "Grain and products of its processing. Method for determination of fat".

The mass fraction of starch was determined by the Ewers polarimetric method according to GOST 10845-98 "Grain and products of its processing. Method for determination of starch".

The mass fraction of fiber was determined according to GOST 13496.2-91 "Feed, compound feed, compound feed raw materials. Method for determination of crude fiber".

The mass fraction of reducing substances was determined according to GOST 5903-89. "Confectionery products. Methods for the determination of sugar".

The activity of hydrogen ions (pH) was determined by the potentiometric method using a pH-meter "Akvilon" pH-420.

Determination of moisture content was carried out using the drying method. The mass fraction of dry substances in the extracts was determined using a refractometer.

The content of minerals (ash) was determined according to GOST 26226-95 "Feed, compound feed, compound feed raw materials. Methods for determination of raw ash".

The mass fraction of phenolic substances was determined by the colorimetric method [2, 5].

To quantitatively determine the composition of phenolic substances in the analyzed sample, we used the HPLC method on a Stayer chromatograph (NPO Akvilon, Russia). The identification of substances was carried out by comparing the retention time and spectral characteristics of the investigated substances with similar characteristics of analytical standards [14].

The fractional composition of carbohydrates was determined by the gravimetric method based on the sequential isolation of fractions of water-soluble polysaccharides, pectin substances, hemicellulose A and B [10, 14, 15].

The qualitative composition of the obtained KOS concentrates was determined using thin layer chromatography (TLC) on 10 × 15 cm SORBFIL plates with CTX-1A silica gel. Mobile phase, n-propanol: ethyl acetate: distilled water, in a ratio of 6: 1: 3 parts by volume. After elution, the plates were treated with a developer, a 50% aqueous solution of sulfuric acid, and dried at a temperature of 120 ± 1 °C for 5 min.

The quantitative determination of CBS was carried out by the spectrophotometric method. For this, sorbent areas corresponding to the zones of the analyzed carbohydrates were scraped off from the SORBFIL plates, after determining the qualitative composition of the CBS, and transferred to glass tubes with stoppers, adding 0.5 cm^3 of aniline phthalate reagent to each sample, followed by heating for an hour at 110° With in a drying cabinet. For the preparation of aniline phthalate reagent, 1.66 g of o-phthalic acid and 0.91 cm^3 of aniline are placed in a 100 cm^3 flask, 48 cm^3 of n-butanol and 4 cm^3 of distilled water are added, brought to the mark with diethyl ether. The resulting solution of the determined carbohydrates is cooled to room temperature, stirring thoroughly, add 4 cm^3 of a mixture of concentrated hydrochloric acid and acetone (in a ratio of 1: 25 parts by volume) and incubate for 1 h in a dark place at room temperature, then centrifuge for 15 min. at 8000 rpm and photometric at 520 nm. The concentration of the determined carbohydrates in the sample was determined using a calibration graph.

The monosaccharide composition of the obtained polysaccharides was determined by hydrolysis of these polysaccharides with a solution at a temperature of 100 °C with a solution of 1 mol/L sulfuric acid [3, 14].

The qualitative composition of monosaccharides was determined by TLC [2].

To determine the effect of the BAS concentrate obtained during the study on laboratory animals (young rats) with a single intake into the body, 5 groups of laboratory animals were created, 5 heads each weighing 200–220 g:

Group 1: 4 females and 1 male received the test preparation. During the experiment, 4x-carbon chloride was injected 4 times intraperitoneally.

Group 2: 4 females and 1 male received IMO 900 isomaltooligosaccharide (BaoLingBao, China). During the experiment, epinephrine hydrochloride was injected 4 times into the tail vein intravenously.

Group 3: 4 females and 1 male received isomaltooligosaccharide IMO 900. During the experiment, 4x-carbon chloride was injected intraperitoneally 4 times.

Group 4: 4 females and 1 male received the test preparation. During the experiment, epinephrine hydrochloride was injected intravenously into the tail vein 4 times.

The nutritional supplement was administered orally. The experiment used clinically healthy young rats sustained on a 12-h fasting diet. The food supplement was administered in the morning on an empty stomach. The introduction was performed using a disposable insulin syringe and a venous catheter. The rats were fixed in an upright position with their heads thrown back. Food and water were given 2 h after administration. The dose of the added additive is a concentrate of 3 ml inside with water per head per day.

Daily for 1 head: crushed grain mixture (barley - 10 g, wheat - 5 g, corn - 5 g, fodder chickpeas - 4 g) - 24 g, vegetables (carrot - 5 g, apple - 5 g) - 10 g.

Additionally, once a week - a raw chicken egg weighing from 40 to 60 g for 5 heads.

Feeding frequency: two times (morning and evening) with free water intake. The account of the completeness of the feed intake was carried out daily. In order to avoid the influence of side factors (fluctuations in ambient temperature, disturbances in feeding and maintenance, etc.), animals of all groups (three experimental and control) were kept in the same conditions.

3 Results

The obtained BAS concentrates, in addition to polyphenols, contain: protein from 0.9%, carbohydrates 91.50%, including XOS with prebiotic properties - 68.50% and ash 6.30% (Table 1).

Table 1. Physicochemical composition of concentrates of biologically active substances obtained from the products of enzymatic hydrolysis of millet husk

Mass fraction of moisture,%	Mass fraction of protein in terms of dry matter,%	Mass fraction of ash in terms of dry matter,%	Mass fraction of carbohydrates		Mass fraction of polyphenols in terms of dry matter,%
			XOS in terms of dry matter,%	Residual carbohydrates in terms of dry matter,%	
27.20	0.90	6.30	68.50	23.00	0.98

To obtain a XOS of food grade, the hydrolyzate must be maximally purified from accompanying substances. Extraction with organic solvents can remove non-carbohydrate components from solution, as well as phenolic compounds and other extractives. The product yield and the degree of purification depend on the type of solvent used: methanol, ethanol, propanol, etc. In our case, purification and fractionation, separation of polyphenols and xylooligosaccharides from the concentrate of biologically active substances were carried out by extraction with ethyl alcohol, the ratio of the hydrolyzate of the BAS concentrate to 70% aqueous ethanol solution was 1: 3 parts by volume. As a result of the effect of ethanol on the hydrolyzate, the concentrate is stratified, the polyphenol compounds dissolve and pass into the ethyl alcohol solution, the xylooligosaccharides are precipitated. Centrifugation of the resulting solution for 25 min at 5000 rpm allows to separate fractions of biologically active substances. The physicochemical composition of the XOS concentrate is presented in Table 2.

Table 2. Physicochemical composition of xylooligosaccharide concentrate, in absolutely dry matter

Indicators		Results,%
Moisture content		7.40
Mass fraction of ash, calculated on dry matter		4.26
Mass fraction of crude protein, calculated on dry matter		3.17
Carbohydrates (total), calculated on dry matter	Xylooligosaccharides	78.29
	Monosaccharides	14.17
Xylooligosaccharide composition, calculated on dry matter	Xylose	10.59
	Xylobiose	11.61
	Xylotriosis	18.44
	Xylotetrosa	18.91
	Xylopentose	14.98
	Other	3.77
Monosaccharide composition, calculated on dry matter	Arabinose	2.35
	Glucose	8.25
	Other monosaccharides	3.56

As a result of the conducted studies of the PF concentrate obtained from millet husk, mainly rutin, chlorogenic, gallic and ferulic acids were identified.

HPLC analysis of extracts of model samples of millet husk fermentalizates showed that phenolic profiles differed from those previously observed in soluble and bound fractions of raw millet grain (Fig. 1).

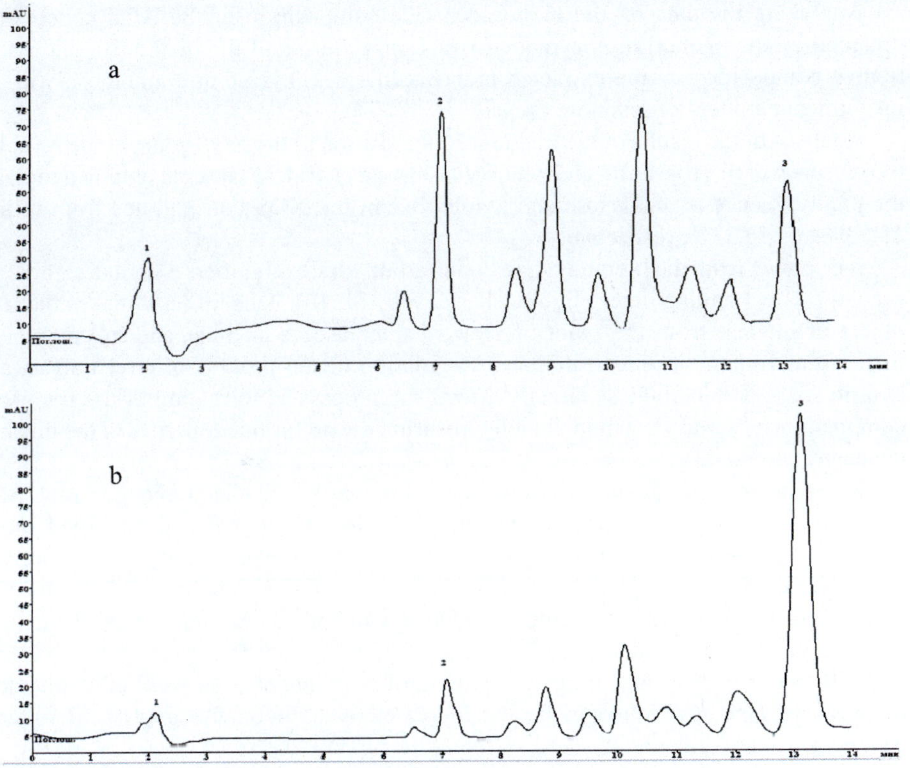

Fig. 1. Chromatogram of a sample of millet husk of the "Saratovskoe yellow" variety (a) and polyphenol concentrate obtained from the products of enzymatic hydrolysis of millet husk (b). No. 1 - gallic acid, No. 2 - chlorogenic acid, No. 3 - ferulic acid.

4 Discussion

The development of an integrated technology for processing millet husk with the use of hydrolytic enzymes to ensure the production of a number of functional ingredients was based on the results of numerous experimental studies. The main stage in the implementation of the task was to conduct research to optimize the key technological parameters of the process of enzymatic processing of husk, development and justification of a complex of operations for processing secondary grain raw materials, optimization of the parameters of the process of enzymatic hydrolysis of millet husk. Hull biomodification or enzymatic hydrolysis is the main technological process for obtaining of functional ingredients [2–4, 15]. This method is based on the extraction of biologically active substances using the selective activity of enzyme preparations together with the effect of temperature, hydromodule, dispersion of raw materials, quality of homogenization and ultrasonic action.

Analyzing the data on the monosaccharide composition of the XOS concentrate obtained by the method of enzymatic extraction, in terms of the qualitative and quantitative composition of the extracted monosaccharides, xylose and arabinose with an insignificant content of mannose prevail.

Analysis of the results of the fractional composition of the XOS concentrate revealed the presence of di-, tri-, tetra- and pentaxylooligosaccharides. The data obtained indicate the predominance of xylotriose and xylotetrose in the XOS concentrates from millet, 15.83% and 16.23%, respectively.

It is known from the literature that xylan carbohydrate oligomers exhibit a significant prebiotic effect among other oligosaccharides [9, 14, 15]. This, in turn, makes them an object of interest from the point of view of their use as an independent component for food and pharmaceutical products. Regulation of the process of enzymolysis and hydrolysis of hemicelluloses directly affects the process of formation of the fractional composition of waste water and, thereby, ensuring the prebiotic properties of the finished concentrates.

In the process of enzymatic hydrolysis, there was a significant change in the fractional composition of the extracted oxycinnamic acids, which make up the polyphenolic compounds of millet husk. This change is associated with the peculiarity of the processes during the enzymatic hydrolysis of raw materials, the release due to the destruction of ether bonds of oligomers, molecules of hydroxycinnamic acids and incomplete extraction of only free polyphenols during the extraction of raw materials with a solution of methanol [5, 9, 10]. So, in the concentrate of polyphenols, the yield of ferulic acid increased by 19%, the yield of gallic acid increased by 2.5%, but the yield of chlorogenic acid decreased by 13%.

At the end of the *in vivo* experiment, blood samples were taken from the animals for general and biochemical analysis. A postmortem examination was performed. As a result of the autopsy, the following pathological picture was observed:

Group 1: the fatness of laboratory animals is above average, the membranes of the brain are moist, shiny, gray-milky; the organs of the abdominal and thoracic cavity are correctly located, the lungs are pink, airy, there are no signs of inflammation, the trachea is unchanged, the esophagus is normal, the stomach is not hyperemic, the mucous walls of pink color do not exfoliate, the intestines are without visible changes, the heart is unchanged, the liver is slightly loose, dark cherry color. The borders of the cortical and medullary layers of the kidneys are well defined, the spleen is not enlarged, with a dense consistency. No visible pathological changes were found, with the exception of the structure of the liver tissue, which is due to artificially induced hepatosis. Due to the action of the tested supplement, the liver retained its functions with a pronounced clinical picture of acute hepatosis that turned into a chronic form.

Group 2: the fatness of laboratory animals is above average, the membranes of the brain are moist, shiny, gray-milky; the organs of the abdominal and thoracic cavity are correctly located, the lungs are pink, airy, there are no signs of inflammation, the trachea is unchanged, the esophagus is normal, the stomach is not hyperemic, the mucous walls of pink color do not exfoliate, the intestines are without visible changes, the heart is hypertrophied, the heart muscle is flabby, the kidneys have punctate hemorrhages, the border of the cortical and medullary layers is smoothed. No visible pathological changes were found except for the structure of the heart muscle and hemorrhages on the surface of the kidneys. In this laboratory animal, degenerative changes in the structure of the heart muscle are observed, which is a consequence of the administration of epinephrine hydrochloride. The use of a commercial drug for artificially induced arrhythmias did not have the desired therapeutic effect.

Group 3: the fatness of laboratory animals is above average, the membranes of the brain are moist, shiny, gray-milky; the organs of the abdominal and thoracic cavity are correctly located, the lungs are pink, airy, there are no signs of inflammation, the trachea is unchanged, the esophagus is normal, the stomach is not hyperemic, the mucous walls of pink color do not exfoliate, the intestines are without visible changes, the heart is unchanged, the liver is loose, dark cherry color, with signs of fatty degeneration. The boundaries of the cortical and medullary layers of the kidneys are poorly expressed, the spleen is not enlarged, and of dense consistency. No visible pathological changes were found, with the exception of the structure of the liver tissue, the presence of fatty degeneration, which is due to artificially induced hepatosis. Despite the use of a commercial supplement, the desired therapeutic effect has not been established, the symptoms of chronic hepatosis are pronounced.

Group 4: the fatness of laboratory animals is above average, the membranes of the brain are moist, shiny, gray-milky; the organs of the abdominal and thoracic cavity are correctly located, the lungs are pink, airy, there are no signs of inflammation, the trachea is unchanged, the esophagus is normal, the stomach is not hyperemic, the mucous walls of pink color do not exfoliate, the intestines are without visible changes, the heart is of normal size, signs of cardiac dystrophy muscles are not observed, the liver is normal, the kidneys are unchanged, the spleen is not enlarged. Against the background of the administration of epinephrine hydrochloride, the test drug had a good therapeutic effect.

5 Conclusion

In this work, a study of the physicochemical properties of biologically active substances from millet husk of the "Saratovskoe yellow" variety, harvest of 2020, - concentrates of KOS and polyphenols was carried out. An analysis is carried out for a number of physicochemical indicators to establish its component composition, and its characteristics are given.

The physicochemical and biological properties of the obtained concentrates of XOS and polyphenolic substances were studied. The concentrate of polyphenols is represented to a greater extent by ferulic acid, up to 33.47% in the concentrate of biologically active substances in terms of dry matter. The XOS concentrate consists mainly of XOS fragments with prebiotic properties - up to 78.29% in terms of dry matter.

In the study of the effect of the BAS concentrate on the organism of laboratory animals, it was revealed that the drug obtained in the study has therapeutic properties in artificially induced hepatosis and arrhythmias.

Thus, as a result of the work carried out, the potential significance of the practical use of millet husk processing products as a source of biologically active substances with antioxidant activity and prebiotic properties with a confirmed therapeutic effect was demonstrated.

Acknowledgments. The authors express their gratitude to the Grant of the President of the Russian Federation for the support of young scientists, doctors of science MD-1551.2020.11.

References

1. Althubiani, A.S., et al.: Chapter 4 - plant-derived prebiotics and its health benefits. In: Khan, M.S.A., Chattopadhyay, I.A.D. (eds.) New Look to Phytomedicine Advancements in Herbal Products as Novel Drug Leads, pp. 63–88. Elsivier Science, Amsterdam (2019)
2. Bityukova, A.V., Amelkina, A.A., Evteev, A.V., Bannikova, A.V.: Development of technology for obtaining phyto-substances from secondary products of grain processing. Tech. Technol. Food Product. **1**(49), 5–13 (2019)
3. Chen, J., Yang, J., Ma, L., Li, J., Shahzad, N., Kim, C.K.: Structure-antioxidant activity relationship of methoxy, phenolic hydroxyl, and carboxylic acid groups of phenolic acids. Sci. Rep. **10**, 2611 (2020)
4. Verma, D.K., Thakur, M.: Phytochemicals - A Global Perspective of Their Role in Nutrition and Health. CRC Press, Boca Raton (2021)
5. Kaprelyants, L., Zhurlova, O.: Technology of wheat and rye bran biotransformation into functional ingredients. Int. Food Res. J. **24**(5), 1975–1979 (2017)
6. Karnjanapratum, S., Kaewthong, P., Takeungwongtrakul, S., Sae-leaw, T., Hong, J.H., Nalinanon, S.: Production of fiber hydrolysate from bamboo shoot with antioxidative properties by enzymatic hydrolysis. Curr. Appl. Sci. Technol. **3**(3), 535–549 (2019)
7. Kaur, R., Ahluwalia, P., Sachdev, A.P., Kaur, A.: Development of gluten-free cereal bar for gluten intolerant population by using quinoa as major ingredient. J. Food Sci. Technol. **55**(9), 3584–3591 (2018)
8. Liu, Y., Sun, Y., Huang, G.: Preparation and antioxidant activities of important traditional plant polysaccharides. Int. J. Biol. Macromol. **111**, 780–786 (2018)
9. Ognyanova, M., Remoroza, C.A., Schols, H.A., Petkova, N.T., Georgiev, Y.N.: Structural study of a pectic polysaccharide fraction isolated from “mountain tea” (Sideritis scardica Griseb.). Carbohyd. Polym. **260**, 117798 (2021)
10. Pigman, W.: The Carbohydrates: Chemistry and Biochemistry. Elsevier Science, Amsterdam (2012)
11. Singh, R.D., Banerjee, J., Arora, A.: Prebiotic potential of oligosaccharides: a focus on xylan derived oligo-saccharides. Bioact. Carbohyd. Dietary Fibre **5**(1), 19–30 (2015)
12. Tringali, C.: Bioactive Compounds from Natural Sources: Isolation, Characterization and Biological Properties. CRC Press, Health & Fitness (2020)
13. Vassileva, M., Flor-Peregrin, E., Malusá, E., Vassilev, N.: Towards better understanding of the interactions and efficient application of plant beneficial prebiotics, probiotics, postbiotics and synbiotics. Front. Plant Sci. **11**, 1068 (2020)
14. Vermerris, W., Nicholson, R.: Phenolic Compound Chemistry. Springer Science & Business: Science Media (2007). https://doi.org/10.1007/978-1-4020-5164-7
15. Xie, J.-H., Jin, M.-L., Morris, G.A., Zha, X.-Q., Chen, H.-Q., Yang, Y.: Advances on bioactive polysaccharides from medicinal plants. Crit. Rev. Food Sci. Nutr. **56**, S60–S84 (2016)

Isolation of Cellulase-Containing Exosomal Vesicles from *Picea Abies* Xylem

Dmitry Chukhchin, Nikita Shutskiy(✉), and Ksenia Vashukova

Northern (Arctic) Federal University, Arkhangelsk, Russia
{n.shutskij,k.bolotova}@narfu.ru

Abstract. Exosomes of woody plants have important physiological roles in the storage and transport of enzymes. The endo-β-1,4-glucanase may be a component of plant exosomes. The aim of the present study was isolation of exosomal proteins from spruce xylem and analysis of their enzymatic activity. Woody samples were harvested from stem of a Norway spruce (*Picea abies* (L.) Karst.). Exosome-containing fraction was obtained using size-exclusion high-performance liquid chromatography. A scanning electron microscopy was used to visualize exosomes on the inner surface of the cell wall of developing tracheids. The protein concentration was determined in the lyophilisate solution by Bradford method. The cellulolytic activity of the exosome-containing fraction was measured based on the protocol for copper number on filter paper. The endoglucanase activity of the exosomal vesicles was 43.5 nmol/(min·mg protein).

Keywords: Exosomal proteins · Enzymatic activity · Xylem · Endoglucanase · Pits

1 Introduction

The study of the composition and mechanisms of various types of plant exosomes is of great interest in plant physiology. Exosomes of woody plants have physiological roles in the storage and transport of enzymes; in the cell communication and nutrient delivery. There are reports of the presence and practical potential of plant vesicles that resemble exosomes in structure and function [1]. The analysis of the existence of exosome-like vesicles in sunflower fluids is shown [11]. Ultracentrifugation of sunflower extracellular fluids allowed the isolation of particles of 50–200 nm with apparent membrane organization. A small GTPase Rab was putatively identified in this vesicular fraction [11].

Rutter and Innes conducted the first study to isolate and purify plant extracellular vesicles from Arabidopsis (*Arabidopsis thaliana*) leaves [12]. The analysis of proteins from the leaves of Arabidopsis showed the presence of endo-β-1,4-glucanase in the microsomal fraction [13]. This study reports for the frst time that endo-β-1,4-glucanase may be a component of plant exosomes.

The plant cellulase complex contains an incomplete pool of cellulases. According to the literature [8, 10, 13], it contains endo-1,4-β-glucanases (EG), cellobiases and

V. Kurchenko et al. (Eds.): ICAETT 2021, LNNS 408, pp. 75–80, 2022.
https://doi.org/10.1007/978-3-030-96641-6_9

glucohydrolases, but cellobiohydrolases are absent [9]. Endo-1,4-β-glucanases (EGs) are enzymes capable of degrading cellulose by randomly breaking down internal bonds in microfibrils. These enzymes can hydrolyze amorphous cellulose to glucooligosaccharides, but cannot hydrolyze crystalline cellulose, xyloglycan, xylan, $(1 \rightarrow 3)(1 \rightarrow 4)$-β-D-glucan and other polysaccharides and oligosaccharides. The presence of EG was shown in those plant tissues in which cellulose formation takes place: differentiating xylem, phloem, cork, young and mature leaves, root and other [13]. In higher plants, EGs play a key role in the cellulose biosynthesis [5, 10, 13]. Overexpression of EG led to an acceleration of plant growth and an increase in biomass yield [13]. Suppression of EG expression caused a decrease in cellulose content, a decrease in leaf growth, and cell wall defects [13]. Several hypotheses have been put forward to explain the participation of plant EGs in cellulose biosynthesis. It was assumed that the degradation of cellulose microfibrils can occur to stop the synthesis of the cellulose chain, the release of synthases from growing microfibrils, the removal of non-crystalline glucan chains, or the elimination of tensile stress, presumably caused by the assembly of long-chain molecules [5, 10, 13]. However, there is no evidence for all of these assumptions.

The presence of EG activity in extracts isolated from the secondary xylem of *Picea abies* was shown using carboxymethylcellulose as a substrate [6]. Some researchers do not identify CMC-ase activity with cellulase activity and express doubts about the correctness of this approach. Therefore, we made an attempt to study the enzymatic hydrolysis of cellulose in a heterogeneous reaction, where the substrate is filter paper; the membrane-bound cellulase is represented by exosomal vesicles, and the reaction is controlled by the number of aldehyde groups in the solid phase. For this control, the method for determining the copper number of cellulose is best suited.

The aim of the present study was isolation of exosomal proteins from spruce xylem and analysis of their enzymatic activity.

2 Materials and Methods

2.1 Plant Materials and Sample Preparation

For this study, we harvested stem samples of a Norway spruce (*Picea abies* (L.) Karst.) growing on the border of its range in the Russia, Arkhangelsk region (64,46° N, 40,94° E). The diameter of the spruce stem was 120 mm. Samples taken in August during the period of active spruce growth and synthesis of exosomes. The stem samples were cut into blocks. The bark and the secondary phloem layer were carefully removed from the blocks. The xylem 100 μm thick was separated, collected and treated by liquid nitrogen. The xylem samples were ground using a mill MЧ-C (Russia) after freezing in liquid nitrogen. At this temperature, cell walls crack in all directions. The xylem material in the form of a powder was stored at −80 °C in freezer (Thermo Fisher Scientific, USA). The xylem extract was prepared by adding 1 ml of 0.1 M acetate buffer (pH 5.0) to 100 mg of xylem material with stirring for 1 h. Sodium azide in the amount of 0.05% was added to the suspension to prevent microbial contamination. The supernatant (xylem extract) was separated using centrifugation (Centrifuge 5804R, Eppendorf, Germany) at 14000 rpm for 1 h to remove all particles larger than 1 μm from the extract (fragments of cell walls, intracellular objects).

2.2 High Performance Liquid Chromatography (HPLC)

Exosome-containing fraction was obtained using size-exclusion high-performance liquid chromatography Dionex Ultimate 3000 (Thermo Fisher Scientific, USA). Preparative column BioSep-SEC-S4000 with a pore size of 50 nm was used. 0.1 M acetate buffer (pH 5.0) with sodium azide was used as an eluent. The flow rate was 10 ml/min. The volume of xylem extract used to obtain the exosome-containing fraction was 2 ml. The HPLC system used the 280 nm UV detector. The fraction corresponding to the first peak characterizing particle sizes more than 50 nm was selected for study as an exosome-containing fraction. The Chromeleon Version 6.80 SR11 Build 3161 (Dionex, USA) was used for HPLC system control. Thus, it was possible to cut off particles smaller than 50 nm.

2.3 Protein Concentration in the Exosome-Containing Fraction

The exosome-containing fraction after HPLC was frozen at −80 °C (Thermo Fisher Scientific freezer, USA) and lyophilized (Labconco FreeZone, USA). Freeze Drying Conditions: vacuum 0.02 mBar; temperature −40 °C. The lyophilisate was dissolved in 1 ml of distilled water. The protein concentration was determined in the lyophilisate solution by Bradford method for protein quantitation [3].

The calibration graph of the dependence of optical density on the protein concentration in solution was built using the Albumin, Bovine (BSA) standard (VWR Life Science, USA) in the concentration range from 2.5 mg/ml to 25.0 mg/ml.

2.4 Enzymatic Activity of the Exosome-Containing Fraction

The enzymatic activity of the exosome-containing fraction was determined using a modified protocol for determining the copper number [4]. 1 g of filter paper was grind and placed in 250 ml flask with 70 ml of 0.1 M acetate buffer with sodium azide. 1 ml of exosome-containing fraction was added to the solution directly after HPLC. Four experimental variants were carried out depending on the duration of the reaction: the mixture was incubated for 0 h, 0.5 h, 1 h or 2 h at 35 °C. All flasks were heated to a boil after incubation. The further procedure for determining the copper number was carried out according to the protocol [4].

2.5 Sample Preparation for Electron Microscopy

Small blocks (10 × 10 × 15 mm) containing some xylem were prepared from the woody stems. An approach to sample preparation associated with preliminary cryomechanical destruction and freeze-drying was used [2]. The samples were treated with liquid nitrogen in a plastic vessel until boiling no longer occurred. Freeze-drying was performed using a Labconco freeze dryer (FreeZone 2.5 L, Labconco Corporation, USA).

2.6 Scanning Electron Microscopy

Most of the sample images were obtained using an SEM Sigma VP Zeiss (Carl Zeiss Microscopy GmbH, Germany). To increase the image contrast of the samples, a gold–palladium coating with a thickness of 5 nm at a ratio of 80:20 was applied to the split surface. For this treatment, we used a Q150T ES (Quorum Technologies Ltd, UK).

3 Results

Exosomes of the secondary xylem were found at the sites of pit formation in the S_1 layer of the secondary fiber wall [8] (Fig. 1).

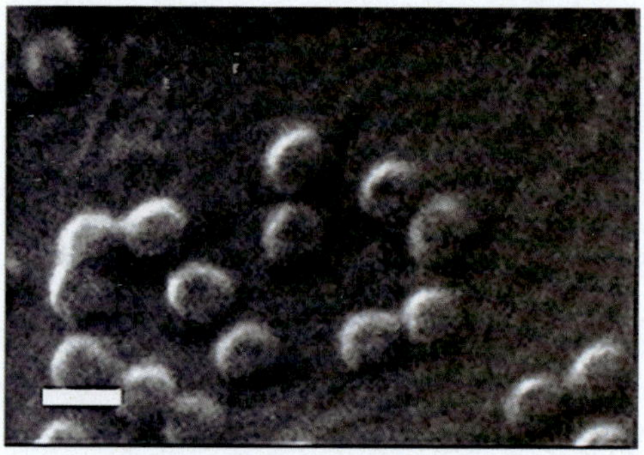

Fig. 1. Exosomes on the inner surface of the cell wall of developing tracheids. Bar 100 nm

There are grounds to assert that endo-1,4-β-glucanases delivered by exosomes participated in pit cavity formation in the S_1 layer of xylary fibres. The exosome-containing fraction outside the fractionation limits was separated from the xylem extract using the HPLC system. The first peak characterizing particle sizes larger than 50 nm was peak from 8 to 8.5 min (Fig. 2, marked red).

The isolation of the fraction in the mentioned range made it possible to remove all particles less than 50 nm – ribosomes, free proteins, tannins and low molecular weight compounds. The Bradford method was used to determine the presence of exosomal proteins in the isolated exosome-containing fraction.

The optical density in the exosome-containing fraction determined by the Bradford method was 0.451 optical density units (o.d.u.). The protein concentration in the fraction was calculated and equal to 18.4 mg/ml. The isolated fraction demonstrates the presence of proteins, presumably exosomal. The next step of the study was to study the cellulolytic activity of the exosome-containing fraction.

The cellulolytic activity of the exosome-containing fraction was measured based on the protocol for copper number. Three experimental variants were carried out depending on the duration of the incubation. The efficiency of cellulose hydrolysis in the presence of the exosome-containing fraction was detected at 0, 0.5, 1 or 2 h (Table 1).

The increase in the copper number occurs under the action of the exosome-containing fraction. The enzyme activity was expressed in international units (1 unit corresponds

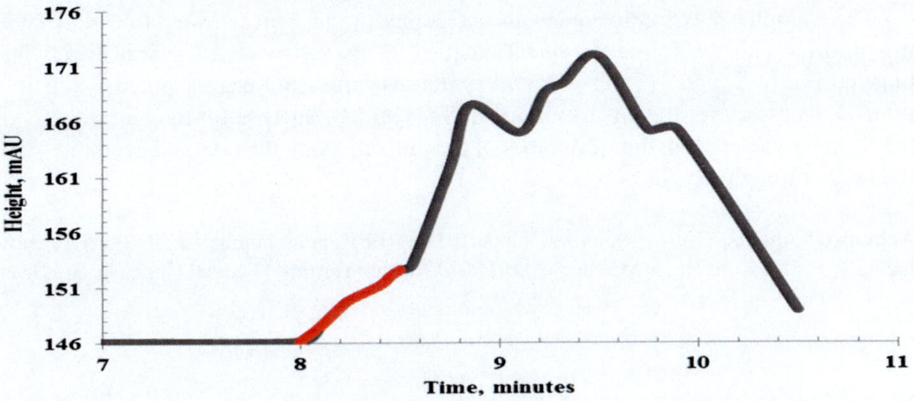

Fig. 2. Chromatogram of the spruce xylem extract with an isolated fraction

Table 1. The enzymatic activity of the exosome-containing fraction measured by copper number

Duration of enzymatic hydrolysis, h	Copper number, g Cu/100 g cellulose
0	1.03 ± 0.02
0.5	1.14 ± 0.02
1	1.21 ± 0.01
2	1.26 ± 0.02

to the formation of 1 μmol of the product in 1 min under the action of enzymes on the substrate).

4 Discussion and Conclusion

The spruce xylem extract fraction was analyzed for the presence of exosomal proteins and their functions. The presence of cellulases in the exosomes isolated from the spruce *Picea abies* xylem was shown.

The endoglucanase activity of the exosomal vesicles on filter paper was 43.5 nmol/(min·mg protein). In comparison with the CMC-ase activity of *P.verriculosum* EG II (40 μmol/(min·mg protein)) [9], it is approximately 1000 times less. This may be due to several reasons: CMC-ase activity is determined in a homogeneous medium, in our case both the substrate and the vesicles are insoluble in water; the vesicle moves along the substrate by Brownian motion, which is many times less than the Brownian motion of molecules; only those enzyme molecules «work» whose catalytic site touch cellulose, and the rest are located in other parts of the vesicle membrane and cannot participate in the reaction, although they contribute to the protein concentration; as a rule, fungi cellulases are more «aggressive» than plant cellulases. EG affects only the amorphous part of cellulose.

The endoglucanases delivered by the exosomes produce local destruction of S_1 layer microfibrils with the formation of pit cavities in the cell walls of xylem fibers. The biosynthesis of S_1 and S_2 layers with the synthases participation is separated by the time interval necessary for the pits formation in S_1 layer [7]. Sucrose for the biosynthesis of the S_2 layer comes from the sieve tubes of the phloem along the rays and penetrates into the fibers through the pits.

Acknowledgments. This research was supported by RFBR, grant number 20–04-00457. Instrumentation of the Core Facility Center «Arktika» of Northern (Arctic) Federal University was used in this work.

References

1. Akuma, P., Okagu, O., Udenigwe, C.: Naturally occurring exosome vesicles as potential delivery vehicle for bioactive compounds. Front. Sustain. Food Syst. **3**, 23 (2019)
2. Bogolitsyn, K.G., Zubov, I.N., Gusakova, M.A., Chukhchin, D.G., Krasikova, A.A.: Juniper wood structure under the microscope. Planta **241**(5), 1231–1239 (2015). https://doi.org/10.1007/s00425-015-2252-1
3. Bradford, M.M.: A rapid and sensitive method for the quantitation of microgram quantities of protein utilizing the principle of protein-dye binding. Anal. Biochem. **72**, 248–254 (1976)
4. Burton, J.O., Rasch, R.H.: The determination of the alpha-cellulose content and copper number of paper. Bur. Stand. J. Res. **6**(4), 603 (1931). https://doi.org/10.6028/jres.006.037
5. Cosgrove, D.J.: Plant cell wall extensibility: connecting plant cell growth with cell wall structure, mechanics, and the action of wall-modifying enzymes. J. Exp. Bot. **67**(2), 463–476 (2016)
6. Chukhchin, D.G., Bolotova, K., Sinelnikov, I., Churilov, D., Novozhilov, E.: Exosomes in the phloem and xylem of woody plants. Planta **251**(1), 1–14 (2019). https://doi.org/10.1007/s00425-019-03315-y
7. Chukhchin, D.G., Vashukova, K., Novozhilov, E.: Bordered pit formation in cell walls of spruce tracheids. Plants **10**(9), 1968 (2021)
8. Hrmova, M., MacGregor, E.A., Biely, P., Stewart, R.J., Fincher, G.B.: Substrate binding and catalytic mechanism of a barley β-D-glucosidase/(1, 4)-β-D-glucan exohydrolase. J. Biol. Chem. **273**(18), 11134–11143 (1998)
9. Korotkova, O.G., et al.: Comparative analysis of the properties of recombinant endo-1,4-β-glucanase ii and its chimeric form with a cellular binding module. Mosc. Univ. Chem. Bull. **60**(5), 296–303 (2019)
10. Minic, Z., Jouanin, L.: Plant glycoside hydrolases involved in cell wall polysaccharide degradation. Plant Physiol. Biochem. **44**(7–9), 435–449 (2006)
11. Regentea, M., Corti-Monzóna, G., Maldonado, A.M., Pinedo, M., Jorrín, J., de la Canal, L.: Vesicular fractions of sunflower apoplastic fluids are associated with potential exosome marker proteins. FEBS Lett. **583**(20), 3363–3366 (2009)
12. Rutter, B.D., Innes, R.W.: Extracellular vesicles isolated from the leaf apoplast carry stress-response proteins. Plant Physiol. **173**(1), 728–741 (2017)
13. Yu, L., Sun, J., Li, L.: PtrCel9A6, an endo-1,4-β-glucanase, is required for cell wall formation during xylem differentiation in Populus. Mol. Plant **6**(6), 1904–1917 (2013)

Composition of Biologically Active Substances of Flower Extracts from a Variety of *Achillea* Species and Their Antioxidant Properties

V. Kurchenko[1(✉)], N. Sushinskaya[1], K. Maiorava[1], E. Tarun[2], and A. Kupriyanov[3]

[1] Belarusian State University, Minsk, Belarus
kurchenko@tut.by
[2] International State Ecological Institute, A.D. Sakharova, Minsk, Belarus
ktarun@tut.by
[3] Kuzbass Botanical Garden of SB RAS, Kemerovo, Russia

Abstract. A comparative study has been carried out on the composition of biologically active substances of methanol extracts of flowers from six species of Achillea of the aboriginal flora of Siberia and Kazakhstan. According to the results of GC-MS analysis, it can be concluded that there are significant differences in the composition and content of mono-, sequi-, triterpene and terpenoid compounds, alkaloids and steroids, which are associated with the growing region and species characteristics of researched species of Achillea. The antioxidant activity of flower extracts of Achillea decreases in the following order: *A. asiatica* Serg. (Kemerovo, Russia), *A. micrantha* Willd. (Kazakhstan), *A. setacea* Waldst. & Kit. (Kazakhstan), *A. asiatica* Serg. (Khakassia, Russia), *A. asiatica* Serg. (Kazakhstan), *A. karatavica* Kamelin (Kazakhstan), *A. biebersteinii* Afan. (Kazakhstan), which is associated with the peculiarities of the composition of biologicaly active substances (BAS). Recommended for introduction for use in pharmaceuticals: *A. asiatica* (Kemerovo), *A. micrantha* (Kazakhstan) and *A. setacea* (Kazakhstan), which contain the largest amount of biologically active substances.

Keywords: GC-MS · Achillea · Terpenoids · Sesquiterpenes · Alkaloids · Steroids

1 Introduction

The genus Yarrow (Achillea) includes more than 100 species, which are mainly distributed in the northern hemisphere [5]. They are widely used in traditional European medicine to treat fever, hypertension, gastrointestinal disorders, anti-haemorrhaging, and heal wounds [2, 4]. Previous studies have shown that biologically active substances (BAS) included in the extracts from various Achillea species have antioxidant, anti-inflammatory, antiseptic, analgesic and antiproliferative activity [1, 7, 8, 10, 12]. Phytochemical research has shown that many species of the genus Achillea contain mono- and sequiterpenes, terpenoids, flavonoids, flavonols, flavones, and other biologically active

V. Kurchenko et al. (Eds.): ICAETT 2021, LNNS 408, pp. 81–91, 2022.
https://doi.org/10.1007/978-3-030-96641-6_10

substances [1, 4, 6, 8, 12]. Studies are under way to expand the objects of phytochemical analysis of Achillea to identify their new forms and species with a high content of biologically active substances, suitable for use in pharmaceuticals [3, 8, 11, 13]. It should be noted that intraspecific differences in the composition of secondary metabolites may be associated with the existence of their chemotype, as well as the region of growth and climatic conditions. In addition, endemic Achillea species growing in various regions of Kazakhstan and Siberia, for which phytochemical studies have not been conducted before, are of great interest for obtaining BAS. In this regard, it seemed appropriate to conduct a comparative study of the composition and content of biologically active substances of extracts from flowers of various Achilles species growing in Siberia and Kazakhstan. In addition, it is important to study the biological activity of the obtained extracts and give their toxic-hygienic assessment.

The purpose of the work was to conduct a comparative analysis of the composition of biologically active substances of extracts from flowers of various species of yarrow and to assess their antioxidant activity.

2 Materials and Methods

The object of the study was methanol extracts of flowers from 5 species of yarrow (see Table 1). Plant samples were collected and identified by employees of the Kuzbass Botanical Garden and deposited in its herbarium.

To prepare the extracts, the flowers were crushed. Substances were extracted from the crushed flowers with methanol in a ratio of 1–10 (m/v). The composition of biologically active substances from flower extracts was studied using a chromato-mass spectrometric system: Agilent 6850 gas chromatograph with Agilent 5975B mass-selective detector (GC-MS). A capillary column DB-5MS (5% PhenylMethylSiloxane, J&W 122-5062) with a length of 30 m and an internal diameter of 0.25 mm was used for analysis. The injector temperature is 300 °C, the interface temperature is 300 °C. Carrier gas is helium, volume flow rate of the carrier gas is 1 ml/min. Temperature program: the initial temperature of the thermostat is 35 °C, the temperature rises at a rate of 5 °C/min to 170 °C (isothermal mode for 7 min); the temperature rises at a rate of 7 °C/min to 280 °C (isothermal mode for 10 min). The analysis time was 59.71 min. The sample volume is 1 μl, the sample input mode with a reset of 70:1. Mass detector parameters: source temperature – 230 °C, quadrupole temperature – 150 °C. Identification of components by mass spectrum using the NIST mass spectrum library. The relative content of individual compounds was determined as a percentage of the mass of extractive substances.

The ORAC (Oxygen Radical Absorption Capacity) fluorimetric method was used to evaluate the antioxidant activity (AOA) of methanol extracts of flowers [9, 14]. The method is based on measuring the decrease in the fluorescence intensity of fluorescein (FL), during its interaction with oxygen radicals. Antioxidants in the reaction medium, interacting with oxygen-containing radicals, slow down free radical oxidation of PL. The AOA of the substances included in the extracts was determined and evaluated by their ability to bind free radicals formed in the Fenton system. Fluorescence measurements were performed on an RF-5301 PC fluorometer («Shimadzu», Japan). The fluorescence intensity was recorded at a wavelength of 514 nm. The excitation wavelength was 490 nm.

Table 1. Species of studied extracts of yarrow flowers, places of their collection and catalogue number in herbariums.

Herbarium catalog number	Plant species	Gathering place
1	2	3
KUZ KEM 16466	Asian yarrow (*Achillea asiatica* Serg. Kemerovo, Russia)	Russia, Kemerovo Region (54.353912° N, 86.77906° E)
KUZ SIB 16332	Asian yarrow (*Achillea asiatica* Serg. Khakassia, Russia)	Russia, Khakassia, Ordzhoniki-dzevsky district, Lake Sulfatnoe (54°58′40" N, 89°36′35" E.)
KUZ KAZ 06526	Asian yarrow (*Achillea asiatica* Serg. Kazakhstan)	Kazakhstan, East Kazakhstan region, Saur ridge (47.43306° N, 85.27296° E, A = 669 m.)
KUZ KAZ 07720	Biberstein's Yarrow (*Achillea biebersteinii* Afan. Kazakhstan)	Kazakhstan, South Kazakhstan region, Karatau ridge (42.86126° N, 69.86825° E)
KUZ KAZ 06524	Small - flowered yarrow (*Achillea micrantha* Willd Kazakhstan)	Kazakhstan, East Kazakhstan region, Kokpektinsky district (43.83109° N, 83.38052° E,A = 427 m)
KUZ KAZ 06521	Bristly yarrow (*Achillea setacea* Waldst. & Kit. Kazakhstan)	Kazakhstan, East Kazakhstan region, Tarbagatai district (47.49057° N, 81.07756° E)
KUZ KAZ 09113	Yarrow karatausky (*Achillea karatavica* Kamelin Kazakhstan)	Kazakhstan, South Kazakhstan region, Karatau ridge (42.86062° N, 69.90652° E)

The calculation of OAO indicators was carried out according to the degree of fluorescence intensity (A, %), calculated by the formula:

$$A = \frac{Fl}{Fl_0} \times 100,$$

where Fl_0 – fluorescence intensity of the control FL sample (FL solution without Fe^{2+}, EDTA, hydrolyzate and H_2O_2), Fl – the fluorescence intensity of the solution after the addition of the extract.

Graphs of the dependence of the fluorescence intensity (A, %) on the content of the flower extract were plotted. According to the obtained equation, the concentration of the sample IC50, corresponding to 50% inhibition of fluorescence, was calculated. Plotting and mathematical processing of the research results were carried out using the computer program «Microsoft Office Excel 2003» (Microsoft Corporation, USA). The results of independent experiments are presented as the arithmetic mean. The reliability of the differences between the data samples was determined by the method of confidence intervals.

3 Results and Discussions

The comparative study of the composition and content of BAS of methanol extracts of 7 yarrow samples collected in various regions of Kazakhstan, Khakassia and the Kemerovo region of the Russian Federation was conducted. According to the results of GC-MS analysis, 87 biologically active substances of different structure were found. The relative abundances of the main BAS in methanol extracts of flowers of the studied yarrow is presented in Table 2. Analysis of the results shows that there are significant differences in the composition and content of secondary metabolites in the flowers of *A. asiatica*, which are associated with the region of growth. Thus, *A. asiatica* (Kemerovo) contains a number of compounds: thujene, trans-geranylgeraniol, osthole, linoleic acid, beta.-amyrin, which are absent in the extract of this species from Khakassia and Kazakhstan. The total content of monoterpene and terpenoid compounds in the extract of this species is 36.4%. A distinctive feature of the composition of the BAS from extract of the form *A. asiatica* (Khakassia) is the presence of limonene oxide, tricosane, camphor, d-oxime, santene, alpha-bisabolol, which are not detected in other forms of this species.

The content of monoterpene and terpenoid compounds in the extract is 49.8%. The extract obtained from *A. asiatica* flowers collected in Kazakhstan is characterized by the presence of o-cymene, beta.-terpineol, alpha.-terpineol, beta.-eudesmol, sesquisabinene hydrate, 1-nonadecene. The content of monoterpene and terpenoid compounds in the extract of this species is 61.0%. Common BAS for extracts of flowers of this species collected in different regions are chamazulene and sitosterol. Guided by chemosystematics, it can be assumed that this species has chemotypes depending on the growing region [11].

The main indicators of the antioxidant activity of the investigated flower extracts in comparative analysis are A_{max} – the fluorescence intensity corresponding to the maximum inhibition of free radicals by the studied extracts, expressed as a percentage, C_{max} – the concentration of the extract at which A_{max} is achieved and IC_{50} – the concentration of the extract at which 50% inhibition of free radicals is achieved. The study showed that, depending on the growing region, there are significant intraspecific differences in the antioxidant activity of *A. asiatica* flower extracts. The minimum IC_{50} index was obtained for the sample of *A. asiatica* (Kemerovo), which indicates its maximum antioxidant activity.

IC_{50} index of the extract from *A. asiatica* flowers collected in the vicinity of Lake Sulfatnoe (Russia, Khakassia) is 4.6 times higher and the village of Karakan (Russia, Kemerovo region) is 7.7 times higher, which indicates their low antioxidant activity. It should be noted that with an increase in the total content of monoterpene and terpenoid compounds in the researched extracts, the antioxidant activity decreases.

A comparative analysis of the composition of secondary metabolites of extracts from flowers of 5 Achillea species collected from various mountain ranges of Kazakhstan showed differences in the composition and total content of monoterpenes and terpenoid compounds. Depending on this indicator, the antioxidant activity of extracts from flowers

Table 2. Relative content of the main biologically active substances in methanol extracts of yarrow flowers and their antioxidant activity

CAS Registry Number Substance,	Relative content of BAS in extracts, %						
	(Kemerovo) A. asiatica	*(Khakassia) A. asiatica*	*(Kazakhstan) A. asiatica*	*(Kazakhstan) A. micrantha*	*(Kazakhstan) A. setacea*	*(Kazakhstan) A. karatavica*	*(Казахстан) A. biebersteinii*
1	2	3	4	5	6	7	8
O-Cymene; 000527-84-4	n.d	n.d	**2.2**	n.d	**1.6**	n.d	**2.3**
4(10)-Thujene; 003387-41-5	**12.9**	n.d	n.d	**2.8**	n.d	**8.9**	n.d
Cineole; 000470-82-6	**6.4**	**2.7**	**13.3**	**2.1**	**10.8**	**11.0**	**0.6**
Beta.-Terpineol; 007299-41-4	n.d	n.d	**2.8**	n.d	n.d	n.d	n.d
(+)-2-Carene; 000554-61-0	n.d	n.d	n.d	n.d	n.d	n.d	**22.9**
(−)-Camphor; 000464-48-2	n.d	**1.2**	**13.6**	**2.4**	**11.6**	**11.3**	n.d
Borneo Camphor; 000507-70-0	n.d	**1.0**	**1.7**	n.d	n.d	n.d	n.d
Camphor,D-Oxime; 005655-61-8	n.d	**1.0**	n.d	n.d	n.d	n.d	n.d
Alpha.-Terpineol; 000098-55-5	n.d	n.d	**7.1**	n.d	**1.3**	**3.9**	n.d

(*continued*)

Table 2. (*continued*)

CAS Registry Number Substance,	Relative content of BAS in extracts, %						
	(Kemerovo) A. asiatica	*(Khakassia) A. asiatica*	*(Kazakhstan) A. asiatica*	*(Kazakhstan) A. micrantha*	*(Kazakhstan) A. setacea*	*(Kazakhstan) A. karatavica*	*(Казахстан) A. biebersteinii*
Limonene Oxide; 006909-30-4	n.d	**2.4**	n.d	n.d	n.d	n.d	n.d
Caryophyllene Oxide 001139-30-6	**1.3**	n.d	**1.9**	**2.4**	**1.4**	n.d	n.d
Beta.-Eudesmol; 000473-15-4	n.d	n.d	**6.8**	**2.6**	**2.1**	**4.8**	n.d
Chamazulene; 000529-05-5	**7.9**	**6.5**	**4.7**	n.d	**10.8**	**23.9**	n.d
Tricosane; 000638-67-5	n.d	**2.3**	n.d	n.d	n.d	n.d	n.d
Trans-Geranylgeraniol; 024034-73-9	**4.1**	n.d	n.d	n.d	n.d	n.d	n.d
2-Methylcyclohexen-1-Yl Triflate; 2000296-75-1	n.d	n.d	n.d	n.d	**1.3**	n.d	n.d
Germacrene D; 023986-74-5	**1.1**	**2.7**	n.d	n.d	**2.1**	n.d	n.d
Lavandulyl Acetate; 025905-14-0	n.d	n.d	n.d	n.d	**6.3**	n.d	n.d

(*continued*)

Table 2. (*continued*)

CAS Registry Number Substance,	Relative content of BAS in extracts, %						
	(*Kemerovo*) *A. asiatica*	(*Khakassia*) *A. asiatica*	(*Kazakhstan*) *A. asiatica*	(*Kazakhstan*) *A. micrantha*	(*Kazakhstan*) *A. setacea*	(*Kazakhstan*) *A. karatavica*	(*Казахстан*) *A. biebersteinii*
3-Keto-Trans-Caryophyllene; 000087-44-5	**2.7**	**2.0**	**2.0**	n.d	**1.8**	n.d	n.d
9-Hydroxynerolidol 3,7,11Trimethyldodeca-1,6,10-Triene-3,9-Diol; 2000281-51-0	n.d	**5.2**	n.d	n.d	n.d	n.d	n.d
Sesquisabinene Hydrate; 2000232-87-7	n.d	n.d	**4.9**	n.d	n.d	n.d	n.d
Santene; 000529-16-8	n.d	**22.4**	n.d	n.d	n.d	n.d	n.d
2-Oxo-3-Phenyl-2h-Pyrane-6-Carboxylic Acid, Ethyl Ester; 2000298-76-9	n.d	**4.0**	n.d	n.d	n.d	n.d	n.d
3-Bromo-1,1,2-Trimethyl-Cyclopropane; 036617-00-2	n.d	n.d	n.d	**14.2**	n.d	n.d	n.d
6-(N-Allylamino)-7methylquinoline -5,8-Dione; 98217-14-2	n.d	n.d	**5.6**	n.d	n.d	n.d	n.d

(*continued*)

Table 2. (*continued*)

CAS Registry Number Substance,	Relative content of BAS in extracts, %						
	(Kemerovo) A. asiatica	*(Khakassia) A. asiatica*	*(Kazakhstan) A. asiatica*	*(Kazakhstan) A. micrantha*	*(Kazakhstan) A. setacea*	*(Kazakhstan) A. karatavica*	*(Казахстан) A. biebersteinii*
Osthole; 000484-12-8	**4.7**	n.d	n.d	n.d	n.d	n.d	n.d
1[(Dimethyl)Methylene]3-[(T-Butyl)Ethynyl]-5,5-Dimethylcyclohex -2-Ene; 2000257-15-9	n.d	**2.3**	n.d	n.d	n.d	n.d	n.d
1-Nonadecene; 018435-45-5	n.d	n.d	**8.1**	n.d	n.d	n.d	n.d
Heptadecane; 000629-78-7	**6.9**	**5.8**	n.d	n.d	**3.2**	**2.1**	n.d
Hexadecanoic Acid; 000057-10-3	**3.7**	**2.9**	**1.4**	n.d	n.d	**1.1**	**1.9**
Cis-Vaccenic Acid; 000506-17-2	n.d	n.d	n.d	n.d	n.d	n.d	**6.1**
1-Nonadecanol; 001454-84-8	n.d	n.d	n.d	n.d	n.d	**1.5**	n.d
Linoleic Acid; 000060-33-3	**4.9**	n.d	n.d	n.d	n.d	**1.5**	n.d
Achillicin; 071616-00-7	n.d	n.d	n.d	**6.4**	n.d	n.d	n.d

(*continued*)

Table 2. (*continued*)

Substance, CAS Registry Number	Relative content of BAS in extracts, %						
	(Kemerovo) A. asiatica	*(Khakassia) A. asiatica*	*(Kazakhstan) A. asiatica*	*(Kazakhstan) A. micrantha*	*(Kazakhstan) A. setacea*	*(Kazakhstan) A. karatavica*	*(Казахстан) A. biebersteinii*
Artemisin; 000481-05-0	**3.9**	n.d	**5.2**	n.d	n.d	n.d	n.d
Alpha-Bisabolol; 000515-69-5	n.d	**2.9**	n.d	n.d	**8.7**	n.d	n.d
.Beta.-Amyrin; 000559-70-6	**4.3**	n.d	n.d	n.d	n.d	n.d	**1.7**
Alpha.-Amyrin; 000638-95-9	n.d	n.d	n.d	n.d	n.d	n.d	**1.3**
5-Eicosene; 074685-30-6	n.d	n.d	n.d	n.d	n.d	n.d	**1.6**
Sitosterol; 000083-47-6	**3.5**	**3.8**	**2.3**	n.d	**2.1**	n.d	**1.3**
Indicators of antioxidant activity of yarrow flower extracts							
Amax, %	83	81	81	88	88	76	87
Cmax, %	0,10	0,1	0,1	0,1	0,1	0,1	0,1
$IC50 \cdot 10^{-4}$, %	**0,47**	**2,14**	**3,63**	**1,3**	**2,0**	**4,5**	**5,0**

decreases according to the IC_{50} index in a number of species *A. micrantha* (12.3%), *A. setacea* (51.3%), *A. asiatica* (61.2%), *A. karatavica* (63.8%).

For *A. biebersteinii* (25.8%), the IC50 index was the maximum, which indicates low antioxidant activity with a minimum number of monoterpene compounds, among which Carene is 22.9%. This species is characterized by the presence in the extract of beta.-amur in, alpha.-amurin, 5-eicosane, sitosterol. There are interspecific differences in the composition of extracts of researched species, represented by fatty acids; sequiterpenes: alpha-bisabolol, artemisin, achillicin; triterpenes: alpha.-amyrin, beta.-amyrin; alkaloids: 5-eicosene and steroids: sitosterol. The extracts of *A. micrantha* and *A. karatavica* lack fatty acids and their derivatives, sesquiterpenes, triterpenes, sitesterol and a number of other compounds. The biological activity of the studied extracts is not a simple sum of the activities of its constituent components. It manifests itself in their joint action. Compounds included in the extracts of the studied Achillea species are characterized by different biological activity. It may be associated with differences in the composition of mono-, sesqui-, triterpene and terpenoid compounds, alkaloids and steroids [1, 4, 12].

Monoterpenes, aldehydes and ketones: O-cymene, 4(10)-thujene, cineole, beta.-terpineol, 2-carene, -camphor, borneo camphor, camphor, D-oxime, alpha.-terpineol, limonene oxide are active against tuberculosis bacillus and a number of pathogenic fungi [3, 9, 10]. These compounds exhibit antiseptic, analgesic, antihistamine and hypotensive properties [7, 8, 10, 12, 13]. They stimulate the gastrointestinal tract. Some substances, in particular terpineol, have high antifungal activity [2].

Terpenoids: camphor, borneo camphor, camphor D-oxime have an exciting effect on the gastrointestinal tract. Camphor and its derivatives affect the central nervous system, calm the work of the heart in shock conditions [11, 12]. The pentacyclic triterpenoid Amyrin exhibits anti-inflammatory properties by inhibiting cytokine production. Sesquiterpenes: Artemisia is an antimalarial, anthelmintic substance, and Achillicin exhibits hemostatic properties [12]. Monocyclic sesquiterpene ethanol Bisabolol has anti-inflammatory antimicrobial properties [11]. Sitosterol has anticancer effects, slowing down the growth of prostate, colon and breast tumors [4, 12].

4 Conclusion

GC-MS analysis of the composition of BAS extracts from flowers of 6 yarrow species of the native flora of Siberia and Kazakhstan showed significant intraspecific differences in the composition and content of secondary metabolites between *A. asiatica* growing in different regions. Differences in the composition of BAS were found between endemic and native Achillea species of mountainous regions of Kazakhstan. The antioxidant activity of extracts from Achillea flowers is determined by the composition of biologically active substances and has significant intra- and interspecific differences. A comparative study of biologically active substances of flowers and their antioxidant activity made it possible to determine species and places of growth of various yarrow plants that are promising for obtaining extracts and application in the food and pharmaceutical industry. Among them are yarrow: *A. asiatica* (Kemerovo), *A. micrantha* (Kazakhstan) and *A. setacea* (Kazakhstan).

References

1. Agar, O.T., Dikmen, M., Turkmenoglu, F.P.: Comparative studies on phenolic composition, antioxidant, wound healing and cytotoxic activities of selected achillea L. species growing in Turkey. Molecules **20**, 17976–18000 (2015). https://doi.org/10.3390/molecules201017976
2. Benedek, B., Kopp, B.: *Achillea millefolium* L. s. l. revisited: recent findings confirm the traditional use. Wien. Med. Wochenschr **157**, 312–314 (2007)
3. Bourgou, S., Bettaieb, I., Megdiche, W.M.: LC-ESI-TOF-MS and GC-MS profiling of *Artemisia herba-alba* and evaluation of its bioactive properties. Food Res. Int. **99**(1), 702–712 (2017). https://doi.org/10.1016/j.foodres.2017.06.009
4. Ivancheva, S., Stancheva, B.: Exudate flavonoid aglycones of *Achillea sp. Sect.* Millefolium and sect. Phytologia Balcanica **2**, 102–105 (1996)
5. Nemeth, E., Bernath, J.: Biological activities of yarrow species (*Achillea spp.*). Curr. Pharm. Des. **14**, 3151–3167 (2008)
6. Nikolova, M., Vitkova, E., Gesheva, E.: Comparison of flavonoid profiles of cultivated plants of *Achillea asplenifolia, Achillea collina* and cultivar "Proa". Biologica Nyssana **4**(1–2), 65–69 (2013)
7. Serdar, G., Sökme, M., Bektaş, E.: Extraction of antioxidative principles of *Achillea biserrata* M. Bieb. and chromatographic analyses. Int. J. Second. Metabol. **2**(2), 3–15 (2015)
8. Tarek, F.E., Gonzalez-Burgos, E., Gomez-Serranillos, M.P.: Chemical characterization of polyphenols of egyptian *Achillea fragrantissima* with *In vitro* antioxidant study. Chiang Mai J. Sci. **45**(2), 897–904 (2018)
9. Tarun, E.I., Zaitseva, M.V., Golovach, T.N.: Influence of peptides of whey proteins of milk on the restoration of the level of fluorescence in the system with activated forms of oxygen. Proc. Belaru. State Univ. **11**, 231–236 (2016). (in Russian)
10. Thoppil, R.J., Harlev, E., Bishayee, A.: Antitumor activities of extracts from selected desert plants against HepG2 human hepatocellular carcinoma cells. Pharm. Biol. **51**, 668–674 (2013)
11. Trendafilova, A., Todorova, M., Duddeck, H.: Flavonoids in flower heads of three *Achillea species* belonging to *Achillea millefolium* group. Chem. Nat. Compd. **43**(2), 212–213 (2007)
12. Valant-Vetschera, K., Wollenweber, E.: Leaf flavonoids of the *Achillea millefolium* group part II: distribution patterns of free aglycones in leaf exudates. Biochem. Syst. Ecol. **16**(7–8), 605–614 (2001)
13. Vitalini, S., Beretta, G., Fico, G.: Phenolic compounds from *Achillea millefolium* L. and their bioactivity. Acta Biochim. **58**, 203–209 (2011)
14. Zharin, V.A., Fedorovich, S.V., Sysa, A.G.: Clinical innovation in food allergy. Mil. Med. **1**(38), 141–143 (2016). (in Russian)

Approaches to the Determination of the Growing Location of Mossy Pine Forests Based on the Spectral Characteristics of Ecologically Dependent Wood Components

Anna Khokh(✉)

Scientific and Practical Center (SPC) of the State Forensic Examination Committee of the Republic of Belarus, Minsk, Belarus

Abstract. In the article, the wood of mossy pine forests (*Pinetum pleuroziosum*) from different areas of Belarus was investigated by spectroscopy in the near infrared region. NIR spectra were recorded using a portable NIR spectrometer MicroNIR OnSite with a diode array detector. The conditions for sample preparation have been optimized and methods have been developed for correcting scattering taking into account the specifics of the samples under study. As a result of experimental studies, 6 homogeneous regions were identified, which are characterized by similar spectral variations. According to known methods, the content of the main biopolymers in the wood of each zone, which may be sensitive to the effect of environmental gradients, was determined. It was found that the differentiation of forest stands is due to the different content of lignin, cellulose and hemicellulose in the wood. The holocellulose content remains unchanged. The research carried out is innovative and can find application in forensic science in solving expert tasks related to determining the location of a tree.

Keywords: Wood · Mossy pine forests · Lignin · Cellulose · Growing place · Spectroscopy

1 Introduction

At present, the method of spectroscopy in the near infrared region (hereinafter referred to as NIR spectroscopy) is rapidly developing, since it is non-destructive, does not require expensive consumables and is characterized by a high speed of analysis [1, 2, 4, 15].

Abroad, this method began to be actively used to study wood, which is a complex organic material, mainly consisting of lignin, cellulose, holo- and hemicellulose. It should be emphasized that the fundamental absorption bands of the vibrational spectra of the molecules of these biopolymers lie in the mid-infrared region, while in the near one there are overtones and composite frequencies, which are mainly due to the vibrations of the bonds of the hydrogen atom with the atoms of other chemical elements.

A wide range of studies in this area is associated with the identification of wood species, including for species included in the International Union for Conservation of

V. Kurchenko et al. (Eds.): ICAETT 2021, LNNS 408, pp. 92–98, 2022.
https://doi.org/10.1007/978-3-030-96641-6_11

Nature (IUCN) Red List of Threatened Species, as well as those whose import is limited on ecological grounds and requires special permission in accordance with the requirements of the Convention on International Trade in Endangered Species of Wild Fauna and Flora (CITES) [7, 9, 13]. There are a number of works concerning the research of chemically and thermally modified wood [6, 8, 10, 14], as well as the assessment of wood quality indicators, including its physical and mechanical properties [3, 5].

A new promising survey area is associated with the use of NIR spectroscopy to prove the facts of illegal felling or non-compliance of timber with those specified in the accompanying documentation in cases of illegal timber trafficking [11, 12]. At the same time, this kind of research has not yet been carried out in Belarus.

2 Materials and Methods

The purpose of the work was to study the possibility of using the NIR spectroscopy method to determine the place of growth of mossy pine forests (*Pinetum pleurozium*).

It is these forest plantations, due to their high productivity, economic value and better quality of the assortments obtained, that most often become the object of violations in the field of forest legislation, including illegal logging.

The working hypothesis was based on the assumption that the differences between the NIR spectra are due to the different contents of the main biopolymers in the wood.

The study used experimental material (drill cores) obtained from 102 temporary sample plots (hereinafter – TSP), laid in mossy pine forests in 2016–2018 throughout Belarus in accordance with the methods accepted in forest taxation. A total of 1.810 Scots pine trees were analysed in aggregate.

In laboratory conditions, drill cores were naturally dried. Their moisture content did not exceed 20%. Sample roughness was minimized.

NIR spectra were recorded using a portable NIR spectrometer MicroNIR OnSite with a diode array detector (VIAVI, USA) in the range 10526–6060 cm^{-1}. The integration time was 10.4 ms, the number of counts was 200, the resolution was 2 cm^{-1}, and the number of scans was 64. At least 20 consecutive measurements were carried out for each core. In some cases, at least 50 measurements to get more information on acceptable variations.

Before carrying out chemical analyzes, the drill cores were dried to an air-dry state, ground in an electric mill, and sieved through a sieve with a pore diameter of 1 mm. The lignin content was determined by the sulfuric acid method modified by Komarov; cellulose – nitrogen-alcohol method; holocellulose – the delignification method, which consists in the treatment of wood with sodium chlorite in acetic acid (Wise's method); the hemicellulose content was calculated as the difference between holocellulose and cellulose.

Statistical processing of research materials was carried out using the statistical packages SPSS v.22.0 and The Unscrambler X (CAMO, USA). The following methods were used: one-dimensional descriptive statistics, the Kruskal-Wallis H-test, and the principal component method.

3 Results

As a result of the studies carried out, it was established that the NIR spectra of wood in mossy pine forests are a poorly differentiated picture of wide absorption bands that are poorly resolved due to mutual overlapping. In this regard, their preliminary processing was used with the calculation of the second-order derivative by the Savitsky-Golay method (using 7 smoothing points), which made it possible to make the signs of separation more explicit.

Further, the principal component method was applied to the entire set of NIR spectra, which made it possible to analyze the relative position of the samples with each of the investigated TSP relative to each other and to assess their homogeneity.

Figure 1 and Fig. 2 show that the NIR spectra of samples with TSP No. 1-102 form six isolated groups (conditional regions), which practically do not overlap and are located in different regions of the model relative to the main components, which indicates the existence of significant differences between them (area No.1 – 34 of TSP, area No. 2 – 4 of TSP, area No. 3 – 24 of TSP, area No. 4 – 31 of TSP, area 5 – 5 of TSP and area 6 – 4 of TSP). The first two principal components account for 87% of the total cumulative variation in the data.

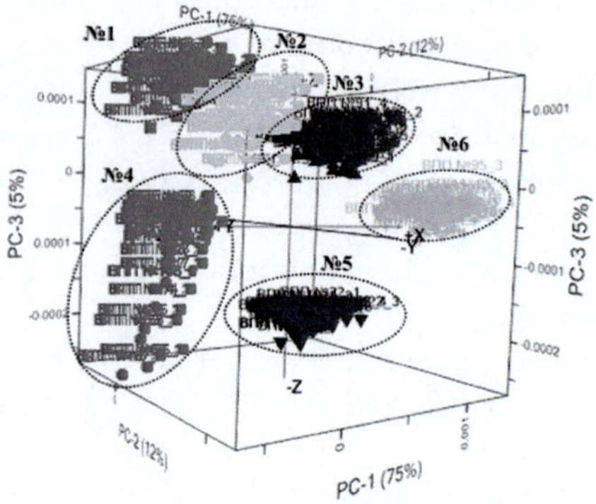

Fig. 1. Plot of RSA accounts for 102 investigated TSP in the space of principal components

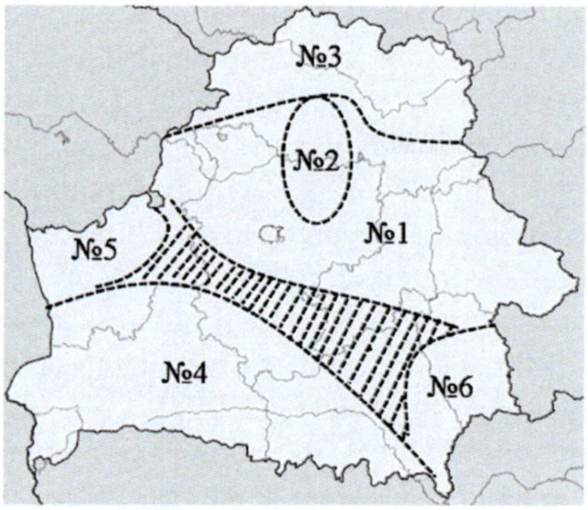

Fig. 2. Zoning of the territory of Belarus based on NIR spectra data

The next stage of the work was to study the content of lignin, cellulose, holo- and hemicellulose in each of the 6 selected areas. For the analysis, 80 drill cores from each area were used. As can be seen from the results of the experimental study (Table 1), the Kruskal-Wallis H-test is statistically significant for lignin, cellulose, and hemicellulose, the significance level being $p < 0.001$. For holocellulose, no statistically significant differences between regions were found ($p > 0.05$).

Further, the data presented in Table 1 were analyzed by the method of cluster analysis. The results are presented on a dendrogram (Fig. 3).

As can be seen from the dendrogram, the investigated areas No. 1-6 can be combined into two territorially larger regions (region No. 1 (conditionally southern) – areas No. 4-6 and region No. 2 (conditionally northern) – areas No. 1-3).

Table 1. Average content of biopolymers recovered from drill cores

Group No	Lignin %	Cellulose %	Holocellulose %	Hemicellulose %
No.1	27.5	45.1	70.4	25.3
SD	0.2	0.8	1.1	1.4
No.2	27.9	44.9	70.3	25.4
SD	0.2	0.4	1.1	1.4
No.3	28.2	45.8	70.3	24.5
SD	0.2	0.8	1.0	1.2
No.4	28.6	43.8	70.4	26.6
SD	0.2	0.8	1.2	1.4
No.5	29.7	44.3	70.3	26
SD	0.1	0.9	1.1	1.3
No.6	30.9	44.2	70.4	26.2
SD	0.2	0.7	1.2	1.5
H-test	434.20	167.71	4.97	52.564
p-value	0.000	0.000	0.42	0.000

Note: SD – the standard deviation

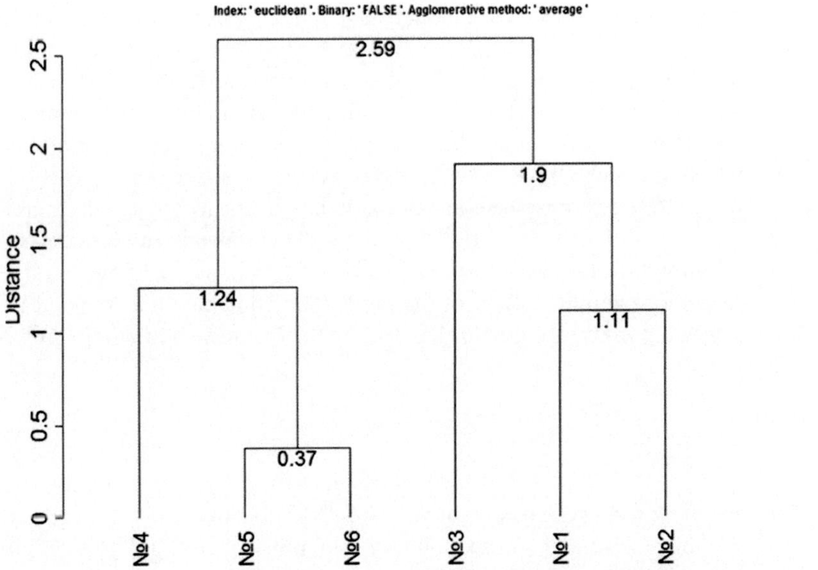

Fig. 3. Vertical dendrogram of the hierarchical classification of 6 areas identified by NIR spectroscopy

In addition, in each of the 6 areas, based on calculations of radial growth, the average width of the annual layer was calculated for each of the 1810 investigated trees, as well as the calculation of the content of late wood in the annual layer (Table 2).

Table 2. Some physical characteristics of mossy pine forests

Group number	% late wood in the annual layer	Average annual layer width, mm	Average age, years
No.1	38	1.53	98
No.2	36	1.51	98
No.3	38	1.38	98
No.4	34	1.80	97
No.5	31	2.16	99
No.6	24	2.13	97
F-value	268.126	191.658	3.091
p-value	0.000	0.000	0.686

According to the results of the study, it was found that for region No. 1 there is a smaller (on average by 20%) content of late wood in the annual layer, while for region No. 2 – large width of the annual layer (on average by 37%). The data obtained adequately characterize the sample, since there were no significant differences in the age of the studied stands ($p > 0.05$), namely, age has the greatest influence on the calculated indicators.

In general, analyzing the results obtained, we can state the following: when moving from south-west to north-east, i.e. with increasing continental climate, the content of lignin increases (on average by 8%) and the content of cellulose and hemicellulose decreases (on average by 3 and 5%, respectively) in mossy pine forests (*Pinetum pleuroziosum*). These circumstances can be explained by the fact that to date has been reliably established the fact of increasing the content of late wood with a decrease in the value of radial growth (total width of the annual layer), and, as we know, in late wood is less lignin and more cellulose than in the early. The content of early wood in mossy pine forests growing in the region No. 1 is 1.43 ± 0.37 (mm), and in the region No. 2 0.92 ± 0.26 (mm).

4 Conclusion

Thus, the work shows the possibility of using NIR spectroscopy as a screening for tracing the origin of pine timber (for example, mossy pine forests). It has been proven that the identification of 6 homogeneous NIR spectra areas is due to differences in the content of lignin, cellulose and hemicellulose, which, in turn, arise due to the different content of early and late wood in the annual layers of the studied stands, i.e. these components are sensitive to the effects of ecological gradients (characteristic of the ecological characteristics of the place of growth). The results obtained, including the currently available electronic database of NIR spectra, can be used for both scientific research in institutions of botanical, ecological and forestry profiles, and forensic botanical examinations in solving expert problems related to the establishment of the place of growth of a tree. In addition to the study of spectral characteristics in the future it is planned to carry out zoning of the territory of Belarus on the basis of the study of morphological and anatomical features of annual layers of mossy pine forests, which will reveal the heterogeneity within the areas No. 1-6 and thus increase the level of reliability of the expert tasks.

Acknowledgments. The author would like to thank Iosif Stanislavovich Tsybovsky, Chief Specialist of BelYurObespechenie, RUE, Candidate of Biological Sciences, for his interest in the work, discussion of the obtained data and valuable advice.

References

1. Beć, K.B., Grabska, J., Huck, C.W.: Near-infrared spectroscopy in bio-applications. Molecules **25**(12), 2948 (2020)
2. Beć, K.B., Grabska, J., Huck, C.W.: Current and future research directions in computer-aided near-infrared spectroscopy: a perspective. Spectrochim. Acta Part A: Mol. Biomol. Spectrosc. **245**, 119625 (2021)
3. Devos, P., Commandre, J.M., Brancheriau, L., Candelier, K., Rousset, P.: Modeling mass loss of biomass by NIR-spectrometry during the torrefaction process. Int. J. Energy Res. **44**(12), 9787–9797 (2020)
4. Huck, C.W.: Bio-applications of NIR spectroscopy. In: Ozaki, Y., Huck, C., Tsuchikawa, S., Engelsen, S.B. (eds.) Near-Infrared Spectroscopy: Theory, Spectral Analysis, Instrumentation, and Applications, pp. 413–435. Springer, Singapore (2021). https://doi.org/10.1007/978-981-15-8648-4_19
5. Kelley, S.S., Rials, T.G., Snell, R., Groom, L.H., Sluiter, A.: Use of near infrared spectroscopy to measure the chemical and mechanical properties of solid wood. Wood Sci. Technol. **38**(4), 257–276 (2004)
6. Lengowski, E.C., Muñiz, G.I.B.D., Klock, U., Nisgoski, S.: Potential use of NIR and visible spectroscopy to analyze chemical properties of thermally treated wood. Maderas. Ciencia y tecnología **20**(4), 627–640 (2018)
7. Nisgoski, S., de Oliveira, A.A., de Muñiz, G.I.B.: Artificial neural network and SIMCA classification in some wood discrimination based on near-infrared spectra. Wood Sci. Technol. **51**(4), 929–942 (2017). https://doi.org/10.1007/s00226-017-0915-8
8. Popescu, C.M., Navi, P., Peña, M.I.P., Popescu, M.C.: Structural changes of wood during hydro-thermal and thermal treatments evaluated through NIR spectroscopy and principal component analysis. Spectrochim. Acta Part A Mol. Biomol. Spectrosc. **191**, 405–412 (2018)
9. dos Santos, J.X., et al.: Discrimination of "Louros" wood from the Brazilian Amazon by near-infrared spectroscopy and machine learning techniques. Eur. J. Wood Wood Prod. **79**(4), 989–998 (2021). https://doi.org/10.1007/s00107-021-01685-3
10. Schwanninger, M., Hinterstoisser, B., Gierlinger, N., Wimmer, R., Hanger, J.: Application of Fourier transform near infrared spectroscopy (FT-NIR) to thermally modified wood. Holz als Roh-und Werkstoff **62**(6), 483–485 (2004)

11. Sierra, V., De Luque, M., Millán, Á.S., Pérez, S., García, E., Luque, L.: NIRWOOD: An EU innovation project to determine species and geographical origin of timber using NIR spectrometry. In: Proceedings of the 28th International Conference on Wood Science and Technology, pp. 7–8. Zagreb, Croatia (2017)
12. Silva, D.C., et al.: Determination of the country of origin of true mahogany (Swietenia macrophylla King) wood in five Latin American countries using handheld NIR devices and multivariate data analysis. Holzforschung **72**(7), 521–530 (2018)
13. Snel, F.A., et al.: Potential field-deployable NIRS identification of seven Dalbergia species listed by CITES. Wood Sci. Technol. **52**(5), 1411–1427 (2018)
14. Todorović, N., Popović, Z., Milić, G., Veizović, M., Popadić, R.: Quality evaluation of heat-treated sessile oak (Quercus petraea L.) wood by colour and FT-NIR spectroscopy. Wood Mater. Sci. Eng. 1–8 (2020). https://doi.org/10.1080/17480272.2020.1847188
15. Zareef, M., et al.: An overview on the applications of typical non-linear algorithms coupled with NIR spectroscopy in food analysis. Food Eng. Rev. **12**, 1–18 (2020)

A Novel Laccase from Basidiomycete *Fomes fomentarius VKPM F-1531*

Sofia V. Kiseleva[1(✉)], Ekaterina A. Cherepanova[2], Elnara R. Nasibullina[1], Lilija Kh. Khalimova[1], and Igor V. Maksimov[2]

[1] Department of Biochemistry and Microbiological Engineering, Ufa State Petroleum Technological University, Ufa, Russia

[2] Institute of Biochemistry and Genetics of the Ufa Federal Research Centre of the Russian Academy of Sciences, Ufa, Russia

Abstract. Laccases are a group of copper-containing phenoloxidases, which activity can stimulated by copper ions. The conditions of obtaining, isolation and purification of extracellular laccase of the basidiomycete *Fomes fomentarius VKPM F-1531* were studied. The optimal concentration of copper in the glucose-peptone nutrient medium and duration of the surface-liquid cultivation for the highest yield of the enzyme were established. The efficient synthesis of fungal laccase by the fungus *F. fomentarius VKPM F-1531* requires the optimal concentration of copper in the nutrient medium of 200 mg/l. The highest enzyme activity was achieved on the 14th day of surface-liquid cultivation on glucose-peptone nutrient medium with optimal $CuSO_4$ supplementation and it was more than 6–6.5 times higher than the one obtained on the conventional glucose-peptone medium. The protein fraction that precipitated in the range from 50% to 70% saturation had a laccase activity for 40 times higher than the initial activity in the culture liquid. The molecular weight of the purified laccase determined by the SDS-PAGE was about 60 kDa.

Keywords: Laccase · *Fomes fomentarius* · Basidiomycete · Copper induction

1 Introduction

Laccases (EC 1.10.3.2) are a group of copper-containing phenoloxidase enzymes capable to catalyze the reduction of an oxygen molecule to two water molecules without formation of hydrogen peroxide. At the same time, a wide range of organic substrates can be oxidized, including lignin and its monomers, as well as inorganic compounds [1–3].

Laccases were found in various types of living organisms, but are the most abundant in xylotrophic basidiomycetes belonging to the group of white rot fungi.The functional role of laccases *in vivo* includes, first of all, the lignin degradation [2], as well as participation in the life cycle of fungi: during the development of the fruitful body, pathogenesis, synthesis of pigments, etc. [3–5].

However, stability (including thermal stability and acid resistance), the ability to work without the activity regeneration stage and the ability to increase activity in the

V. Kurchenko et al. (Eds.): ICAETT 2021, LNNS 408, pp. 99–105, 2022.
https://doi.org/10.1007/978-3-030-96641-6_12

presence of redox mediators to indirectly oxidize wide range of organic and inorganic substrates make this enzyme suitable and attractive for industrial applications [1–3]. At the same time, the practical application of laccases are annually expanding. Nowadays they possess a great biotechnological potential and can be used: for biopulping and biobleaching, transformation of colorants in the textile industry; for obtaining antitumor and anti-inflammatory drugs; for creating biochips and biosensors to detecte chemical compounds; for the production of adhesives; for design new lignocellulosic constructions and technical materials based in the woodworking and chemical industries; for wastewater and soil treatment etc. [1–3].

It has been reported that copper ions are widely known inducers for laccase synthesis, and its production is regulated at transcriptional level [5, 6]. There are several studies on laccase synthesis stimulation by optimal copper concentrations in the nutrient media [5–12]. For example, it has been shown that the optimal concentration of copper for the synthesis of laccases for micromycete *Trichoderma harzianum WL1* and the actinomycete *Streptomyces psammoticus* is in the medium is 2 mM; for ascomycete *Paraconiothyrium variabile* - 10 mM [7–9].

Fungus *Fomes fomentarius VKPM F-1531* is an effective producer of extracellular phenol oxidases during the liquid-state fermentation [11]. The current investigation focuses on the studying the effect of various copper concentrations on production and activity of laccases secreted by basidiomycete *Fomes fomentarius*, as well as to obtain a highly active purified laccase and to study a number of its biochemical properties.

2 Materials and Methods

2.1 Sulfate-Polyacrylamide Gel Electrophoresis (SDS-PAGE) Technique

The strain of the fungus *F. fomentarius* was isolated into pure culture from the fruiting body of *F. fomentarius* (tinder fungus), which has grown on decayed wood samples, and further deposited at All-Russian National Collection of Industrial Microorganisms.

Basidiomycete was cultivated under liquid-state conditions on a modified glucose-peptone medium (HPS) with the media compositions of, g/l: glucose - 20.0; peptone - 3.0; $KH_2PO4*12H_2O$ - 1.0; $K_2HPO_4*2H_2O$ 1.0; $MgSO_4*7H_2O$ 0.25; $CuSO_4*7H_2O$–0.2 2 at a temperature of 22 ± 1 °C for 14 days under static conditions. The initial pH was 6.1 ± 0.1. Each flask was inoculated with 2 agar plugs (1 × 1 cm) cut from the growing 7–9 day old wort agar culture of *F. fomentarius VKPMF-1531*.

The effect of copper ions on the laccase activity of the *Fomes fomentarius VKPM F-1531* was studied using the liquid-state cultivation on glucose-peptone medium (HPS) in the presence of copper ions (CuSO4) in the concentration range 0–300 mg/l in 50 mg/l steps. The solutions of copper sulfate and glucose were added aseptically into the medium after sterilization at the start of the cultivation.

The fungus was cultivated under static conditions without aeration and stirring; temperature during the first three days was 30 ± 1 °C, the next day - 22 ± 1 °C, without lighting (in the dark) for 18 days.

2.2 Enzyme Assay and Protein Content Determination

Laccase activity (U/ml) was determined on a PE-5400 UV spectrophotometer according to the oxidation rate of 10 mM pyrocatechol (ε 740 mM/cm^{-1}) in 0.1 M Na-acetate buffer pH 5.4 at a temperature of 22 ± 1° C and wavelength 410 nm. The unit of activity was determined as the amount of the enzyme catalyzing the oxidation of 1 μmol of the substrate for 1 min. Protein concentration was determined by the Bradford method using bovine serum albumin as a standard.

2.3 Isolation and Purification of the Extracellular Fungal Laccase

Fungal mycelium was removed and 14-day culture liquid of *F. fomentarius* F-1531 obtained by centrifugation for 60 min at 6000 g in a Hettich Universal 320R centrifuge (Germany) was subjected to a two-step ammonium sulfate fractionation.

At first, to remove ballast proteins, ammonium sulfate was added to the supernatant up to 50% saturation and kept at +4 °C overnight, than it was centrifuged for 30 min at 12000 g and the precipitate has been taken away. In the supernatant, the concentration of ammonium sulfate was increased to 70%; after centrifugation at 12000 rpm for 20 min, the precipitate was collected. The resulting fraction was dissolved in a small volume of 0.1 M Na-acetate buffer, pH 5.4, and was dialyzed against 0.02 M Na-acetate buffer. Protein concentration and laccase activity were analyzed in all fractions.

2.4 Molecular Weight Analysis Using Sodium Dodecyl Sulfate-Poly Acrylamide Gel Electrophoresis (SDS-PAGE) Technique

Denaturing electrophoresis was carried out on a 12% polyacrylamide gel according to the method of Laemmli on a Mini-PROTEAN device (BioRad, USA). The mixture of marker proteins Precision Plus Protein ™ All Blue Prestained Protein Standards (BioRad, USA) was used as molecular weight standards. Proteins were visualized by staining with Coomassie (brilliant blue R-250) and gel has been photographed.

The data presented are the average of the results of three parallel experiments carried out at least in three chemical replicates. For evaluation of the results we used the standard error method using Microsoft Office Excel 2010.

3 Results

3.1 Influence of Copper Ions on Laccase Activity of *F. fomentarius* F-1531

As a result of studying the dynamics of the laccase activity of *F. fomentarius F-1531* Culture liquid, which has been obtained in the presence of copper ions in the concentration range of 0–300 mg/l on a liquid HPS medium, the maximum activity was reached on day 14 in the presence of Cu^{2+} at a concentration of 200 mg/l (Fig. 1).

The laccase activity in the culture liquid on the HPS medium with $CuSO_4$ (200 mg/L) was in 6 times higher than the activity on a medium without copper sulfate (Fig. 2).

As a result, the optimal concentration of copper ions to stimulate the biosynthesis of laccase by the fungus *F. fomentarius F-1531* was its content in the HPS nutrient medium of 200 mg/l. In this case, the maximum level of laccase activity (14–17 U/ml) was reached on 14 day.

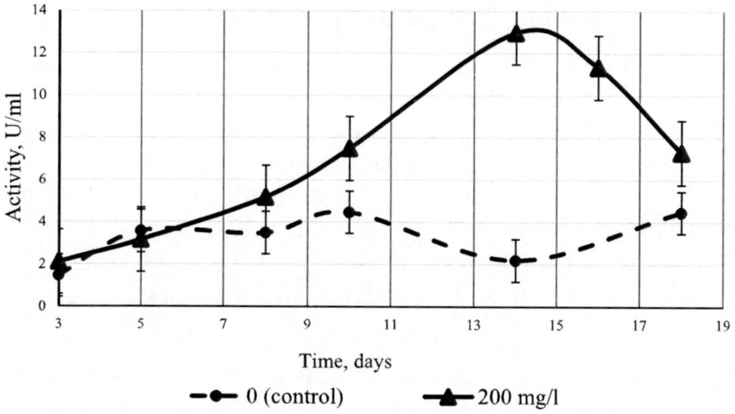

Fig. 1. Time-course of laccase production by *F. fomentarius F-1531* in a liquid HPS medium for 18 days: in the presence of $CuSO_4$ (200 mg/l) and without CuSO4 (control).

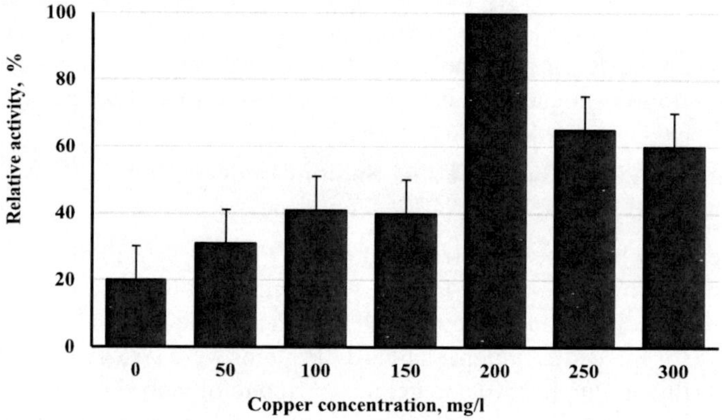

Fig. 2. Profile of phenoloxidase activity in 14-day culture liquid of fungus *F. fomentarius F-1531* in the presence of various concentrations of $CuSO_4$ in culture medium

3.2 Purification *F. fomentarius F-1531* Laccase at the Optimum Concentration of Copper in the Nutrient Medium

Isolation and purification of laccase from 14-day copper supplemented culture broth (200 mg/l Cu^{2+}) of the fungus *F. fomentarius F-1531* was carried out by the ammonium sulfate precipitation method.

As a result of SDS-PAGE electrophoresis and following staining gel by Coomassie R-250, in the purified sample was detected a major laccase protein with a mobility corresponding to a molecular weight of about 60 kDa (Fig. 3).

The protein fraction that precipitated in the range from 50% to 70% saturation had a laccase activity of up to 560 U/ml (for pyrocatechol), which is about for 40 times higher than the initial activity in the culture liquid.

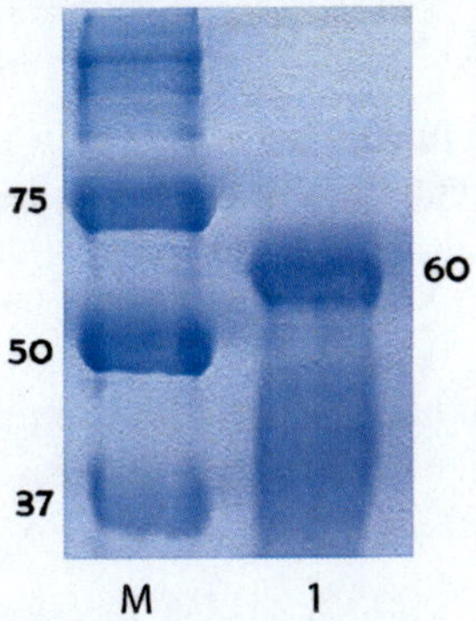

Fig. 3. Electrophoregram of purified enzymes from *F. fomentarius F-1531*: M - protein standards (mass, kDa); 1 - purified enzymes

4 Discussion

To increase the yield of oxidases, including laccases, during the microbial growth, using various inducers, which promote effective biosynthesis are processes that has been reported [1–4]. Copper ions in culture media can serve as inducers, including for the xylotrophic basidiomycete *Fomes fomentarius* [13].

The study of oxidase activity during the *F. fomentarius* liquid-state fermentation with various copper concentrations (0–300 mg/l) for 18 days was found out the time-course of oxidase production and indicated that the maximum laccase activity was reached with $CuSO_4$ (200 mg/l) at 14 day.

The efficiency of oxidase enzymes biosynthesis in the presence of copper was in 6–6.5 times higher than in a medium without copper, as shown on Figs. 1 and 2. This phenomen indicates stimulation of copper-containing oxidase enzymes synthesis in culture broth of fungus *F. fomentarius* [10].

The enzymes were isolated from cultural liquid and purified by ammonium sulfate precipitation. The investigation of salting out the enzymes by the stepwise addition of ammonium sulfate between 0–70% saturation (with the interval of saturation 10–15%) showed that the fraction precipitated between 50–70% of saturation has the highest laccase activity.

The molecular weight of the purified laccase determined by the SDS-PAGE was about 60 kDa which corresponds with the range of 50–90 kDa for the majority of the fungal laccases according to the data from previous studies conducted by other authors [1–4, 13–15]. For example, the purified laccase of the fungus *Fomes fomentarius* obtained under the conditions of solid-state fermentation had a molecular weight of 51 kDa [13].

The combination of the results of gel electrophoresis, an absence of tyrosinase activity, inhibition of the enzymatic activity by EDTA and the capability to oxidize a wide range of phenol derivatives indicates on the laccase nature of the isolated enzyme with high probability [13, 15].

5 Conclusions

Thus, the efficient synthesis of fungal laccase by the fungus *F. fomentarius VKPM F-1531* requires the optimal concentration of copper in the nutrient medium of 200 mg/l. The highest enzyme activity was achieved on the 14th day of surface-liquid cultivation on HPS medium with optimal $CuSO_4$ supplementation and it was more than 6–6.5 times higher than the one obtained on the conventional glucose-peptone medium.

Using the ammonium sulfate fractionation method made it possible to concentrate the enzyme with molecular weight of 60 kDa and laccase activity being 40 times higher than in the original cultural liquid.

Acknowledgments. The study was carried out using the equipment of the Centre for Shared Access "Agidel" and "KODINK" at the Ufa Federal Research Centre of Russian Academy of Sciences. The study was supported by the Foundation for Assistance to Small Innovative Enterprises in Science and Technology (Innovation Promotion Fund) within the UMNIK-2020 competition (contract No. 16276ГУ/ 2021 (dated May 18, 2021)).

References

1. Upadhyay, P., Shrivastava, R., Kumar, P.: Bioprospecting and biotechnological applications of fungal laccase. Biotech. **6**, 1–12 (2016)
2. Moiseenko, K.V., Maloshenok, L.G., Vasina, D.V., Bruskin, S.A., Tyazhelova, T.V., Koroleva, O.V.: Laccase multigene families in Agaricomycetes. J. Basic. Microbiol. **56**, 1392–1397(2016)
3. Kües, U., Rühl, M.: Multiple multi-copper oxidase gene families in basidiomycetes-what for? Curr Genomics **12**(2), 72–94 (2011)
4. Langfelder, K., Streibel, M., Jahn, B., Haase, G., Brakhage, A.A.: Biosynthesis of fungal melanins and their importance for human pathogenic fungi. Fungal. Genet. Biol. **38**(2), 143–158 (2003)
5. Saparrat, M., Balatti, P.A., Martínez, M.J., Jurado, M.: Differential regulation of laccase gene expression in *Coriolopsis rigida* LPSC No. 232. Fungal Biol. **114**, 999–1006 (2010)
6. Sadhasivam, S., Savitha, S., Swaminathan, K., Lin, F.-H.: Production, purification and characterization of mid-redox potential laccase from a newly isolated *Trichoderma harzianum WL1*. Process Biochem. **43**(7), 736–742 (2008)
7. Forootanfar, H., Faramarzi, M.A., Shahverdi, A.R., Yazdi, M.T.: Purification and biochemical characterization of extracellular laccase from the ascomycete *Paraconiothyrium variabile*. Bioresour. Technol. **102**(2), 1808–1814 (2011)
8. Niladevi, K.N., Jacob, N., Prema, P.: Evidence for a halotolerant-alkaline laccase in *Streptomyces psammoticus*: purification and characterization. Proc. Biochem **43**(6), 654–660 (2008)
9. Collins, P.J., Dobson, A.: Regulation of laccase gene transcription in *Trametes versicolor*. Appl. Environ. Microbiol. **63**, 3444–3450 (1997)
10. Bisswanger H.: Prakticheskaja jenzimologija. [Practical Enzymology], 328 p. *BINOM Laboratorija znanij*, Moscow (2015)
11. Shakirova, A.A., Kovalenko, S.V., Khalimova, L.Kh.: Research of influence of the polyunsaturated fatty acids isopropyl esters on the growth and phenoloxidase activity of fungus Fomes fomentarius E-14. Bashkir Chem. J. **26**(1), 85–90 (2019)

12. Ponomareva, E.G., Kupryashina, M.A., Vetchinkina, E.P., Nikitina, V.E.: Extracellular Lac case Activity of Bacteria of the Genera Azospirillum and Niveispirillum. Biol. Bull. Russ. Acad. Sci. Izvestiia Akademii nauk **46**(2), 161–167 (2019)
13. Neifar, M., Jaouani, A., Ellouze-Ghorbel, R., Ellouze-Chaabouni, S.: Purification, characterization and decolourization ability of *Fomes fomentarius* laccase produced in solid medium. J. Mol. Cat. B: Enzym. **64**, 68–74 (2010)
14. Giardina, P., Faraco, V., Pezzela, C., Piscitelli, A., Vanhule, S., Sannia, G.: Laccases: a never-ending story. Cell. Mol. Life Sci. **67**(3), 369–385 (2010)
15. Ezike, T.C., Ezugwu, A.L., Udeh, J.O., Eze, S.O.O., Chilaka, F.C.: Purification and characterisation of new laccase from Trametes polyzona WRF03. Biotechnol. Rep. (Amst) **28** (2020)

Biocompatible Materials Based on Modified Starch and Chitosan with High Mechanical Properties

Ekaterina Kachalova(✉), Kristina Apryatina, and Larisa Smirnova

National Research Lobachevsky State University of Nizhny Novgorod, Nizhny Novgorod, Russia
katerina_k1997@mail.ru

Abstract. Biocompatible film materials based on alkylated chitosan (CTS) and grafted copolymer of starch (PS) with acrylamide (AA) have been obtained by watering from a homogeneous solution of modified polysaccharides. The optimum ratio of PS and grafted AA chains has been found to ensure solubility of the product in the pH range from 2 to 12 and compatibility with CTS in solution. Physical and mechanical properties depend on the ratio, modified CTS and PS. With a mass ratio of the components in the film of 1:1, the breaking stress was 65 MPa, with an elongation of 11%. According to the X-ray phase analysis the films have amorphous structure. High proliferation of human fibroblasts of the hTERT BJ-5ta cell line indicates the presence of biocompatibility and high adhesion of cells to the surface of the film material. The resulting composition is promising as packaging materials and products for medical and biological purposes.

Keywords: Starch · Chitosan · Modification · Film materials · Mechanical strength · Biocompatibility

1 Introduction

Currently, the problem of environmental protection in connection with the emissions of enterprises, pollution of water bodies and the accumulation of unrecyclable waste is urgent [1]. According to the United Nations report, about two billion tons of household waste is generated per year, one-sixth of which are synthetic polymeric materials, which is approximately 300 million tons/year [2]. Among them ¼ part falls on packing materials, the leading position in which takes polyethylene, polypropylene [3].

Today only 20% of polymer waste is recycled, 25% is incinerated and 55% is discarded. Limited and improper collection of plastic waste has led to the visible accumulation of polymer residues in the environment, where they suffer from systematic fragmentation caused mainly by UV radiation and mechanical abrasion, partially transforming into microplastics [4–6].

In order to make the synthetic polymers biodegradable various additives have been developed working on the principle of oxo-degradation, but their work requires certain

V. Kurchenko et al. (Eds.): ICAETT 2021, LNNS 408, pp. 106–113, 2022.
https://doi.org/10.1007/978-3-030-96641-6_13

conditions (light, temperature, humidity), which are not available in many landfills, where waste is buried and covered with layers of sand. Even under these conditions, these additives would not provide biodegradation of polymeric materials, but would turn it into microplastic (fragments of any polymeric material, size < 5 mm) [7].

Gradually microplastics can be found everywhere in various environmental matrices, food, drinking water and even in the human body [8–11]. The effect of microplastics on the human body has not yet been fully studied, but it cannot be called positive. The authors [12] investigated the effects of microplastics on marine life and provided evidence that microplastics adsorb various toxic substances on their surface, emphasizing the increased bioavailability of these chemicals after ingestion, as well as toxicological consequences due to the sensitivity of some molecular and cellular systems.

One of the most important issues in this field is the rational use of plastic containers, their recycling, and, as the best option, replacement with biodegradable materials.

In the light of the above, the aim of the work was to obtain a composition of a PS-AA copolymer soluble in the whole pH range of the medium with CTS and to study their properties.

2 Materials and Methods

Potato starch (PS) with Molecular Mass = $1.5 * 10^5$ (JSC "Verkhovichesky starch plant"), acrylamide (chemically pure), chitosan with a molecular weight of $0.8 * 10^5$ to $3 * 10^5$ and a degree of deacetylation from 75 to 85% (JSC "Bioprogress"), acetic acid (98%, chemical pure), hydrochloric acid (36%, chemical pure), toluene diisocyanate (chemical pure), heptaldehyde (HpA) (99%, Sigma Aldrich), ammonium persulfate (chemical pure).

2.1 Synthesis of Graft Copolymer PS-AA

The grafted AA polymerization on PS was carried out in a three-neck round-bottom flask placed in a thermostat with controlled temperature under continuous stirring. The reaction temperature was 60 °C. First, the required amount of PS was loaded in 0.5% NaOH alkaline solution, (pH ~ 10.5). The system was incubated for 15 min to establish the desired temperature of the solution. Then AA dissolved in NaOH solution (pH = 10.5) was added, stirred and when the temperature reached 60 °C an initiator was added. The synthesis was carried out for 2.5 h.

The concentration of the initiator was varied from $1 * 10^{-3}$ to $0.25 * 10^{-3}$ mol * L^{-1}.

$Ce(SO_4)_2*4H_2O$, $(NH_4)_2S_2O_8$, $[Co(NH_3)_5Cl]Cl_2$ could be used as initiator systems.

All the resulting synthesis products using different types of initiating systems, with different ratios of components PS:AA precipitated by selective double resuspension from water with ethanol. This ensures the removal of unreacted AA. The depth of transformation was determined by the griviometric method. In the process of synthesis samples were taken, purified from the monomer with ethanol by multiple resuspension, the precipitate was dried to constant weight and weighed. The yield of the synthesis product was determined in relation to the theoretically possible copolymer yield, assuming 100% conversion of AA to PAA, taking into account the mass of PS and PAA.

2.2 Obtaining Film Samples

The grafted PS-AA copolymer was dissolved in 3% aqueous acid CTS solution. After complete dissolution, the samples were centrifuged to remove air bubbles, and then films were obtained on a XiamenTMAX-TMH film filling machine on a lavsan substrate. The films were dried under conditions of uniform solvent evaporation at 40 °C and atmospheric pressure until constant weight.

2.3 Study of Mechanical Properties

Physical and mechanical characteristics (tensile strength and elongation at break) of the material were determined on a Roell/Zwick Z005 tensile machine. Tests were conducted at a tensile speed of 10 mm/min on samples 60 ± 5 μm thick in the form of rectangles 15 mm wide. At least 10 samples were measured for each film composition.

2.4 Study of Biocompatibility of Films Based on Modified PS and CTS

Adhesion, cytotoxicity, and cell growth on the surface of the films during cultivation of human fibroblasts of the hTERT BJ-5ta cell line were studied. The films of material, after sterilization by autoclaving at 110 °C, were placed in the wells of a cell culture plate and filled with 500 μl of DMEM medium. Cells were seeded on the film surface at a density of $1.6 * 10^5/cm^2$ and cultured for 24 h. Cell visualization and evaluation of cell viability were assessed by fluorescent microscopy. A $2 * 10^{-4}$ wt.% acridine orange solution in phosphate buffer was used as a dye for staining fibroblasts. This dye interacts selectively with DNA and RNA located in the cell nucleus and mitochondria, respectively, by intercalation or electrostatic attraction. This makes it possible to assess the overall state of the cells-activity, proliferation and apoptosis. Microanalysis of the films was performed on an OlympusIX71 inverter microscope (Japan/Germany) using a "green" filter (510–555 nm emission, 460–495 nm excitation), which allows visualization of the green color of the nucleus of living cells.

2.5 X-ray Phase Analysis

X-ray phase analysis of the samples was performed on an X-ray diffractometer "Bruker D8 Discover", using CuKα radiation. Diffractograms were recorded for the angular range of 10–60° at a diffraction angle of 2θ in a symmetrical geometry with a 0.6 mm slit on the primary beam and a LynxEye linear position-sensitive detector.

3 Results and Discussion

3.1 Synthesis of Grafted Copolymer PS:AA

The combination of CTS and PS in solution limits the difference in pH value of the medium at which solutions of individual polysaccharides are formed ($pH < 6$ for CTS, $pH > 9$ for PS). To solve this problem, modification of PS by graft polymerization with

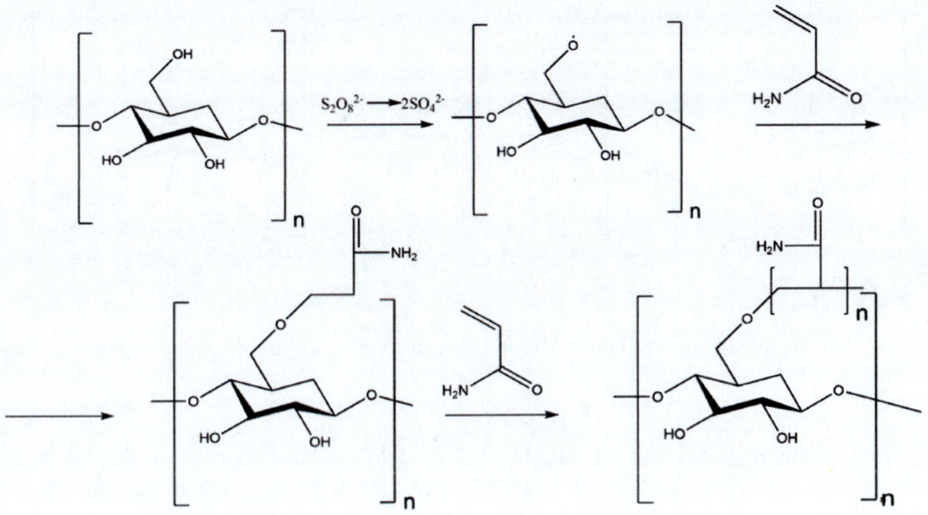

Fig. 1. Scheme of AA polymerization by PS in the presence of persulfate [13].

AA was carried out. The resulting PS derivative is soluble in the entire pH range of the medium.

Based on the literature data the grafted AA polymerization on PS can be represented by the following scheme (Fig. 1) [13].

Since the work was aimed at obtaining products biocompatible and soluble in a wide range of pH values of the medium, the compounds with a high PS content in the copolymer composition are of the greatest interest. The optimal ratio of 1:2 (PS:AA) by mass in the loading mixture has been revealed by experiments. It was found that the copolymer is soluble in the pH range from of 2 to 12, which makes it possible to combine PS derivatives with CTS in solution.

3.2 Synthesis of Film Samples

As noted above, the grafted copolymer PS:AA (1:2) is soluble in a wide range of media pH values, this allowed to obtain a homogeneous system when mixing with polysaccharide solutions. However, the viscosity of the obtained solution (3% sol. CTS, 3% sol. grafted copolymer PS:AA (1:2)) was very low and did not allow obtaining high-quality film samples.

It should be noted that films based on pure CTS are fragile, so CTS was modified with HpA to form Schiff bases (Fig. 2). The introduction of alkyl substituents into the structure of glucosamine links, which can have a plasticizing effect, is fundamental.

To obtain more concentrated solutions of the modified polysaccharide mixture, the PS derivative was precipitated after synthesis, excess solvent and precipitant were removed, and the wet grafted copolymer was introduced into the HpA-treated CTS solution under continuous stirring until a true solution was formed. The dissolution stage was faster and less energy-consuming.

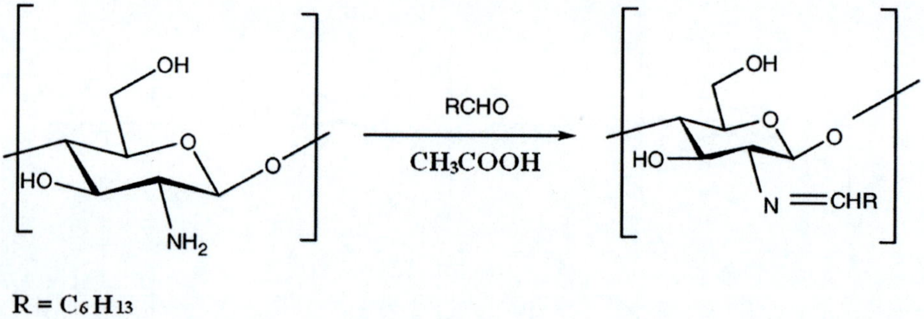

Fig. 2. Scheme of the nucleophilic addition reaction to the CTS amino group of HpA to form Schiff base [14].

Homogeneous solutions of alkylated CTS and grafted copolymer PS:AA were obtained at different ratios of 1:1, 1:2, 1:3 respectively in terms of weight of dry components.

From the obtained solutions under conditions of uniform evaporation of the solvent at T = 40 °C. The physical and mechanical properties were determined. The average test results of five samples in each series are presented in Table 1, the error of measurement is 7%.

Table 1. Physical and mechanical characteristics of films

$CTS_{(10\%HpA)}$:($PS:AA_{1:2}$)1:1, with a ratio by mass	Tensile strength, σ (MPa)	Elongation, ε (%)
1:1	65.5 ± 4.5	11.88 ± 0.83
1:2	43.5 ± 3.5	5.96 ± 0.42
1:3	33.5 ± 2.5	1.68 ± 0.12

Table 1 shows that the strength and strain value significantly depend on the ratio of CTS and grafted copolymer. The best indicators are realized at the same mass ratio of PS:AA and CTS, the ultimate tensile strength of 65 MPa, with a strain of 11%. For comparison, the tensile strength of film samples made of pure-unmodified CTS is 15.5 MPa (ε = 1.51%), getting films made of pure PS is brittle, their strength quantitative characteristics cannot be obtained.

The high strength characteristics can be attributed to a change in the structure of the original polymers and the plasticizing effect of the alkyl fragments of the side substituents in CTS. The structure of initial polymers-pure CTS and PS, as well as film samples based on $CTS_{(10\%HpA)}$:($PS:AA_{1:2}$) 1:1 (Fig. 3) was determined by X-ray phase analysis method.

As can be seen from the diffractograms, the reference PS (Fig. 3, curve 1) has many peaks, it is close to polycrystalline and has rather large crystallites. Initial CTS (Fig. 3, curve 2) is a polymer with a high degree of crystallinity. Increased crystallinity negatively

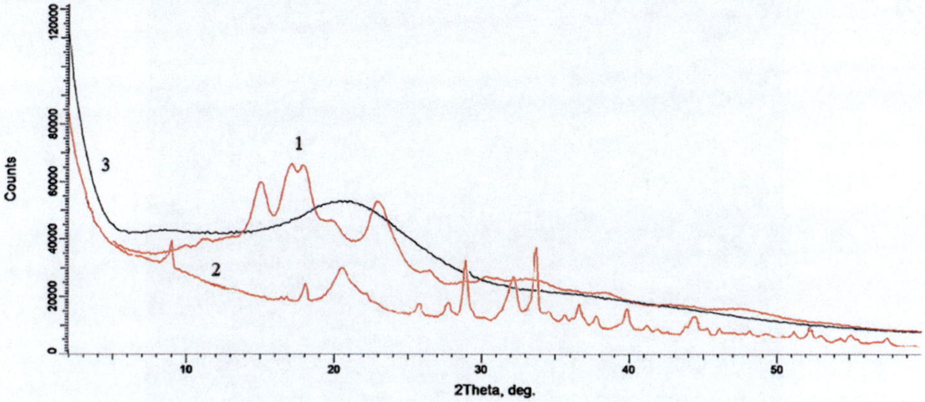

Fig. 3. Diffractograms of samples: 1) PS; 2) CTS; 3) $CTS_{(10\%HpA)}$:($PS:AA_{1:2}$) 1:1

affects such properties of film samples as strength and elasticity. An amorphous halo is observed in the spectrum of the sample based on CTS_{HpA}:PS:AA, on the basis of which we can conclude that it has an amorphous structure. The estimate of the size of coherent scattering regions is on the order of 1 nm.

The biocompatibility, adhesion, and cytotoxicity of the samples based on the obtained composition were studied by culturing human fibroblasts of the hTERT BJ-5ta cell line (Fig. 4) and compared with the pure chitosan sample (Fig. 5).

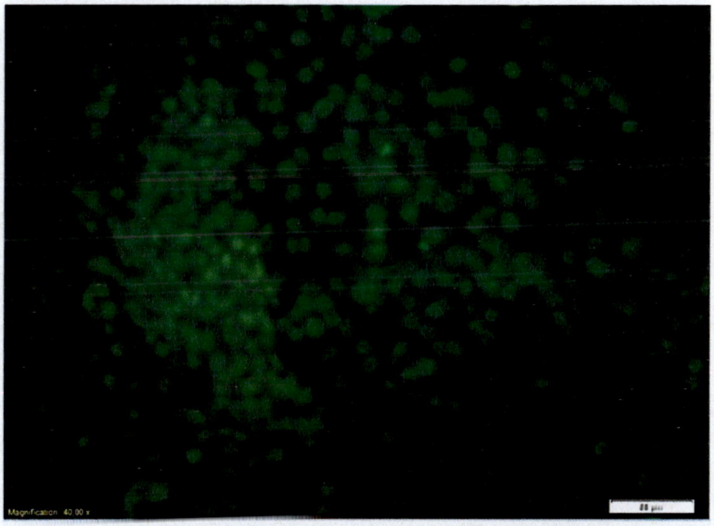

Fig. 4. Film based on $CTS_{(10\%HpA)}$:($PS:AA_{1:2}$) 1:1 + toluene diisocyanate, populated with fibroblasts, after 24 h

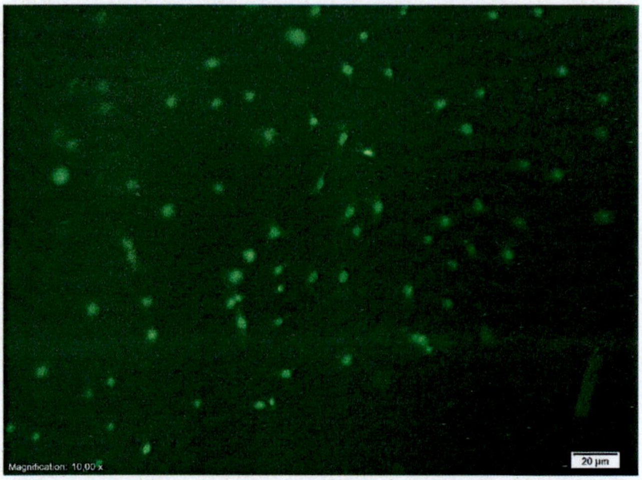

Fig. 5. Film based on pure CTS, populated with fibroblasts, after 24 h

Compared to the pure CTS film sample, the analysis showed high cell viability on the surface of the films. Active cell growth and division, indicating a high degree of adhesion, as well as biocompatibility and absence of cytotoxicity.

4 Conclusion

The resulting composition based on modified PS and its composite with CTS, subject to its optimization, is promising as a packaging material, film biocompatible medical devices-tissue-substituting materials in traumatology for wound surface closure.

Acknowledgments. We acknowledge to PhD in Physics and Mathematics, Yunin P., for testing samples by X-ray diffraction analysis.

This work was supported by the Fund for Assistance to the Development of Small Forms of Enterprises in the Scientific and Technical Sphere (Fund for Assistance to Innovation), within the framework of the UMNIK program (contract code 0064457).

This work has been performed with the support of Research and Educational Center of Nizhny Novgorod region in the framework of the agreement No. 16-11-2021/48.

References

1. Chen, Y., Leng, Y., Liu, X., Wang, J.: Microplastic pollution in vegetable farmlands of suburb Wuhan, central China. Environ. Pollut. **257**, 113449 (2020)
2. Al Rayaan, M.: Recent advancements of thermochemical conversion of plastic waste to biofuel-A review. Clean. Eng. Technol. **2**, 100062 (2021)
3. Geyer, R., Jambeck, J.R., Law, K.L.: Production, use, and fate of all plastics ever made. Sci. Adv. **3**(7), e1700782 (2017)
4. Su, L., Cai, H., Kolandhasamy, P.: Using the Asian clam as an indicator of microplastic pollution in freshwater ecosystems. Environ. Pollut. **234**, 347–355 (2018)

5. Wang, W., Ge, J., Yu, X., Li, H.: Environmental fate and impacts of microplastics in soil ecosystems: Progress and perspective. Sci. Total Environ. **708**, 134841 (2020)
6. Yang, L., Zhang, Y., Kang, S.: Microplastics in soil: a review on methods, occurrence, sources, and potential risk. Sci. Total Environ. **780**, 146546 (2021)
7. Akdogan, Z., Guven, B.: Microplastics in the environment: a critical review of current understanding and identification of future research needs. Environ. Pollut. **254**, 113011 (2019)
8. Curren, E., Leaw, C.P., Lim, P.T., Leong, S.Y.: Evidence of marine microplastics in commercially harvested seafood. Front. Bioeng. Biotechnol. **8**, 1390 (2020)
9. Guo, J.J.: Source, migration and toxicology of microplastics in soil. Environ. Int. **146**, 105263 (2020)
10. Ragusa, A.: Plasticenta: first evidence of microplastics in human placenta. Environ. Int. **146**, 106274 (2021)
11. Wang, C., Zhao, J., Xing, B.: Environmental source, fate, and toxicity of microplastics. J. Hazardous Mater. **407**, 124357 (2021)
12. Alfaro-Núñez, A.: Microplastic pollution in seawater and marine organisms across the Tropical Eastern Pacific and Galápagos. Res. Square **11**(1), 1–8 (2020)
13. Wang, S.: Synthesis and characterization of starch-poly(methyl acrylate) graft copolymers using horseradish peroxidase. Carbohyd. Polym. **136**, 1010–1016 (2016)
14. Yu, C., Kecen, X., Xiaosai, Q.: Grafting modification of Chitosan. Biopolymer Grafting Synth. Prop. **1**, 295–364 (2018)

Products of Hydrolysis of Bleached Hardwood Kraft Pulp by Carbohydrate-Active Enzymes

Aleksandr Shevchenko[1](✉), Andrey Aksenov[1], Irina Tyshkunova[2], Danil Falev[1], and Evgeniy Toptunov[1]

[1] Northern (Arctic) Federal University, Arkhangelsk, Russia
[2] Institute of Macromolecular Compounds of the Russian Academy of Sciences, St. Petersburg, Russia

Abstract. Enzymatic technologies for processing cellulose-containing raw materials are of great importance for the development of industry. Monomers of sugars and non-hydrolyzed residue are formed in the destruction process of biopolymers. Monosaccharides can be used for assimilation by microorganisms to produce organic acids. The non-hydrolyzed residue is modified cellulose with great potential value for use in various industries. Conversion of kraft pulp, with a concentration of 10%, was carried out with economically justified dosages of biocatalysts based on an enzyme complex from the fungus *Penicillium verruculosum*. The maximum degree of hydrolysis was 68% for reducing sugars (glucose, xylose, mannose, xylobiose and oligosaccharides). The possibility of obtaining cryogel from biomodified hardwood pulp was investigated in the paper. The article shows the possibility of controlled destruction of kraft pulp with a change in the dosage of enzyme preparations.

Keywords: Kraft pulp · Hydrolysis · Biomodified cellulose · Carbohydrate-active enzymes · Cryogel · Aerogel · Hardwood · *Penicillium verruculosum*

1 Introduction

In recent years, the prevailing trend is to improve and reduce the cost of biological processes for the conversion of lignocellulosic materials [32]. The increasing environmental problems associated, with the possible depletion of resources, have led to a special practical interest in the development and use of a complex of natural polysaccharides, as a potential source of carbohydrates for microbiological synthesis, and biomodified cellulose for use in industry [15, 24].

The high content of polysaccharides in wood pulp, which is comparable to grain, offers great prospects within the framework of an integrated processing approach. To date, carbohydrate products of hydrolysis of cellulose-containing materials are the most promising nutrient media for the microbiological synthesis of various kinds of organic acids and the production of bioethanol [15, 25, 29], and due to their abundance and

V. Kurchenko et al. (Eds.): ICAETT 2021, LNNS 408, pp. 114–123, 2022.
https://doi.org/10.1007/978-3-030-96641-6_14

stability in various media, plant cellulose and cellulose nanomaterials are attracted as an alternative to synthetic materials, especially for fillers and amplifiers of composite materials [7, 10]. Hardwood and softwood surpasses many other non-wood materials of plant origin in terms of the potential production of sugars [19, 20].

The development of carbohydrate-active enzymatic technologies for the conversion of plant materials allows the introduction of alternative, non-food sources of carbon for industrial applications [16, 20, 22, 30]. Increasing production efficiency, reducing the cost of carbon source [20, 35] and improving the quality of finished products can be achieved by developing technologies that play a crucial role in ensuring key biotechnological processes [21, 25].

Cheap and readily available substrates for microbiological synthesis, with low competing value in food quality [8], can become an alternative source of carbon. There are various chemical pretreatment technologies of wood. These include such as acid treatment, alkaline and kraft pulping. The kraft pulping is a cost-effective and efficient method of preparing wood for the production of bleached pulp [1, 20]. For industrial fermentation, substrates are selected based on the geographical location and the availability of raw materials [13]. For the synthesis of amino acids in Europe, molasses is used, as well as in China and South America, whereas for food resources, corn syrup is used in North America and grain syrup in Russia [2].

The main requirements for the composition of the complex carbohydrate-active enzymes for the most complete hydrolysis of cellulose include the presence of high endoglucanase and cellobiohydralase activity. In addition to all this, the complex must have a sufficiently high betaglucanase and cellobiase activity in order to convert the intermediate hydrolysis product – cellobiose disaccharide – into glucose monosaccharide, since cellobiose accumulating in the medium exhibits an inhibitory effect on the rate of enzymatic hydrolysis [37]. An example of high catalytic activity is the *Penicillium verruculosum* carbohydrases complex [5, 20].

The economic component of work consists in using not only the sugars of the hydrolyzate of plant raw materials, but also another conversion product - non-hydrolyzable insoluble cellulose-containing biomaterials - biomodified cellulose [34]. The materials obtained in this way can be used in various technologies, including nanotechnology [30, 33]. Using this enzymatic treatment, it is possible to improve the properties of not only traditional types of viscose pulp, but also paper and non-wood pulps. One of these areas is the production of powdered cellulose and bioaerogels/cryogels. These products are used in various fields of industry, including those for which the use of cellulose fibers is not typical: drug production, food industry, in insulating materials, electrical devices and for energy storage [14, 34].

Bioaerogels are dry, ultralight porous materials with a high specific surface area obtained by supercritical drying [4, 12]. This generation of aerogels appeared at the beginning of the twenty-first century from biomass, based on polysaccharides, and therefore they are called bioaerogels. Bioaerogels have a low density of 0.05–0.2 g/cm^3 and a fairly high specific surface area from 200 to 600 m^2/g [3]. Polysaccharide aerogels have properties similar to synthetic polymers and inorganic aerogels, they can also have heat-insulating properties [9, 23, 27], and when compared with brittle flint aerogels, bioaerogels do not collapse under compression, their plastic deformation reaches 80% of the deformation before the destruction of the pore wall [26, 27].

The advantage of aerogels is their high efficiency, and the disadvantage is the high price. The widespread development of new products, in particular such as bioaerogels, is combined with the principles of eco-friendly use and processing of natural resources [26].

Another type of porous materials is cryogels. Cryogelation is a new, developing promising direction in the field of obtaining polysaccharide materials for biomedical purposes [7]. The formation of a unique, macroporous structure during the lyophilization of cellulose solutions opens up the possibility of obtaining potential biomaterials (cryogel scaffolds) for regenerative biomedicine and tissue engineering [36].

An important aspect of the conversion of the kraft pulp is the regulation of the degree of biomodification of the structure of plant fibers. In the process of such selective and controlled exposure, the polysaccharide complex of plant origin acquires new specified properties. For deep controlled bioconversion, the selection of effective carbohydrate-active enzymes is crucial, which include a whole complex of glycosylhydrolases with a different mechanism of breaking in macromolecules. From the above, it can be argued that it is necessary to solve the problem of studying and developing methods for converting kraft pulp to the specified characteristics. The solution to the problem is an optimal balance of biomodified cellulose and sugars in the hydrolysate. Therefore, it is envisaged to obtain the optimal content of highly concentrated hydrolysates and use of modified cellulose for possible use in the production of bioaerogels, cryogels and nanocellulose.

2 Materials and Methods

2.1 Purpose and Objectives of the Study

The aim of the work was to compare different levels of enzymatic degradation of hardwood kraft pulp on the characteristics of hydrolysates and biomodified cellulose. For this purpose, experiments of various dosages of enzymes B1-221-151 and F10 obtained by recombinant strains of *Penicillium verruculosum* were carried out and processed to obtain conversion products of the necessary parameters. The enzyme dosage was adjusted so that the total activity was 5 and 10 filter paper units (FPU)/g of dry pulp.

2.2 Substrates and Enzymes

The substrates included industrial samples of hardwood pulp obtained from a mixture of aspen and birch in a 1:1 ratio at the Arkhangelsk Pulp and Paper Mill, which is part of Pulp Mill Holding GmbH. The composition of bleached kraft pulp was determined earlier [22]. Hardwood pulp contained 74.5% cellulose and 21.5% xylane. The study used complex cellulolytic enzyme preparations (B1-221-151 and F10) produced by recombinant strains of *Penicillium verruculosum*. The B1 complex contained a mixture of endoglucanase, cellobiohydrolase, xylanase (B1-221-151), and F10 contained cellobiase (F10) [5].

2.3 Enzymatic Hydrolysis

Enzymatic hydrolysis with a concentration of 10% was carried out on (in) a laboratory bioreactor Biostat A Plus (Sartorius) at 50 °C with constant stirring (200–300 rpm) for 48 h. The pH was maintained at 5.0 using an acid-sodium buffer. The cellulose residue after enzymatic hydrolysis was separated from the sugar solution by centrifugation at 4200 rpm for 15 min. The supernatant was collected and analyzed for the content of various sachets. The biomodified cellulose was thoroughly washed with distilled water to separate the soluble sugars. Washing was carried out by sequentially repeating the cycles of mixing residues and centrifugation. After each cycle, the supernatant was carefully drained and a fresh portion of distilled water was added. The biomodified cellulose was further analyzed to calculate the degree of hydrolysis, fractionation and determination of the degree of crystallinity. To determine the degree of polymerization, a viscometric method was used, with cadmium ethylenediamine with a cellulose concentration of 0.1% was used as a solvent (GOST 25438-82). According to the obtained results of the characteristic viscosity, the average degree of polymerization is determined.

2.4 Determining Glucose and Reducing Sugars, Inversion

Glucose and reducing sugars in hydrolysates were determined by the Shomody–Nelson method [18]. Concentrations of minor sugars and disaccharides (cellobioses, xylobioses) were determined using the LC-20 isolation system (Shim-pack ISA-07/S2504 column, Shimazu, Kyoto, Japan) with post column derivatization and fluorometric determination. Inversion of disaccharides and oligosaccharides is carried out to simple sugars. The effect is aimed at determining the unresolved parts of polysaccharides. H_2SO_4 was added to the samples of bleached kraft cellulose hydrolysates to a concentration of 4%, and incubated for 20 min at a temperature of 100 °C. The resulting mixture was cooled and neutralized to a neutral rastover using 25% NaOH. The analysis of the sugar content was carried out by the Shomody–Nelson method.

2.5 Design Cryogels

The biomodified cellulose obtained from experiment at 10 FPU/g dosage of enzymes was washed and dried on freeze drying. Cryogels were made from residue after 94-h enzymatic hydrolysis of kraft pulp. The concentration of the residue was 5%, and phosphoric acid 85% (H_3PO_4) was used as a solvent. The dissolution time was 27 h at a temperature of 24 ± 2 °C. Distilled water was used as a precipitant of cellulose from phosphoric acid solution. The regenerated cellulose was washed with distilled water to a neutral reaction with phenolphthalein. After all manipulations, the resulting gel was frozen and freeze-dried.

3 Results

3.1 Soluble Hydrolysis Products

As a result of enzymatic hydrolysis of kraft pulp with a dosage of 5 FPU/g enzymes, the following composition of the hydrolysate was determined (Fig. 1A). To construct the diagram, data on sugars, disaccharides and oligosaccharides after inversion were used after chromatographic analysis, where the sum of all sugars in the hydrolysate was taken as an integer value. The diagram shows that the dominant monosaccharide in the hydrolysate was glucose, its concentration reached 40 g/L. A great contribution to the content of hemicellulose residues was made by disaccharide – xylobiose, its concentration after 48 h of enzymatic destruction was 15 g/L. Other sugars were in smaller amounts: xylose 6%, mannose 1% and other sugars 7%, which included arabinose, galactose, disaccharides, oligosaccharides, etc.

Enzymatic hydrolysis with a higher dosage of enzymes (10 FPU/g) showed other results (Fig. 1B). Glucose, 61%, with a concentration of about 50 g/L, remained the predominant monosaccharide in the hydrolysate. Xylose in this case contributes more to the total sugar content than xylobiose. Their concentrations reached 9,0 and 8.3 g/L, respectively, after 48 h of hydrolysis. In this experiment, there was an increase in the total content of mannose to 3%, and of other sugars to 11%.

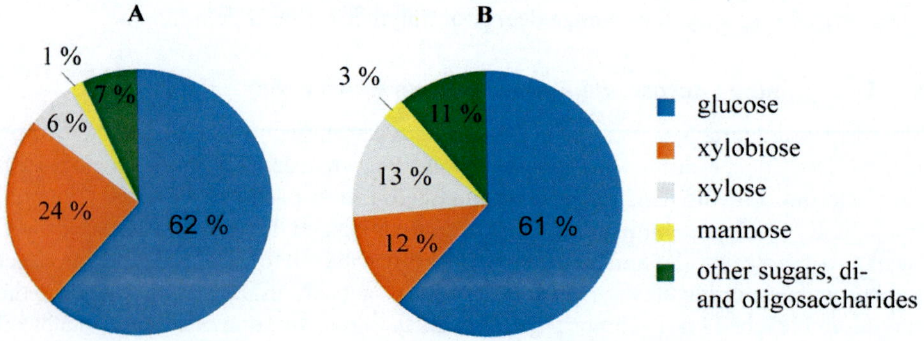

Fig. 1. Sugar content after 48 h of enzymatic hydrolysis with 5 (A) and 10 (B) dosage of enzymes FPU/g

3.2 Degree of Hydrolysis of Polysaccharides

In all experiments, the degree of conversion of the glucan chains of kraft pulp was investigated. Thus, enzymatic hydrolysis with an enzyme dosage of 5 FPU/g reached 50% of the degree of destruction of glucans in 48 h, hydrolysis with 10 FPU/g reached the level of destruction of almost 55% (Fig. 2B). This result was 10% higher than in the other experiment. The yield of dry glucose from the kraft pulp had a different level throughout the hydrolysis time. There for the 10 FPU/g experiment had an advantage of 5 to 8% in terms of glucan conversion rate throughout the entire hydrolysis time. The biggest difference was observed in the first 6 h of the experiment, which could may be due to the enzymes of the xylanase group, which could increase the availability of cellulose fibers for the action of cellulases because of a higher dosage.

The degree of xylan destruction was determined by the sum of the content of xylobiose disaccharide and xylose monosaccharide in the hydrolysate (Fig. 2A). For the experiment with 10 FPU/g dosage, a large jump in xylan conversion was observed, which reached the level of 60% already in the first 6 h of hydrolysis, while the experiment with 5 FPU/g dosage showed a result of 46%. The more interesting was the result after 24 h of hydrolysis. The levels of xylan destruction were almost the same, 66 and 62%, respectively. After 48 h, the destruction degree with a dosage of enzymes 5 FPU/g became more than 10% higher compared to the experiment with a dosage of 10 FPU/g enzymes.

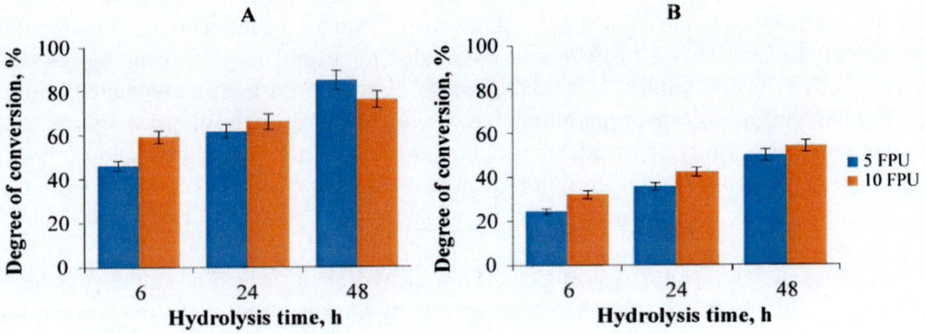

Fig. 2. Dependence of the degree of hydrolysis of xylan (A) and glucans (B) on time

To determine the optimal method of using modified cellulose, the chemical composition of modified cellulose was studied. The modified cellulose after enzymatic hydrolysis 5 FPU/g dosage of enzymes had a polysaccharide composition at the level of 64% cellulose and 36% other polymers. For the experiment with the dosage of enzymes of 10 FPU/g, the composition was 61% cellulose and 39% other polymers. A decrease of the polymerization degree of cellulose was noted during enzymatic hydrolysis. After 48 h of hydrolysis at 10 FPU/g enzyme dosage, the degree of cellulose polymerization decreased by about 2.5 times (up to 300) compared to the initial value (Fig. 3).

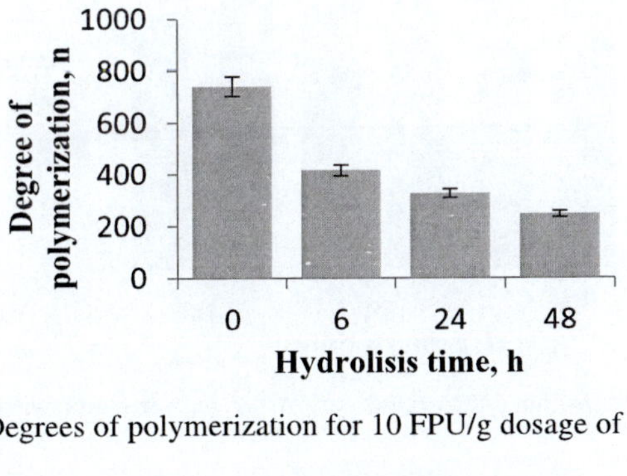

Fig. 3. Degrees of polymerization for 10 FPU/g dosage of enzymes

3.3 Design Cryogels

Cryogel was obtained from modified cellulose after 94 h of enzymatic hydrolysis. The cryogel yield was 68%, cryogel volume - 4.5 cm^3, volume shrinkage - 10%, volume variation - 110%, density - 0.038 g/cm^3, porosity - 97.2%, the swelling - 6098%. These characteristics were determined according to the study [36].

4 Discussion

4.1 Soluble Products and Modified Cellulose

The key sugar formed in the process of enzymatic destruction of kraft pulp is glucose. The value of its concentration in the solution plays a major role for the possibility of its further use for the synthesis of organic acids using industrial producers. The maximum dosage of enzymes was 10 FPU/g and this is due to several factors. This dosage was used in other similar studies [20, 31]. The use of such a dosage is associated with a fairly high cost of enzyme preparations. The use of higher dosages of enzymes can lead to economic instability. To study the regulated biocatalytic destruction and compare the degree of hydrolysis, concentration of sugars, disaccharides and oligosaccharides, a dosage of the same an enzyme complex from the fungus *Penicillium verruculosum* of 5 FPU/g was taken.

Figure 4 shows the ratio of products at different hydrolysis time. The amount of sugars after enzymatic hydrolysis and inversion increased with the time of the experiment for both experiments, but at different rates. Inversion of sugars gives additional 10 and 12% of reducing sugars for 5 and 10 FPU/g dosage of enzymes. The accumulation of sugars, di- and oligosaccharides after 6 h occurs at different rates, which in consequence may affect the composition of the biomodified cellulose at the end of the experiment.

After 48 h of enzymatic conversion, it can be seen that quantity the final products are presented with a significant difference in the ratio for experiments of 5 and 10 FPU/g. The amount of biomodified cellulose in the 5 FPU/g experiment after 48 h of hydrolysis was 2.1 times higher compared to the experiment with a higher dosage of enzymes.

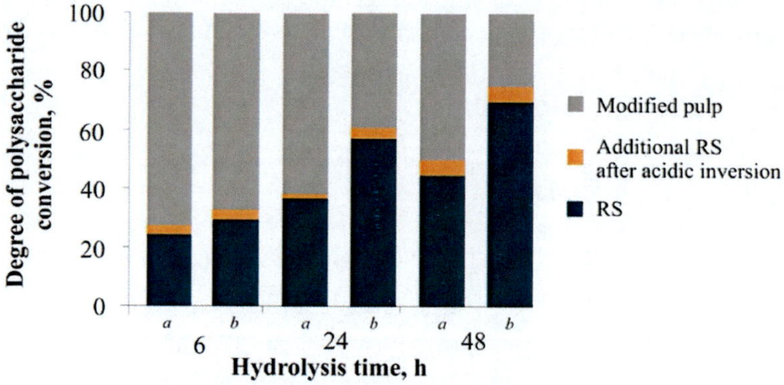

Fig. 4. Dependence of product yield on duration for experiments with 5(a) and 10(b) FPU/g dosage of enzymes

Figure 4 shows the ratio of products at different times of hydrolysis, which makes it possible to vary the degree of hydrolysis kraft pulp based on specific local tasks. It is possible to obtain all the necessary products, such as modified cellulose of a given characteristic and composition, and hydrolysate with high concentration of sugars. Thus, for a small level of cellulose modification, at the initial stage of conversion, in the first 6 h, the concentration of kraft pulp can be increased to 30% to obtain the amount of mono- and disaccharides, as for deep conversion after 48 h (70–80%), but with a concentration of kraft pulp of 5%. As a result the degree of destruction of the cellulose complex and hemicelluloses can be varied without loss in the economic component with the maintenance of high sugar level.

The method of pretreatment with xylanase can be used for additional optimization of the conversion of cellulose-containing materials and regulation of final products. The need in it is caused by an increase in the rate of the enzymatic reaction and the removal of part of the hemicelluloses with an increased yield of xylan sugars at the first stage. A high effect of glucan hydrolysis is possible due to an increase in the available surface of cellulose microfibrils. In addition, the further destruction of xylan increases the total content of monosaccharides in the hydrolysate, assimilable by microorganisms.

4.2 Possibility of Obtaining Cryogel from Biomodified Cellulose

Various biomass sources can be used to produce aerogels and cryogels. Special attention was paid to the ecological and biomedical application of bioaerogels and cryogels. Lignocellulose raw materials, hemicelluloses, cellulose of various types, alginate, etc. were tested from natural polymers to produce cryogels [4]. This work was aimed at studying the possibility of obtaining cryogels from modified cellulose after enzymatic hydrolysis of kraft pulp. The modified cellulose from at the experiment at 10 FPU/g enzyme dosage was treated using phosphoric acid to obtain a cryogel. The possibility of obtaining cryogels from modified cellulose containing both cellulose and hemicellulose has been proven. Cryogel samples were obtained with good quality indicators of porous materials and sent for further study as absorbents and substrate for cultivation of microscopic fungi.

5 Conclusion

Bleached kraft pulp is a good substrate for enzymatic hydrolysis. Using the high efficiency of active carbohydrate enzymes, it is possible to obtain a high yield of sugars with a small dosage. The least studied issue is the further processing of the resulting modified cellulose.

Research on the complex conversion of bleached kraft pulp using *Penicillium verruculosum* with different dosages was continued in our work. It has been shown that different doses of enzyme preparations change the final composition of the modified cellulose while maintaining the total glucose level in the hydrolysate. The substrate concentration of 10% and the enzyme consumption of 10 FPU/g give more acceptable results for the industrial use of sugar solution. Deeper and faster destruction of glucans was noted for the experiment 10 FPU/g dosage of enzymes, while a higher level of destruction of xylan was observed in experiment 5 FPU/g dosage of enzymes. The data presented in the work for the first time the possibility of obtaining products with high added value from biomodified kraft pulp - cryogel with good quality indicators of porous materials.

Acknowledgments. This study was funded by the Russian Foundation for Basic Research (project no. 19-33-60014). The authors wish to thank The Federal Research Centre "Fundamentals of Biotechnology" of the Russian Academy of Sciences (laboratory of Professor A.P. Sinitsyn). The study was (carried out) using the equipment of the Core Facility Center 'Arktika' of the Northern (Arctic) Federal University.

References

1. Akim, G.L., Belodubrovskii, G.B., Burov, A.V.: Tekhnologiya tsellyulozno-bumazhnogo proizvodstva. Pulp Paper Technol. 1 (2003)
2. Anusree, M., Nampoothiri, K.M.: Biosynthesis, recovery and purification of l-lysine from jackfruit seed (JFS) hydrolysate by Corynebacterium glutamicum DM 1729. Biocatal. Agric. Biotechnol. **4**(4), 506–513 (2015)
3. Budtova, T.: Cellulose II aerogels: a review. Cellulose **26**(1), 81–121 (2019)
4. Budtova, T., et al.: Biorefinery approach for aerogels. Polymers **12**(12), 2779 (2020)
5. Chekushina, A., Dotsenko, G., Sinitsyn, A.: Comparing the economy of plant material bioconversion processes using biocatalysts based on Trichoderma and Penicillium verruculosum enzyme preparations. Catal. Ind. **5**, 98–104 (2013)
6. Bakhshpour, M., Idil, N., Perçin, I., Denizli, A.: Biomedical applications of polymeric cryogels. Appl. Sci. **9**(3), 553 (2019)
7. Chen, Z., Liu, G., Zhang, J., Bao, J.: A preliminary study on L-lysine fermentation from lignocellulose feedstock and techno-economic evaluation. Biores. Technol. **271**, 196–201 (2019)
8. Gopinath, V., Murali, A., Dhar, K.S., Nampoothiri, K.M.: Corynebacterium glutamicum as a potent biocatalyst for the bioconversion of pentose sugars to value-added products. Appl. Microbiol. Biotechnol. **93**(1), 95–106 (2012)
9. Groult, S., Budtova, T.: Thermal conductivity/structure correlations in thermal super-insulating pectin aerogels. Carbohyd. Polym. **196**, 73–81 (2018)
10. Henríquez-Gallegos, S., Albornoz-Palma, G., Andrade, A., Soto, C., Pereira, M.: Impact of the enzyme charge on the production and morphological features of cellulose nanofibrils. Polymers **13**(19), 3238 (2021)
11. Himmel, M.E., Adney, W.S., Baker, J.O., Nieves, R.A., Thomas, S.R.: Cellulases: structure, function, and applications. In: Handbook on Bioethanol, pp. 143–161. Routledge (2018)
12. Kistler, S.S.: Coherent expanded aerogels and jellies. Nature **127**(3211), 741 (1931)
13. Kjeldsen, K.R.: Optimization of an industrial L-lysine producing Corynebacterium glutamicum strain (Doctoral dissertation, Technical University of Denmark (DTU)) (2009)
14. Köse, K., Mavlan, M., Youngblood, J.P.: Applications and impact of nanocellulose based adsorbents. Cellulose **27**(6), 2967–2990 (2020). https://doi.org/10.1007/s10570-020-03011-1
15. Leuchtenberger, W., Huthmacher, K., Drauz, K.: Biotechnological production of amino acids and derivatives: current status and prospects. Appl. Microbiol. Biotechnol. **69**(1), 1–8 (2005). https://doi.org/10.1007/s00253-005-0155-y
16. Li, M., et al.: A two-step conversion of corn stover into furfural and levulinic acid in a water/gamma-valerolactone system. BioResources **11**(4), 8239–8256 (2016)
17. Liebner, F., et al.: Shaped hemocompatible aerogels from cellulose phosphates: preparation and properties (2012)
18. Nelson, N.: A photometric adaptation of the Somogyi method for the determination of glucose. J. Biol. Chem. **153**, 375–380 (1944)
19. Novozhilov, E.V., et al.: Application of complex biocatalysts based on recombinant Penicillium verruculosum enzyme preparations in the hydrolysis of semichemical hardwood pulp. Catal. Ind. **6**(4), 348–354 (2014). https://doi.org/10.1134/S2070050414040138

20. Novozhilov, E.V., et al.: Biocatalytic conversion of kraft pulp using cellulase complex of Penicillium verruculosum. Catal. Ind. **8**(1), 95–100 (2016). https://doi.org/10.1134/S2070050416010098
21. Okafor, N.: Modern Industrial Microbiology and Biotechnology. CRC Press, Boca Raton (2016)
22. Pavlov, I.N.: A setup for studying the biocatalytic conversion of products from the processing of nonwood raw materials. Catal. Ind. **6**(4), 355–360 (2014). https://doi.org/10.1134/S207005041404014X
23. Plappert, S.F., Nedelec, J.M., Rennhofer, H., Lichtenegger, H.C., Liebner, F.W.: Strain hardening and pore size harmonization by uniaxial densification: a facile approach toward superinsulating aerogels from nematic nanofibrillated 2, 3-dicarboxyl cellulose. Chem. Mater. **29**(16), 6630–6641 (2017)
24. Rahikainen, J., Mikander, S.: Inhibition of enzymatic hydrolysis by residual lignins from softwood–study of enzyme binding and inactivation on lignin-rich surface. Biotechnol. Bioeng. J. **12**(8), 823–834 (2011)
25. Razak, M.A., Viswanath, B.: Optimization of fermentation upstream parameters and immobilization of Corynebacterium glutamicum MH 20-22 B cells to enhance the production of L-lysine. 3 Biotech. **5**(4), 531–540 (2015)
26. Rudaz, C., Courson, R., Bonnet, L., Calas-Etienne, S., Sallée, H., Budtova, T.: Aeropectin: fully biomass-based mechanically strong and thermal superinsulating aerogel. Biomacromol **15**(6), 2188–2195 (2014)
27. Sescousse, R., Gavillon, R., Budtova, T.: Aerocellulose from cellulose–ionic liquid solutions: preparation, properties and comparison with cellulose–NaOH and cellulose–NMMO routes. Carbohyd. Polym. **83**(4), 1766–1774 (2011)
28. Song, A., et al.: Gel polymer electrolyte based on polyethylene glycol composite lignocellulose matrix with higher comprehensive performances. Electrochim. Acta **247**, 505–515 (2017)
29. Stafford, W., et al.: Forestry biorefineries. Renew. Energy **154**, 461–475 (2020)
30. Sweeney, M.D., Xu, F.: Biomass converting enzymes as industrial biocatalysts for fuels and chemicals: recent developments. Catalysts **2**(2), 244–263 (2012)
31. Tenkanen, M., Tamminen, T., Hortling, B.: Investigation of lignin-carbohydrate complexes in kraft pulps by selective enzymatic treatments. Appl. Microbiol. Biotechnol. **51**(2), 241–248 (1999)
32. Tiwari, R., Nain, L., Labrou, N.E., Shukla, P.: Bioprospecting of functional cellulases from metagenome for second generation biofuel production: a review. Crit. Rev. Microbiol. **44**(2), 244–257 (2018)
33. Torlopov, M.A., Tarabukin, D.V., Frolova, S.V., Shcherbakova, T.P., Volodin, V.V.: Enzymatic hydrolysis of powder celluloses obtained by various methods. Chem. Plant Raw Mater. **3**, 69–76 (2007)
34. Tyshkunova, I.V.: Dissertation of Candidate of technical sciences/Biocatalytic conversion of sulfate cellulose using a complex of PenicilliumVerruculosum (2018)
35. Tyshkunova, I.V., Luzhkova, O.V., Aksenov, A.S., Novozhilov, E.V.: Production of highly concentrated glucose hydrolysates from technical cellulose. New achievements in chemistry and chemical technology of plant raw materials, p. 385 (2017)
36. Tyshkunova, I.V., Chukhchin, D.G., Gofman, I.V., Poshina, D.N., Skorik, Y.A.: Cellulose cryogels prepared by regeneration from phosphoric acid solutions. Cellulose **28**(8), 4975–4989 (2021). https://doi.org/10.1007/s10570-021-03851-5
37. Zhang, Q., Han, Y., Xiao, H.: Microbial α-amylase: a biomolecular overview. Process Biochem. **53**, 88–101 (2017)

Biologically Active Substances of Extracts from Various Wood-Decay Fungi and Their Antioxidant Activity

V. Kurchenko[1(✉)], N. Sushinskaya[1], K. Maiorava[1], E. Tarun[2], and I. Kiseleva[3]

[1] Belarusian State University, Minsk, Belarus
kurchenko@tut.by
[2] International State Ecological Institute, A.D. Sakharova, Minsk, Belarus
ktarun@tut.by
[3] Ural Federal University named after the First President of Russia B.N. Yeltsin, Yekaterinburg, Russia
irina.kiseleva@urfu.ru

Abstract. Using GC-MS, the composition of biologically active substances (BAS) was investigated and the antioxidant activity (AOA) of methanol extracts of 10 samples of 5 species of timber fungi collected in the vicinity of Yekaterinburg (Russian Federation) and Fanipol (Republic of Belarus) was studied. All the samples studied are dominated by aliphatic compounds: saturated hydrocarbons, fatty acids, fatty acid amides. In addition, they contain various phenolic compounds and arabitol. The relative content of unsaturated fatty acids decreases in the series of investigated extracts: *F. fomentarius*, Fanipol – 76,62%; F. *pinicola*, Yekaterinburg – 74,6%; *I. obliquus*, Yekaterinburg – 69,91%; *I. obliquus*, Fanipol – 64,57%; *G. applanatum*, Fanipol – 63,61%; *F. fomentarius*, Yekaterinburg – 55,18%; G. applanatum, Екатеринбург – 46,94%; *P. betulinus*, Yekaterinburg – 33,92%; *P. betulinus*, Fanipol – 27,52%; *F. pinicola*, Fanipol – 25,0%. The composition of BAS obtained methanol extracts depends on the species and the place of collection of fruit bodies of xylotrophic fungi. It should be noted that the antioxidant properties of extracts of the studied species of fungi depend on the composition of BAS. The minimum $IC_{50} \cdot 10^{-4}$ corresponds to the maximum antioxidant activity, which decreases in the series: *G. applanatum*, Fanipol – 5,5; *F. fomentarius*, Yekaterinburg – 6,03; *G. applanatum*, Yekaterinburg – 6,76; *F. pinicola*, Yekaterinburg – 7,94; *I. obliquus*, Fanipol – 10,0; *P. betulinus*, Yekaterinburg – 10,0; *F. fomentarius*, Fanipol – 12,6; *F. pinicola*, Fanipol –12,6; *P.betulinus*, Fanipol – 13,8; I. obliquus, Yekaterinburg – 15,8; *F. fomentarius*, Yekaterinburg – 23,4. Studies have shown that the composition of BAS extracts from the fruit bodies of wood-decay fungi and their AOA depends on the type of decay that they cause, the region of growth and have intraspecific differences.

Keywords: Antioxidant activity · Extracts of timber fungus · Fluorescein · Arbitol · 9,12-octadecadienoic acid · 9-octadecenamide

V. Kurchenko et al. (Eds.): ICAETT 2021, LNNS 408, pp. 124–135, 2022.
https://doi.org/10.1007/978-3-030-96641-6_15

1 Introduction

Among the 16 thousand species of higher basidiomycetes known to science, about 200 can be attributed to the number of medicinal plant. The Aphyllophorales is an obsolete order of fungi in the Basidiomycota that is of particular interest. These xylotrophic fungi hydrolyze cellulose, hemicellulose and lignin of wood due to the release of various enzymes. They are divided into lignin-destroying and cellulose-destroying. Fungi that cause brown rot mainly destroy cellulose and hemicellulose, and white rot is formed by xylotrophic fungi that hydrolyze cellulose, hemicellulose and lignin [11]. Various secondary metabolites are formed in the fruit bodies of aphyllophoroid hymenomycetes, depending on the type of rot and the type of decomposed wood [2, 4, 9, 13]. Their fruiting bodies are an accessible, easily recoverable raw material base for the production of biologically active substances (BAS).

The composition of BAS obtained from fruit bodies depends on the type of destructible wood, the region of growth, climatic conditions and may have intraspecific differences [6]. Extracts from the fruit bodies of aphyllophoroid fungi contain polysaccharides [8, 12], terpenoids, polyphenolic compounds [1, 3, 4, 7, 9, 10], including flavonoids, tannins [9, 13], steroids, sterols, nucleotides, fatty acids, vitamins. Already, representatives of this group of fungi have found wide application in the food and pharmaceutical industry of Southeast Asian countries: *Ganoderma applanatum, Ganoderma lucidum, Lentinus edodes, Fomitopsis officinalis, Grifola frondosa, Fomes fomentarius* и *Inonotus obliquus.* To date, treatment with drugs based on timber fungi is recognized by the official medicine of the USA, Canada, Israel, New Zealand, Germany, etc. [6]. In this regard, it is relevant to conduct comparative studies of the composition, content and biological activity of secondary metabolites from fruit bodies of various taxonomic groups of aphyllophoroid fungi growing in various regions of Belarus and Russia.

2 Materials and Methods

The object of the study was extracts from the fruit bodies of various types of fungi, presented in Table 1.

Substances were extracted from crushed fruit bodies with methanol in a ratio of 1:10 (m/v). The composition of biologically active substances of extracts from the fruit bodies of a number of wood-destroying fungi was carried out using a chromato-mass spectrometric system: Agilent 6850 gas chromatograph with Agilent 5975B mass-selective detector (GC-MS). A capillary column DB-5MS (5% PhenylMethylSiloxane, J&W 122–5062) with a length of 30 m and an internal diameter of 0.25 mm was used for analysis. The injector temperature is 300 °C, the interface temperature is 300 °C. Carrier gas is helium, volume flow rate of the carrier gas is 1 ml/min. Temperature program: the initial temperature of the thermostat is 35 °C, the temperature rises at a rate of 5 °C/min to 170 °C (isothermal mode for 7 min); the temperature rises at a rate of 7 °C/min to 280 °C (isothermal mode for 10 min). The analysis time was 59.71 min. The sample volume is 1 μl, the sample input mode with a reset of 70:1. Mass detector parameters: source temperature – 230 °C, quadrupole temperature – 150 °C. Identification of components by mass spectrum using the NIST mass spectrum library. The relative content of individual compounds was determined as a percentage of the mass of extractive substances.

Table 1. Types of the investigated extracts of wood-decay fungi, places of their collection and growing substrate

Species	Growing location and growing substrate
Brown-rot fungi	
Red-belted conk (Red-banded polypore) (*Fomitopsis pinicola* (Sw.) P. Karst.)	Russian Federation, Yekaterinburg, pine tree (56°85'25" N, 60°43'19"E)
(*F. pinicola*, RF, Yekaterinburg) (*F. pinicola*, RB, Fanipol)	Republic of Belarus, Fanipol, spruce (53°46'59"N, 27°22'20"E)
White-rot fungi	
Hoof fungus (*Fomes fomentarius* (L.) Fr.) (*F. fomentarius*, RF, Yekaterinburg) (*F. fomentarius*, RB, Fanipol)	Russian Federation, Yekaterinburg, birch (56°50'54"N 60°39'02"E) Republic of Belarus, Fanipol, birch (53°45'12"N 27°21'11"E)
Birch tinder fungus (chaga) (*Inonotus obliquus* (Ach. ex Pers.) Pilát) (*I.obliquus*, RF, Yekaterinburg) (*I.obliquus*, RB, Fanipol)	Russian Federation, Yekaterinburg, birch (56°54'14"N 60°26'41"E) Republic of Belarus, Fanipol, birch (53°44'47"N 27°19'42"E)
Birch polypore (*Piptoporus betulinus* (Bull.) P. Karst.) (*P.betulinus*, RF, Yekaterinburg) (*P. betulinus*, RB, Fanipol)	Russian Federation, Yekaterinburg, birch (56°50'49"N 60°22'53"E) Republic of Belarus, Fanipol, birch (53°45'17"N 27°21'03"E)
Artist's bracket (*Ganoderma applanatum* (Pers.) Pat.) (*G. applanatum*, RF, Yekaterinburg) (*G. applanatum*, RB, Fanipol)	Russian Federation, Yekaterinburg, poplar (56°47'55"N, 60°40'37"E) Republic of Belarus, Fanipol, aspen (53°45'06"N 27°20'41"E)

Determination of the antioxidant activity of inclusion complexes.

The ORAC (Oxygen Radical Absorption Capacity) fluorimetric method was used to evaluate the antioxidant activity (AOA) of methanol extracts from the fruit bodies of wood-destroying fungi [5, 14]. The method is based on measuring the decrease in the fluorescence intensity of fluorescein (FL), during its interaction with oxygen radicals. Antioxidants in the reaction medium, interacting with oxygen-containing radicals, slow down free radical oxidation of PL. The AOA of the substances included in the extracts was determined and evaluated by their ability to bind free radicals formed in the Fenton system. Fluorescence measurements were performed on an RF-5301 PC fluorometer («Shimadzu», Japan). The fluorescence intensity was recorded at a wavelength of 514 nm. The excitation wavelength was 490 nm.

The calculation of OAO indicators was carried out according to the degree of fluorescence intensity (A, %), calculated by the formula:

$$A = \frac{Fl}{Fl_0} \times 100,$$

где Fl_0 – fluorescence intensity of the control FL sample (FL solution without Fe^{2+}, EDTA, hydrolyzate and H_2O_2), Fl – the fluorescence intensity of the solution after the addition of the extract.

Graphs of the dependence of the fluorescence intensity (A, %) on the content of extracts from the fruit bodies of wood-destroying fungi were plotted. According to the obtained equation, the concentration of the sample IC_{50}, corresponding to 50% inhibition of fluorescence, was calculated. Plotting and mathematical processing of the research results were carried out using the computer program «Microsoft Office Excel 2003» (Microsoft Corporation, USA). The results of independent experiments are presented as the arithmetic mean. The reliability of the differences between the data samples was determined by the method of confidence intervals.

3 Results and Discussions

Among the species of fungi that cause brown rot, the most common is red-banded polypore (*F. Pinicola* (Sw.) P. Karst.). It is found in stands of all vegetation zones from the forest-steppe zone to the northern taiga subzone. The main destroyer of coniferous wood, which allows this species to be attributed to the group of xylotrophs rather than facultative parasites. depending on the place of growth and the substrate, fruit bodies contain various biologically active substances (Table 2).

Among the detected substances, aliphatic compounds dominate: saturated hydrocarbons, fatty acids, fatty acid amides. In the composition of the *F. pinicola* extract collected in the vicinity of Fanipol and the vicinity of Yekaterinburg, the content of fatty acids was 44.9% and 53.16%. respectively. The total content of unsaturated fatty acids and fatty acid amides was 25.0% for the red-banded polypore from Fanipol, and 74.6% for the red-banded polypore from Yekaterinburg. The antioxidant activity of the investigated extracts differs significantly. This is due to the difference in the composition of biologically active substances. In the extract of *F. pinicola*, collected in the vicinity of Yekaterinburg, the antioxidant activity is significantly higher, which is probably due to the presence of a large amount of unsaturated fatty acids.

Among the fungi that cause white rot are widespread: *F. fomentarius, I.obliquus, P.betulinus, G. applanatum.*

The birch polypore (*Piptoporus betulinus* (Bull.) P. Karst.) has annual fruiting bodies. It is a necrotrophic parasite on weakened birches, and will cause brown rot and eventually death, being one of the most common fungi visible on dead birches. One of the most common types of xylotrophs. The analysis of the results presented in Table 3 shows that there are no significant differences in the composition and content of biologically active substances in the fruit bodies of birch polypore, depending on the place of collection.

Among the identified substances in both objects, arbidol dominates, the content of which is about 50%. In addition, the composition of BAS revealed: saturated hydrocarbons, fatty acids, fatty acid amides.

Table 2. Composition of biologically active substances of methanol extracts of red-banded polypore (*Fomitopsis pinicola*) and their antioxidant activity

Retention time, Min	Substance name	Relative content, %	
		F. pinicola, RB, Fanipol	*F. pinicola*, RF, Yekaterinburg
1	2	3	4
9.989	2-cyclopenten-1-one	2.69	n. d.
10.262	Undecane	3.06	1.28
12.411	(+)-(s)-pantolacton	1.15	n. d.
15.283	5-methoxypyrrolidin-2-	1.32	n. d.
15.822	Methenamine	1.15	n. d.
20.521	d-allose	n. d.	4.06
22.112	pentadecanoic acid	1.25	1.39
22.607	hexadecanoic acid, methyl ester	2.00	n. d.
23.197	tetradecanoic acid	9.55	3.87
23.361	9-hexadecenoic acid	n. d.	1.09
24.642	8-octadecenoic acid, methyl ester	1.71	n. d.
24.648	9-octadecenoic acid, methyl ester, (e)	n. d.	2.23
24.801	9,12-octadecadienoic acid (z,z)-, methyl ester	n. d.	3.18
24.807	dibutyl phthalate	7.10	n. d.
25.295	(z)-11-octadecenoic acid	3.45	13.26
25.422	9,12-octadecadienoic acid (z,z)-	**19.84**	**28.14**
26.411	Octadecanamide	n. d.	3.10
28.383	9-octadecenamide	n. d.	**23.63**
28.402	Tetradecanal	1.67	n. d.
28.573	z,e-7,11-hexadecadien-1-yl acetate	n. d.	3.04
29.937	phenol, 2,2'-methylenebis[6-(1,1-dimethylethyl)-4-methyl-	2.31	n. d.
31.801	bis(2-ethylhexyl) terephthalate	8.83	n. d.
32.492	Squalene	n. d.	1.75
39.283	2-({[4-(diethylamino)phenyl] methyl idene}amino)benzoic acid	n. d.	**5.57**
Indicators of antioxidant activity of extracts of timber fungi			
Amax, %		93	91
Cmax, %		0,1	1
IC50•10^{-4}, %		**12,6**	**7,94**

Table 3. Composition of biologically active substances of methanol extracts of birch polypore (*Piptoporus betulinus*) and their antioxidant activity

Retention time, min	Substance name	Relative content, %	
		P.betulinus, RF, Yekaterinburg	*P. betulinus*, RB, Fanipol
1	2	3	4
5.436	ethanol, 2-(dimethylamino)	n. d.	1.00
9.812	malonic acid	n. d.	1.16
21.358	l-arabinitol	**50.23**	**40.16**
22.119	pentadecanoic acid	1.03	n. d.
23.209	hexadecanoic acid	3.49	2.88
24.649	9-octadecenoic acid (z)-, methyl ester	0.71	1.45
24.801	9,12-octadecadienoic acid, methyl ester	0.69	2.42
25.295	(z)-11-octadecenoic acid	8.86	7.95
25.460	9,12-octadecadienoic acid (z,z)	**10.02**	**18.14**
26.418	n-tetradecanoic acid amide	0.94	0.35
28.377	9-octadecenamide, (z)-	7.24	3.96
28.580	methyl 9,12-heptadecadienoate	0.97	n. d.
30.641	parahigginic acid	n. d.	1.23
Indicators of antioxidant activity of extracts of timber fungi			
Amax, %		94	95
Cmax, %		1	1
IC50•10^{-4}, %		**10**	**13,8**

In the composition of birch polypore extract collected in the vicinity of Fanipol and the vicinity of Yekaterinburg, the content of fatty acids was 25.74% and 33.19%. accordingly. The total content of unsaturated fatty acids and fatty acid amides was 27.52% for birch polypore from Fanipol, and 33.92% for birch polypore from Yekaterinburg. The antioxidant activity of extracts from birch polypore fruit bodies collected in the vicinity of Fanipol and the vicinity of Yekaterinburg is low and does not differ significantly between the objects of the study. This is due to the low content of BAS, which can potentially exhibit antioxidant properties. In the extract obtained from birch polypore collected in the vicinity of Yekaterinburg, the antioxidant activity is significantly higher, which is probably due to the presence of a large amount of unsaturated fatty acids.

The hoof fungus (*Fomes fomentarius* (L.) Fr.) mainly affects the middle, most valuable part of the trunk of many hardwoods. The rot is central, corrosive, rotten-fibrous.

The decayed wood shows black lines in the lightly coloured decayed areas. It is found on almost all hardwoods of the forest zone of Russia. The infestation of stands of birch, aspen, linden at the age of stagnation can reach 50% or more. Food specialization – from facultative saprophytism to saprophytism. The composition of biologically active substances of methanol extracts of this hoof fungus is presented in Table 4.

Table 4. Composition of biologically active substances of methanol extracts of hoof fungus (*Fomes fomentarius*) and their antioxidant activity

Retention time, Min	Substance name	Relative content, %	
		F. fomentarius, RB, Fanipol	*F. fomentarius*, RF, Yekaterinburg
1	2	3	4
10.262	Hendecane	n. d.	1.52
10.268	undecane	1.35	n. d.
13.020	n-docosanene	n. d.	1.06
13.812	1-decene, 4-methyl	n. d.	1.10
16.165	heptadecane, 8-methyl	n. d.	0.88
22.607	hexadecanoicacid	n. d.	0.91
23.197	n-hexadecanoic acid	2.63	n. d.
23.209	hexadecanoic acid tetradecanoic acid	n. d.	13.20
24.642	cis-13-octadecenoic acid, methyl ester	n. d.	1.43
24.80	dibutyl phthalate	n. d.	1.84
24.801	9,12-octadecadienoic acid, methyl	1.99	n. d.
25.251	octadecenoic acid	n. d.	9.73
25.422	9,12-octadecadienoic acid	5.23	6.96
25.650	1,19-eicosadiene	n. d.	0.85
26.405	tetradecanamide	n. d.	5.02
26.411	Hexadecanamide	7.34	n. d.
28.383	9-octadecenamide, (z)-	**58.99**	**37.06**
28.573	z,e-7,11-hexadecadien-1-yl acetate	10.41	n. d.
Indicators of antioxidant activity of extracts of timber fungi			
Amax, %		94	86
Cmax, %		0,1	0,1
IC50•10^{-4}, %		**12,6**	**23,4**

The analysis of the results presented in Table 4 shows that there is a difference in the composition and content of biologically active substances in the fruit bodies of the hoof fungus collected on birch, depending on the place of collection. Among the detected substances, aliphatic compounds dominate: saturated hydrocarbons, fatty acids, fatty acid amides. In the extracts of the hoof fungus collected on birch in the vicinity of Fanipol and Yekaterinburg, the content of fatty acids was 9.85% and 34.07%, respectively. The content of fatty acid amides in these objects was 66.33% and 42.08%, respectively. The total content of unsaturated fatty acids and amides of unsaturated fatty acids was 76.62% and 55.18%, respectively. The antioxidant activity of the studied extracts from the fruit bodies of the hoof fungus is low and depends on the place of collection.

The birch tinder fungus (chaga) *Inonotus obliquus* (Ach. ex Pers.) Pilat. affects live birch trees mainly of older age. It is found on weakened old-aged birches, less often on alder throughout the forest zone of Russia. A typical facultative saprotroph. It is the subject of preparation for the needs of pharmacology. Chaga has medicinal properties, normalizes the activity of the intestinal tract, disinfects wounds, relieves inflammation in burns, abscesses, etc., in the field it is used for brewing tea. The composition of biologically active substances of methanol extracts of the birch tinder fungus, depending on the place of collection, has significant differences in the composition of fatty acids (Table 5).

Thus, the content of fatty acid amides in the extract of birch tinder fungus (chaga) collected in the vicinity from Yekaterinburg was 68.13%, and chaga from Fanipol – 47.93%. The total content of unsaturated fatty acids and amides of unsaturated fatty acids was 64.57% for chaga from Fanipol, and 69.91% for chaga from Yekaterinburg. The antioxidant activity of extracts from the fruit bodies of the chaga collected in the vicinity of Fanipol and the vicinity of Yekaterinburg is low and does not differ significantly. This is due to the low content of BAS, which can potentially exhibit antioxidant properties.

The artist's bracket (*Ganoderma applanatum* (Pers.) Pat.) is found on dead wood and deciduous litter, especially on birch and aspen trees, in forests of all vegetation zones, causing a rot of heartwood in a variety of trees. Much less often it can settle on the shaft of coniferous species. The species is widespread in the forests of the Russian Plain. The composition of biologically active substances of methanol extracts of artist's bracket is presented in Table 6.

The analysis of the results presented in Table 6 shows that there are no significant differences in the composition and content of biologically active substances in the fruit bodies of artist's bracket depending on the place of collection. Among the identified substances, aliphatic compounds dominate: saturated hydrocarbons, fatty acids, amides of unsaturated fatty acids. In the composition of the extract of the artist's bracket collected in the vicinity of Fanipol and the vicinity of Yekaterinburg, the content of fatty acids was 43.66% and 51.63%, respectively. The total content of unsaturated fatty acids and amides of unsaturated fatty acids was 63.61% for the artist's bracket from Fanipol, and 46.94% for the artist's bracket from Yekaterinburg. The antioxidant activity of extracts from the fruit bodies of bracket fungus collected in the vicinity of Fanipol and Yekaterinburg is high. This is due to the content of BAS, which can potentially exhibit antioxidant properties.

Table 5. Composition of biologically active substances of methanol extracts of birch tinder fungus (chaga) (*Inonotus obliquus)* and their antioxidant activity

Retention time, Min	Substance name	Relative content, %	
		I.obliquus, RB, Fanipol	*I.obliquus*, RF, Yekaterinburg
1	2	3	4
10.262	Undecane	3.72	5.54
17.693	Syringol	6.26	n. d.
18.111	Alloaromadendrene	1.86	n. d.
18.638	2,4-di-tert-butylphenol	n. d.	3.03
18.967	dodecanoic acid, 1-methylethyl ester	n. d.	2.12
22.607	hexadecanoic acid, methyl ester	1.40	n. d.
24.160	alpha-oxy-propio-syringone	1.88	n. d.
24.642	6-octadecenoic acid, methyl ester	n. d.	1.78
24.649	9-octadecenoic acid, methyl ester	5.71	n. d.
24.801	9,12-octadecadienoic acid (z,z)-methyl ester	**18.05**	n. d.
26.411	Octadecanamide	4.55	7.28
28.371	9-octadecenamide	**39.51**	**60.85**
28.567	9,12-tetradecadien-1-ol, acetate,(e,z)	n. d.	7.69
39.283	2-({[4-(diethylamino) phenyl] methylidene}amino)benzoic acid	**13.07**	n. d.
Indicators of antioxidant activity of extracts of timber fungi			
Amax, %		89	91
Cmax, %		0,1	1
IC50•10^{-4}, %		**10**	**15,8**

Table 6. Composition of biologically active substances of methanol extracts of artist's bracket (*Ganoderma applanatum)* and their antioxidant activity

Retention time, min	Substance name	Relative content, %	
		G. applanatum, RB, Fanipol	*G. applanatum*, RF, Yekaterinburg
1	2	3	4
10.097	1-butanamine, 3-methyl-n-(3-methylbutylidene)-	n. d.	1.09
10.268	Undecane	1.29	n. d.
11.168	Glycerin	n. d.	2.79
12.855	2-methyl-4-oxopentanoic acid	1.43	0.76
14.003	2,3-dihydro-3,5-dihydroxy-6-methyl-4h-pyran-4-one	n. d.	1.32
20.153	ethanone, 1-(2,5-dihydroxyphenyl	1.67	n. d.
22.112	pentadecanoic acid	1.00	1.38
23.209	hexadecanoic acid	4.57	5.16
23.361	cis-9-hexadecenoic acid	2.31	4.68
24.648	9-octadecenoic acid, methyl ester	2.64	0.53
24.801	9,12-octadecadienoic acid, methyl ester	2.62	1.33
25.333	(z)-11-octadecenoic acid	**12.05**	**12.62**
25.447	9,12-octadecadienoic acid (z,z)	**17.26**	**27.78**
26.411	Hexadecanamide	2.54	n. d.
28.377	9-octadecenamide, (z)-	**29.04**	n. d.
Indicators of antioxidant activity of extracts of timber fungi			
Amax, %		95	98
Cmax, %		0,1	1
$IC50{\bullet}10^{-4}$, %		**5,5**	**6,76**

4 Conclusion

The conducted research has shown that the fruit bodies of wood-decay fungus are a valuable raw material for the production of biologically active substances. Using GC-MS, it was shown that, depending on the place of collection and the species of the studied methanol extracts of fruit bodies of 5 species of timber fungi, the composition of BAS differs significantly. The highest relative content in the investigated extracts is saturated and unsaturated fatty acids, as well as their amides. The content of unsaturated fatty acids decreases in the series of studied samples: *F. fomentarius*, Fanipol; *F. pinicola*, Yekaterinburg; *I. obliquus*, Yekaterinburg; *I. obliquus*, Fanipol; *G. applanatum*, Fanipol; *F. fomentarius*, Yekaterinburg; *G. applanatum*, Yekaterinburg; *P. betulinus*, Yekaterinburg; *P. betulinus*, Fanipol; *F. pinicola*, Fanipol.

The composition of the investigated extracts also contains phenolic compounds and sugars. Depending on the place of collection of fruit bodies of fungi and their species, the composition of BAS varies significantly. The composition of the BAS of the studied extracts is determined by their AOA. Extracts with a minimum value of $IC_{50} \times 10^{-4}$ have the greatest AOA. The maximum AOA decreases in the series: *G. applanatum*, Fanipol; *F. fomentarius*, Yekaterinburg; *G. applanatum*, Yekaterinburg; *F. pinicola*, Yekaterinburg; *I. obliquus*, Fanipol; *P. betulinus*, Yekaterinburg; *F. Fomentarius*, Fanipol; *F. pinicola*, Fanipol; *P.betulinus*, Fanipol; *I. obliquus*, Yekaterinburg; *F. fomentarius*, Yekaterinburg. Studies have shown that the composition of BAS extracts from the fruit bodies of timber fungi and their AOA depends on the type of wood being destroyed, the region of growth and have intraspecific differences.

References

1. Ahmad, M.F.: Ganoderma lucidum: persuasive biologically active constituents and their health endorsement. Biomed. Pharmacother **107**, 507–519 (2018)
2. Angelini, P., et al.: Overview of the biological activities of a methanol extract from Wild Red Belt Conk, fomitopsis pinicola (Agaricomycetes), fruiting bodies from Central Italy. Int. J. Med. Mushrooms **20**(11), 1047–1063 (2018)
3. Ćilerdžić, J., et al.: Pleurotus ostreatus and laetiporus sulphureus (Agaricomycetes): possible agents against Alzheimer and Parkinson diseases. Int. J. Med. Mushrooms **21**(3), 275–289 (2019)
4. Duru, K., et al.: The pharmacological potential and possible molecular mechanisms of action of Inonotus obliquus from preclinical studies. Phytother. Res. **33**(8), 1966–1980 (2019)
5. Ehlenfeldt, M.K., et al.: Oxygen radical absorbance capacity (ORAC) and phenolic and anthocyanin concentrations in fruit and leaf tissues of highbush blueberry. J. Agric. Food Chem. **49**, 2222–2227 (2001)
6. Ermoshin, A.A., et al.: Antioxidant activity and chemical composition of extracts from fruiting bodies of xylotrophic fungi growing on birch. J. Sib. Fed. Univ. Biol. **14**(3), 339–353 (2021). https://doi.org/10.17516/1997-1389-0354
7. Glumac, M., Pejin, B., Karaman, M., Mojović, M., Matavulj, M.: Lignicolous fungi hydrodistilled extracts may represent a promising source of natural phenolics. Nat. Prod. Res. **31**(1), 104–107 (2017)
8. Hao, L., Shen, Z., Shen, Z., Tao, R., Jia, S.: Characterization and antioxidant activities of extracellular and intracellular polysaccharides from Fomitopsis pinicola. Carbohydr. Polym. **141**, 54–59 (2016)
9. Khadhri, A., Aouadhi, C., Aschi-Smiti, S.: Screening of bioactive compounds of medicinal mushrooms collected on Tunisian Territory. Int. J. Med. Mushrooms **19**(2), 127–135 (2017)
10. Knežević, A., et al.: Antifungal, ntioxidative, and genoprotective properties of extracts from the blushing bracket mushroom, Daedaleopsis confragosa (Agaricomycetes). Int. J. Med. Mushrooms **19**(6), 509–520 (2017)
11. Kosolapov, D.A.: Aphyllophoroid fungi of the middle taiga forests of the European North-East of Russia. Ural Branch of the Russian Academy of Sciences, Yekaterinburg, p. 232 (2008)
12. Li, Z.: Chaga medicinal mushroom, inonotus obliquus (agaricomycetes) polysaccharides suppress tacrine-induced apoptosis by ROS-scavenging and mitochondrial pathway in HepG2 cells. Int. J. Med. Mushrooms **21**(6), 583–593 (2019)

13. Prendecka, M., et al.: Stimulation of the activity of a novel tannase produced in white-rot fungi phellinus pini, fomes fomentarius, and tyromyces pubescens by medium supplementation. Biotechnol. Appl. Biochem. **63**(5), 652–658 (2016)
14. Wang, C., et al.: Effects of polysaccharides from Inonotus obliquus and its chromium (III) complex on advanced glycation end-products formation, α-amylase, α-glucosidase activity and H2O2-induced oxidative damage in hepatic L02 cells. Food Chem. Toxicol. **116**(Pt B), 335–345 (2018)

Novel Method of Preparation Biocompatible Porous Materials Based on Modified Chitosan with High Mechanical Properties

Ivan Lednev(✉), Larisa Smirnova, Vitaliy Korokin, and Diana Fukina

National Research Lobachevsky State University of Nizhny Novgorod, Nizhny Novgorod, Russia

ildnv235@gmail.com

Abstract. A new method to obtain three-dimensional porous matrixes based on chitosan by treating its aqueous solutions with hydrocarbonates has been developed. During the reaction, pore formation occurs simultaneously, due to the release of carbon dioxide and the transition from the salt form of chitosan into the basic form, leading to the formation of a three-dimensional porous matrix. The structure is characterized by an interconnected pore system, with average pore size about 100–150 μm and average porosity about 82–85%. The material is biodegradable, biocompatible, exhibits high cell adhesion to fibroblasts and promotes their proliferation.

Keywords: Chitosan · Three-dimensional matrix · Pores · Biomaterial

1 Introduction

Porous structures based on biocompatible and biodegradable polymers are widely used as components in the manufacture of scaffolds, hemostatic sponges or adsorbents [1]. Chitosan is a product of chitin deacylation, which is a biocompatible and biodegradable polymer, and a promising material for these purposes. Chitosan is insoluble at neutral and alkaline pH values, but forms water-soluble salts with a number of inorganic and organic acids. The protonation of the amino groups of chitosan in acidic media, causes its dissolution, due to the repulsion of charges [2]. Chitosan-based materials can be manufactured in various forms, such as films, hydrogels, nonwoven materials and three-dimensional porous structures and they are widely used in medicine, cosmetics, agriculture, and water purification [3]. To form these structures, the polymer is primarily dissolved in an aqueous acid solution, as a result of which, chitosan is converted into a salt form [4].

A well-known method to form porous matrices of three-dimensional structure from chitosan is based on freezing the solution with further lyophilization, to remove water. Afterwards, the chitosan sponge is converted from the salt form to the basic form and lyophilized again. This process is energy- and time-consuming [5].

Recently, a method based on the use of supercritical carbon dioxide, which consists in dissolving chitosan in carbonic acid under high pressure, after which the pressure is

V. Kurchenko et al. (Eds.): ICAETT 2021, LNNS 408, pp. 136–143, 2022.
https://doi.org/10.1007/978-3-030-96641-6_16

brought to normal value, and the carbonic acid decomposes with the formation of water and carbon dioxide, resulting in the formation of a porous material, has attracted much attention. This method is environmentally friendly and allows obtaining highly porous structures, but requires specialized equipment [6].

Thus, the development of new methods of obtaining three-dimensional porous structure remains relevant [7].

The aim of the work was to develop a scientific basis for a highly efficient method of obtaining materials of three-dimensional porous structure based on chitosan, with controlled pore size.

2 Materials and Methods

Chitosan with a molecular weight of to $3 * 10^5$ and a degree of deacetylation 80% ("Bio-progress"), lactic acid (80%, chemically pure, "Reakhim"), ammonium bicarbonate (chemically pure, "Reakhim"), heptaldehyde (99%, Sigma Aldrich).

2.1 Preparation of Porous Samples

Chitosan (3 wt%) was dissolved in aqueous lactic acid solutions (1 wt%). After complete dissolution of the polymer under stirring, the foaming agent was added, resulting in formation of a three-dimensional porous matrix, due to carbon dioxide release with simultaneous transition of chitosan from the salt form to the basic one. Similarly, samples based on Schiff base-chitosan obtained by treatment of polysaccharide with heptaldehyde were obtained.

2.2 Preparation of Film Samples

Chitosan and Schiff base-chitosan solution samples were centrifuged to remove air bubbles, and then films were obtained on a XiamenTMAX-TMH film filling machine on a polyester substrate. The films were dried under conditions of uniform solvent evaporation at 30 °C and atmospheric pressure until constant weight.

2.3 Study of Mechanical Properties

Physical and mechanical characteristics (tensile strength and elongation at break) of the material were determined on a Roell/Zwick Z005 tensile machine. Tests were conducted at a tensile speed of 10 mm/min on samples 60 ± 5 μm thick in the form of rectangles 15 mm wide. At least 10 samples were measured for each film composition.

2.4 Study of Biocompatibility of Films Based on Modified PS and CTS

To study the biocompatibility of the material, films were obtained by molding from a solution under conditions of uniform evaporation of the solvent at t° = + 25 °C. The films were examined for adhesion, cytotoxicity, and cell growth on its surface by culturing human fibroblasts of the hTERT BJ-5ta cell line. The films were placed in the wells of a 6-well plate for cell cultivation and filled with 500 μl of DMEM medium. Cells were seeded on the film surface at a density of $1.6 \times 10^5/cm^2$ and cultured for 24 h. Cell visualization and assessment of cell viability were assessed by fluorescent microscopy. A 2×10^{-4} wt.% acridine orange solution in phosphate buffer was used as a dye for staining fibroblasts. This dye selectively interacts with DNA and RNA located in the cell nucleus and mitochondria by intercalation or electrostatic attraction, respectively. This makes it possible to assess the overall state of the cells - activity, proliferation and apoptosis. Microsampling of the films was performed on an OlympusIX71 inverter microscope using a "green" filter (510–555 nm emission, 460–495 nm excitation), which allows visualization of the green color of the living cell nucleus.

2.5 Study on Microstructure

Electron microscopic studies of the microstructure of the samples were carried out on a JEOL JSM-IT300LV scanning electron microscope at high vacuum in the mode of registration of secondary electrons at an accelerating voltage of 20 kV.

The porosity, size, and pore distribution of the samples were determined using a Pascal 140 and 440 mercury intrusion porometer. The use of a unique ultramacroporous dilatometer makes it possible to analyze pores ranging in size from 3.6 nm to 1200 μm. For the measurements, the samples were cut into small pieces (100–200 mg).

3 Results and Discussion

3.1 Preparation of Porous Samples

The novelty of the developed method of formation of the three-dimensional structure consists in the fact that the formation of three-dimensional porous structures based on chitosan and its derivatives is performed from aqueous solutions of chitosan in organic acids, by reaction with hydrocarbonates, which in interaction with acid generate carbon dioxide, serving as a pore-forming agent. At the same time, the salt form of chitosan, which makes it difficult to form a three-dimensional structure due to the homonymous charges on the protonated amino groups, is converted into the basic form. The mechanism of the process is based on the exchange reaction between the hydrocarbonate anion and the salt form of chitosan, in which the equilibrium is shifted toward the formation of reaction products. As a result, the salt form of chitosan transforms into the basic form accompanied by the formation of a three-dimensional structure stabilized by intermolecular and intramolecular hydrogen bonds of chitosan chains, with simultaneous release of carbon dioxide as a pore-forming agent (Fig. 1).

The obtained sponges were repeatedly washed with water to neutral pH values and dried under vacuum until constant weight. Microphotograph (Fig. 2) show microstructure of the sponge prepared from pure chitosan solution, and sponge made from solution of chitosan modified with heptaldehyde shown in (Fig. 3).

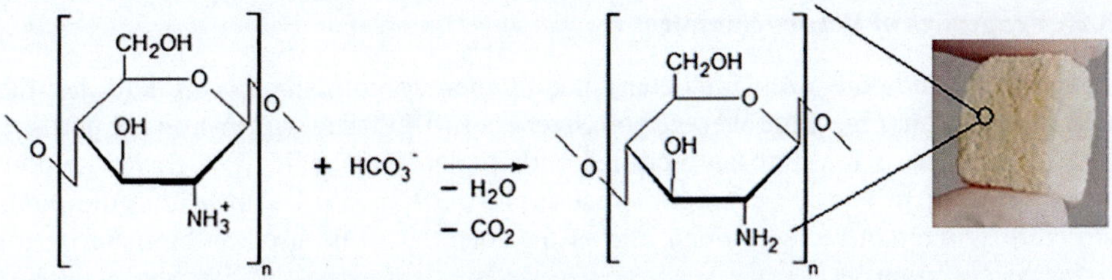

Fig. 1. Scheme of the reaction between chitosan and foaming agent.

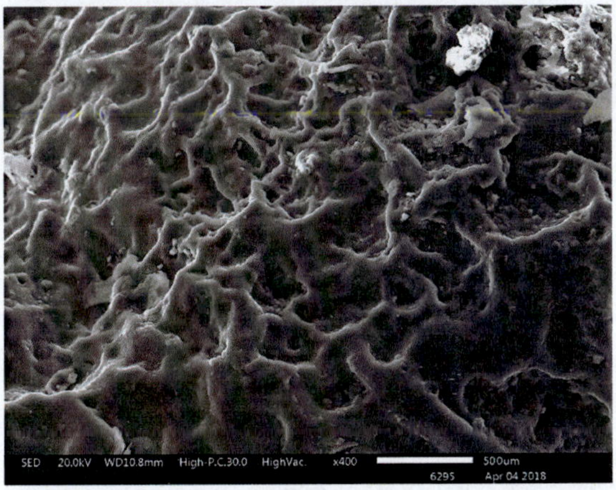

Fig. 2. SEM image of sponge from pure chitosan

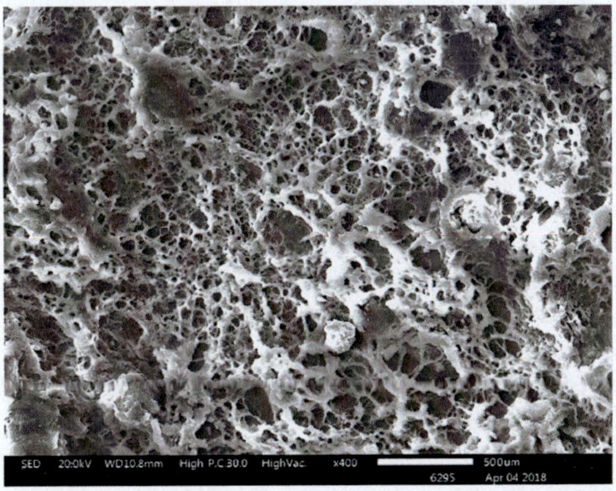

Fig. 3. SEM image of sponge from chitosan modified with heptaldehyde

3.2 Properties of Porous Samples

The porosity and pore size were determined by mercury intrusion porosimetry For the original chitosan (Fig. 4) the average pore size was 150–200 μm with an average porosity of 82.5%, whereas for chitosan modified with heptaldehyde (Fig. 5) the average pore size was 100 μm and the average porosity 84.3%. It should be noted that the pores structure is interconnected, which allows the material to be used as biomaterial for tissue regeneration.

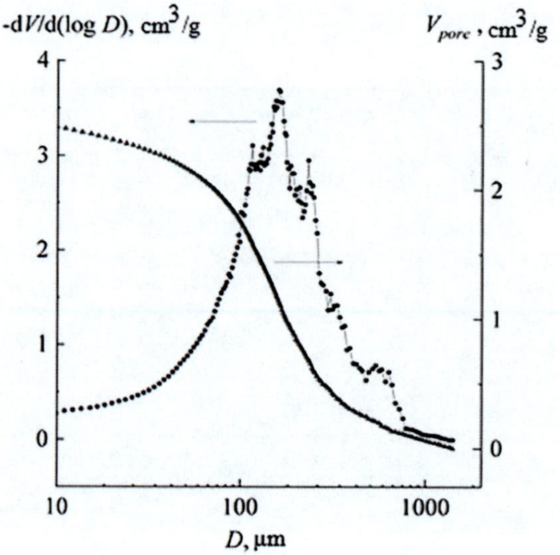

Fig. 4. The pore size distribution of sponge from pure chitosan

Materials used in tissue engineering must have mechanical properties comparable to those of the tissue to be replaced. The mechanical properties of chitosan films and its derivative modified with heptaldehyde were measured. The results of the mechanical properties test are presented in (Table 1).

The table shows that chitosan modified by treatment with 10% heptaldehyde had high mechanical properties, which is apparently due to the plasticizing effect of alkyl fragments. Experiments were performed on the synthesis with more heptaldehyde, but the result was phase separation of the system, which precluded the formation of films.

In view of the application of the material in a living organism, the question of studying biocompatibility was obvious.

The extraordinary value of the material would increase dramatically if, along with increased strength, transparency and biodegradability, it became a matrix that provides adhesion and growth of fibroblasts, thereby accelerating the process of wound healing, ensuring tissue cell proliferation. In this regard, in-vitro film studies were conducted on the adhesion and proliferation of fibroblast cells as connective tissue precursors. (Fig. 6) shows a photograph of the pure chitosan film and photo of the chitosan with 10% heptaldehyde-based film (Fig. 7) after 24 h of seeding with fibroblast cells.

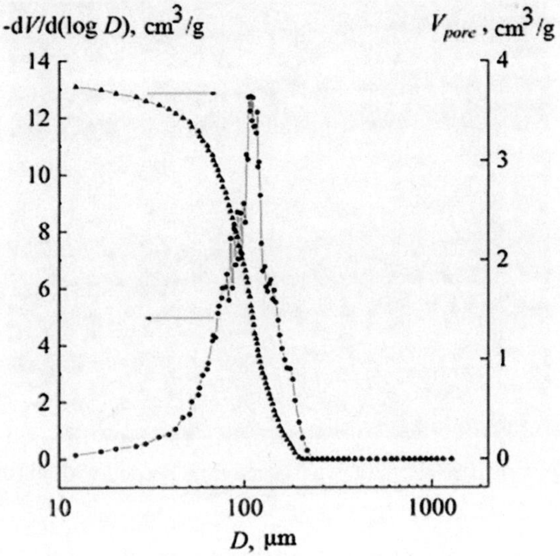

Fig. 5. The pore size distribution of sponge from chitosan modified by heptaldehyde.

Table 1. Mechanical characteristics of chitosan films and its derivatives

Composition	Tensile strength, σ (MPa)	Elongation, ε (%)
Pure chitosan	15.5 ± 1.7	1.51 ± 0.2
Chitosan + 2% heptaldehyde	52.4 ± 3.5	6.9 ± 0.5
Chitosan + 5% heptaldehyde	68.1 ± 4.8	8.1 ± 0.6
Chitosan + 10% heptaldehyde	99.7 ± 6.1	13.8 ± 1.1

Fig. 6. Pure chitosan-based film seeded with fibroblasts after 24 h

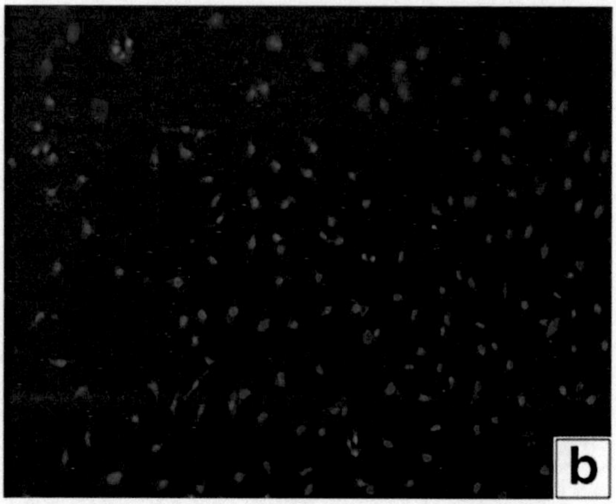

Fig. 7. Chitosan with 10% heptaldehyde-based film, seeded with fibroblasts after 24 h

It can be seen from these photos that the film based on modified chitosan has a more uniform distribution of fibroblasts on the surface, as well as more active growth and cell division. These properties make it possible to use this material as a promising matrix for tissue regeneration.

4 Conclusion

Thus, a new method of obtaining three-dimensional porous objects based on chitosan has been developed. By varying the nature of the components it is possible to change the microstructure of the samples, which expands the possible field of application of the material. The obtained matrix is also biodegradable, biocompatible, hypoallergenic and with good cell adhesion. Based on the results obtained, we can assume the prospects of using this approach as a competitor to the method based on the use of supercritical carbon dioxide.

Acknowledgments. This work was supported by the Fund for Assistance to the Development of Small Forms of Enterprises in the Scientific and Technical Sphere (Fund for Assistance to Innovation), within the framework of the UMNIK program (contract code 0064501).

References

1. Begum, H., Ahmed, M.S., Jeon, S.: New approach for porous chitosan-graphene matrix preparation through enhanced amidation for synergic detection of dopamine and uric acid. ACS Omega **2**(6), 3043–3054 (2017)
2. Berretta, J., Bumgardner, J.D., Jennings, J.A.: Lyophilized chitosan sponges. B Chitosan Based Biomater. **1**, 239–253 (2017). 5; cc.
3. Caro León, F.J., Lizardi-Mendoza, J., Argüelles-Monal, W., Carvajal-Millan, E., López Franco, Y.L., Goycoolea, F.M.: Supercritical CO 2 dried chitosan nanoparticles: production and characterization. RSC Adv. **7**(49), 30879–30885 (2017)

4. Dresvyanina, E.N., Dobrovol'skaya, I.P., Smirnov, V.E., Popova, E.N., Vlasova, E.N., Yudin, V.E.: Thermal properties of salt and base forms of chitosan. Polymer Sci. Ser. A **60**(2), 179–183 (2018)
5. Kou, S.(Gabriel), Peters, L.M., Mucalo, M.R.: Chitosan: a review of sources and preparation methods. Int. J. Biol. Macromol. **169**, 85–94 (2021)
6. Ma, J., Zhong, L., Peng, X., Xu, Y., Sun, R.: Functional chitosan-based materials for biological applications. Curr. Med. Chem. **27**(28), 4660–4672 (2020)
7. Tsurkan, D., et al.: Extreme biomimetics: designing of the first nanostructured 3D spongin-atacamite composite and its application. Adv. Mater. **33**(30), 2101682 (2021)

Effect of Nanosized Forms of Iron-Containing Compounds on the Growth Characteristics of Raphanus Sativus Seeds

Blinov Andrey Vladimirovich[1(✉)], Gvozdenko Alexey Alexeevich[1], Slyadneva Kristina Sergeevna[1], Kravtsov Alexander Alexandrovich[1,2], and Rajabov Uktamjon Ikromjon Ugli[1]

[1] North Caucasus Federal University, Stavropol, Russian Federation
[2] Federal Research Center The Southern Scientific Centre of the Russian Academy of Sciences, Rostov-on-Don, Russian Federation

Abstract. This work presented the experimental data on the change in the characteristics of seed germination of Raphanus sativus radish after their treatment with sols of iron-containing compounds nanoparticles. For research, samples of nanoparticles of the following compounds were synthesized: $Fe(OH)_3$, $KFe[Fe(CN)_6]$, $FePO_4$. The iron concentration was $5.66 \cdot 10^{-2}$ mol/l in all samples. The micelle charge and average hydrodynamic radius were determined by dynamic light scattering and electroacoustic spectroscopy, respectively. The largest average hydrodynamic radius of particles, according to the results of the data obtained, was found in the $FePO_4$ sample, and the smallest – in $Fe(OH)_3$. It was found that all iron-containing compounds have a positive micelle charge. The seeds of Raphanus sativus radish were treated for 10 min, and then they were placed in a thermostat at room temperature for 3 days. A control sample of Raphanus sativus radish seeds was treated in distilled water. The following parameters of seed germination were estimated in the study: germination percentage, germination ability, and degree of seed damage. The negative influence of sols on seed germination was noted. The most inhibitory effect was exerted by a sample of $KFe[Fe(CN)_6]$ nanoparticles, and the least - by a $Fe(OH)_3$ sample. Deterioration is caused by the direct toxicity of nanoparticles and their induction of the development of molds on the seed substrate.

Keywords: Raphanus sativus · Nanoparticles · Radish seeds · Iron-containing compounds · Toxicity

1 Introduction

Nanoparticles of iron-containing compounds have a great potential application in various fields of science and technology, such as biomedicine, medicine, agriculture [1, 5, 11, 19]. Iron-containing nanoparticles are a promising material for plant cultivation, which is due to the properties of the trace element iron - a functional component of enzymatic systems that plays an important role in the formation of chlorophyll, oxidative and

V. Kurchenko et al. (Eds.): ICAETT 2021, LNNS 408, pp. 144–150, 2022.
https://doi.org/10.1007/978-3-030-96641-6_17

energy processes in the body. With a lack of iron in plants, physiological processes are disturbed, chlorosis of the youngest leaves occurs. In agriculture, iron in various forms, in particular, Prussian blue nanoparticles, can be used as an agent capable of removing toxic compounds such as cesium and its compounds from plants [18].

There is a study [20], which shows a stimulating effect on the formation and growth of shoots and the development of the root system, describes the ability of iron nanoparticles in low concentrations from 40 to 80 μmol/L to improve the germination ability of plants. It was found that iron nanoparticles are able to penetrate into the shell of peanut seeds and increase water absorption in order to stimulate seed germination [14]. The influence of colloidal forms of iron and selenium was also studied [10]. It was revealed that the seeds of agricultural crops after treatment with colloidal forms of iron showed 100% germination, however, a negative effect was also shown associated with the suppression of plant growth at an iron concentration of 10 mg/L. In another work [9] it was shown that in the absence of pronounced cytotoxicity of iron nanoparticles, iron nanoparticles penetrate into the epidermis of the root and settle in the cells; the antioxidant activity of plants increased. It is known that the use of iron-containing nanoparticles in various forms can facilitate the assimilation of phosphorus by plants at a low application rate in the composition of mineral complexes [7]. There are many works on the effect of various nanoparticles on seed germination [4, 17]. There is an opinion that the use of nanoparticles of metals and other elements can increase the productivity of agricultural crops [3, 15, 16, 21].

Some works have shown the negative effect of iron-containing nanoparticles on the formation and growth of structural elements of plants. Thus, iron nanoparticles are capable of causing cytogenetic damage to wheat, and this study also showed that iron nanoparticles promote the formation of reactive oxygen species in plant tissues [12]. Iron nanoparticles in high concentrations are of great interest to scientists [13], which is associated with concerns about the environmental behavior and potential of the environmental effect of iron nanoparticles. In the environment, iron-containing nanoparticles can be physically, chemically, or biologically affected by factors such as pH, ionic action, dissolved oxygen species, etc. The toxicity of iron-containing nanoparticles depends on their properties, size and composition of the compound in which these particles are located. The main stress for plants and bacteria is oxidative stress caused by reactive oxygen species. At a concentration of zero-valence nanosized iron of 250 mg/L, an inhibitory effect is observed on samples of ryegrass, barley, and flax [6]. There are data showing the effect of iron compounds on the absorption of arsenic by the roots and shoots of rice seedlings [2].

2 Materials and Methods

This work aimed to study the effect of iron containing compounds' nanoparticles on the growth of Raphanus sativus radish seeds.

Research objectives:

1) synthesize samples of iron-containing compounds' nanoparticles ($Fe(OH)_3$, $KFe[Fe(CN)_6]$, $FePO_4$) and study their size and particle charge;

2) to study the effect of nanoparticles of iron-containing compounds ($Fe(OH)_3$, $KFe[Fe(CN)_6]$, $FePO_4$) on the germination percentage, germination ability, and the degree of damage to Raphanus sativus radish seeds.

The sols of iron (III) hydroxide, iron (III) phosphate, and Prussian blue were selected for the study. The iron concentration in all samples was $5.66 \cdot 10^{-2}$ mol/l.

Iron hydroxide nanoparticles were synthesized by hydrolysis of a precursor at 100 °C. Iron phosphate nanoparticles were obtained as a result of the exchange reaction of sodium dihydrogen phosphate (NaH_2PO_4) and iron (III) chloride ($FeCl_3$) with an excess of iron chloride. $KFe[Fe(CN)_6]$ nanoparticles were obtained by mixing $FeCl_3$ and $K_4[Fe(CN)_6]$ ($FeCl_3$ was in excess in all samples).

The seeds of Raphanus sativus radish were treated for 10 min. Then they were placed in a thermostat for 3 days at room temperature. In the control experiment, the seeds of Raphanus sativus radish were germinated after preliminary treatment in distilled water. The number of seeds in the samples was 100 pcs. The studies were carried out in triplicate.

The main parameters of seed germination, which are given attention, are the germination percentage, germination ability and the degree of seed damage. The studies were carried out following GOST 12038-84 "Seeds of agricultural crops. Methods of determination of the germination" [10].

The hydrodynamic radius of nanoparticles was determined by dynamic light scattering on a Photocor-Complex setting (Antek-97, Russia). Computer processing of the research results was carried out using the DynaLS computer software [13].

The micelle charge was investigated by electroacoustic spectroscopy on a DT 1202 spectrometer (Dispersion Technology Inc., USA).

3 Results and Discussion

At the first stage of research, the obtained sols of iron-containing compounds' nanoparticles were investigated by dynamic light scattering and electroacoustic spectroscopy. Figure 1 and Table 1 present the data obtained.

It was found that the largest average hydrodynamic radius of particles is observed in the $FePO_4$ sample, and the smallest in the $Fe(OH)_3$ sample. It is shown that in all samples of iron-containing compounds particles have a positive charge, which indicates that the potential-forming layer is formed by Fe^{3+} ions. Schemes of the micelles' structure are presented below:

$$\{m(Fe(OH)_3)\ nFe^{3+} \cdot 3(n-x)Cl^-\}^{3x+} \cdot 3xCl^-$$
$$\{m(FePO_4)\ nFe^{3+} \cdot 3(n-x)Cl^-\}^{3x+} \cdot 3xCl^-$$
$$\{m(KFe[Fe(CN)_6])\ nFe^{3+} \cdot 3(n-x)Cl^-\}^{3x+} \cdot 3xCl^-$$

At the next stage, we studied the effect of iron-containing compounds' nanoparticles on the growth parameters of Raphanus sativus radish seeds. The data is presented in Table 2.

All the presented samples of iron-containing compounds' sols had an inhibitory effect on the seeds of Raphanus sativus radish, which may be associated with the direct

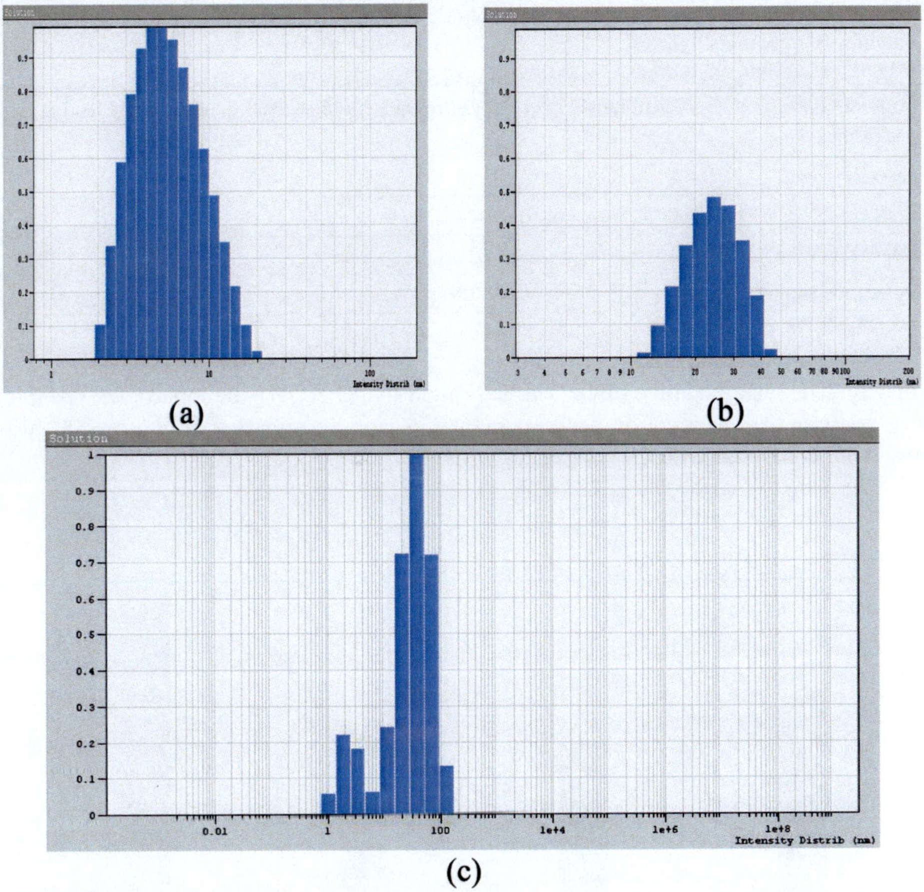

Fig. 1. Histograms of the distribution of the hydrodynamic radius of iron-containing compounds' nanoparticles: $Fe(OH)_3$ (a), $KFe[Fe(CN)_6]$ (b), $FePO_4$ (c)

Table 1. Characteristics of iron-containing compounds' nanoparticles

Iron-containing compound	The average hydrodynamic radius of nanoparticles in sol, nm	Particles' charge
$Fe(OH)_3$	5	+
$KFe[Fe(CN)_6]$	25	+
$FePO_4$	38	+

toxicity of nanoparticles or their induction of the development of molds on the seed substrate. It was found that the greatest negative impact on the germination percentage and germination ability (decrease in germination percentage by 3.85 times and germination ability by 2.3 times, compared with the reference sample) had a $KFe[Fe(CN)_6]$ sample

Table 2. Growth indices of Raphanus sativus radish seeds treated with nanoparticles of iron-containing compounds

Iron-containing compound	Germination percentage, %	Germination ability, %	Degree of seed damage, %
Fe(OH)3	54	58	6
FePO4	44	44	24
KFe[Fe(CN)6]	20	34	4
Reference sample	77	78	0

(Fig. 2). The least negative effect was exhibited by a $Fe(OH)_3$ nanoparticles' sample; a decrease in the germination percentage by 23% and germination ability by 20% was observed, compared with the reference sample.

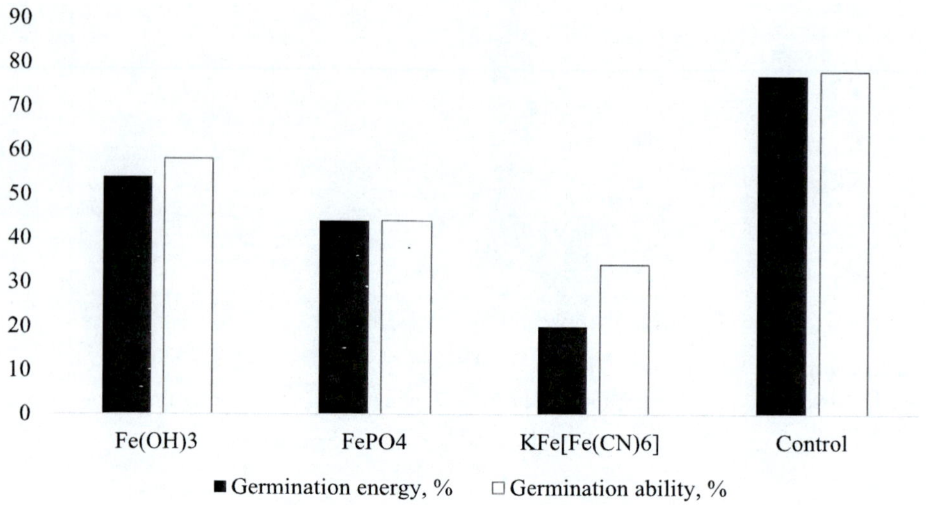

Fig. 2. Comparison of germination percentage and germination ability

It is important to note that the degree of damage to the seeds of Raphanus sativus radish by mold spores was not the same. According to GOST 12038-84 [10], the degree of damage to the seeds of Raphanus sativus radish in the KFe[Fe $(CN)_6$] sample corresponds to a low degree of damage (0–5%), and in the samples of $FePO_4$ and $Fe(OH)_3$, the degree of damage can be classified as "Medium degree of damage" (5–25%). Figure 3 shows a comparison of samples treated with $FePO_4$ and $KFe[Fe(CN)_6]$ sols.

(*a*) (*b*)

Fig. 3. Photographs of germinated seeds of Raphanus sativus radish before the germination percentage measurement; (a) - sample treated with $KFe[Fe(CN)_6]$ nanoparticles; (b) sample treated with FePO4 nanoparticles

4 Conclusion

The studies have shown a strong inhibitory effect of $Fe(OH)_3$, $KFe[Fe(CN)_6]$ and $FePO_4$ nanoparticles on the germination of Raphanus sativus radish seeds, which may be associated with the toxicity of the nanoparticles or with the promotion of molds development on the seed substrate. It has been shown that the use of $KFe[Fe(CN)_6]$ nanoparticles leads to a decrease in the germination percentage by 3.85 times and the germination ability by 2.3 times (in comparison with the reference sample).

Acknowledgments. The work was carried out using the Center for Collective Use of the North Caucasus Federal University (NCFU) with the financial support of the Ministry of Education and Science of the Russian Federation (project No. RF-2296.61321X0029, agreement No. 075-15-2021-687).

References

1. Ali, A., et al.: Synthesis, characterization, applications, and challenges of iron oxide nanoparticles. Nanotechnol. Sci. Appl. **9**, 49–67 (2016)
2. Rahman, M.A., Rahman, M.M., Kadohashi, K., Maki, T., Hasegawa, H.: Effect of external iron and arsenic species on chelant-enhanced iron bioavailability and arsenic uptake in rice (Oryza sativa L.). Chemosphere **84**(4), 439–445 (2011)
3. Barabanov, P.V., Gerasimov, A.V., Blinov, A.V., Kravtsov, A.A., Kravtsov, V.A.: Influence of nanosilver on the efficiency of Pisum sativum crops germination. Ecotoxicol. Environ. Saf. **147**, 715–719 (2018)
4. Blinov, A.V., et al.: Study of stabilization of selenium nanoparticles by polysaccharides. J. Hygienic Eng. Des. **34**, 209–216 (2021)
5. Bouafia, A., Laouini, S.E.: Plant-mediated synthesis of iron oxide nanoparticles and evaluation of the antimicrobial activity: a review. Mini-Rev. Org. Chem. **18**(6), 725–734 (2021)

6. El-Temsah, Y.S., Joner, E.J.: Impact of Fe and Ag nanoparticles on seed germination and differences in bioavailability during exposure in aqueous suspension and soil. Environ. Toxicol. **27**(1), 42–49 (2010)
7. Folmanis, G.E., Fedotov, M.A.: Dispersion methods of preparation of nanosized biological agents for vegetable crops. J. Phys. Conf. Ser. **1431**(1), 1–5 (2019)
8. GOST R 12038-84: Agricultural Seeds. Germination Determination Methods, Moscow (1985). (in Russian)
9. Iannone, M.F., Groppa, M.D., de Sousa, M.E., Fernández van Raap, M.B., Benavides, M.P.: Impact of magnetite iron oxide nanoparticles on wheat (Triticum aestivum L.) development: evaluation of oxidative damage. Environ. Exp. Bot. **131**, 77–88 (2016)
10. Joseph, S., et al.: Effects of enriched biochars containing magnetic iron nanoparticles on mycorrhizal colonisation, plant growth, nutrient uptake and soil quality improvement. Pedosphere **25**(5), 749–760 (2015)
11. Kharisov, B.I., Rasika Dias, H.V., Kharissova, O.V., Manuel Jiménez-Pérez, V., Olvera Pérez, B., Muñoz Flores, B.: Iron-containing nanomaterials: synthesis, properties, and environmental applications. RSC Adv. **2**(25), 9325 (2012)
12. Korotkova, A., Sizova, E., Lebedev, S., Kosyan, D., Rusakova, E.: Influence of iron nanoparticles on induction of oxidative damage in Triticum vulgare. Ecol. Environ. Conserv. **21**, 101–111 (2015)
13. Lei, C., Sun, Y., Tsang, D.C.W., Lin, D.: Environmental transformations and ecological effects of iron-based nanoparticles. Environ. Pollut. **232**, 10–30 (2018)
14. Li, X., Yang, Y., Gao, B., Zhang, M.: Stimulation of peanut seedling development and growth by zero-valent iron nanoparticles at low concentrations. PLoS ONE **10**(4), e0122884 (2015)
15. Liu, R., Lal, R.: Potentials of engineered nanoparticles as fertilizers for increasing agronomic productions. Sci. Total Environ. **514**, 131–139 (2015)
16. Rui, M., et al.: Iron oxide nanoparticles as a potential iron fertilizer for peanut (Arachis hypogaea). Front. Plant Sci. **7**, 815 (2016)
17. Siddiqui, S.A., et al.: Effect of selenium nanoparticles on germination of Hordéum Vulgáre barley seeds. Coatings **11**(7), 862 (2021)
18. Thammawong, C., Opaprakasit, P., Tangboriboonrat, P., Sreearunothai, P.: Prussian blue-coated magnetic nanoparticles for removal of cesium from contaminated environment. J. Nanopart. Res. **15**(6) (2013). Article number: 1689. https://doi.org/10.1007/s11051-013-1689-z
19. Vanin, A.F.: Prospects of using magnetic nanoparticles to potentiate the anticarcinogenic action of dinitrosyl iron complexes with thiol ligands. Biophysics **56**(5), 832–835 (2011). https://doi.org/10.1134/S0006350911050228
20. Yasmeen, F., Razzaq, A., Naveed Iqbal, M., Jhanzab, H.M.: Effect of silver, copper and iron nanoparticles on wheat germination. Int. J. Biosci. **6**(4), 112–117 (2015)
21. Xie, Y., et al.: The interactions between nanoscale zero-valent iron and microbes in the subsurface environment: a review. J. Hazard. Mater. **321**, 390–407 (2017)

Study of the Possibility of Application of Acoustic Spectroscopy in Dairy Products

Kaishev Vladimir Grigorievich[1,2], Khramtsov Andrey Georgievich[3], Blinov Andrey Vladimirovich[3], Gvozdenko Alexey Alexeevich[3](✉), and Maglakelidze David Guramievich[3]

[1] Russian Academy of Sciences, Moscow, Russian Federation
[2] Stavropol State Agrarian University, Stavropol, Russian Federation
kaishev@stagau.ru
[3] North Caucasus Federal University, Stavropol, Russian Federation
akhramtcov@ncfu.ru

Abstract. In this paper, we have studied the possibility of using acoustic and electroacoustic spectroscopy in the dairy industry to study processes occurring at the nanoscale. Studies of the dispersed phase of milk and the process of milk fermentation were carried out. The studies were carried out by the method of acoustic and electroacoustic spectroscopy on the DT-1202 spectrometer. Measurements of attenuation spectra, particle size of the dispersed phase, pH, ξ-potential, electrical conductivity were carried out. It is shown that the samples of pasteurized milk have a bimodal particle size distribution. The sample contains 2 fractions of particles. The first with an average diameter of about 65 nm, which corresponds to casein. And the second – 300 nm, corresponding to the fat fraction of milk. It was found that in the study of samples with a low content of the fat fraction (1.5%), a statistical error occurs, which is associated with a significant predominance of the number of casein micelles in comparison with fat globules. As a result of a comprehensive study of the milk fermentation process, it was found that 8 h after the addition of the starter culture, a transformation from free (sol) to structure-bound (gel) is observed due to loss of surface charge and coagulation, followed by coalescence of casein micelles with the formation of a gel structure protein clot. The data obtained indicate the possibility of using the method of acoustic and electroacoustic spectroscopy for the study and control of various processes occurring in milk and dairy products at the nanoscale.

Keywords: Acoustic and electroacoustic spectroscopy · Spectrometer DT-1202 · Dairy industry · Milk · Dispersed phase · Protein · Fat content · Fermentation

1 Introduction

From the point of view of colloidal chemistry, milk and dairy products can be considered as complex natural colloidal systems, where the dispersion medium is represented by water with salts, vitamins, hormones, etc. dissolved in it, and the dispersed phase consists of casein micelles (protein fraction) and milk fats [1, 11, 12, 17]. It is important to note

V. Kurchenko et al. (Eds.): ICAETT 2021, LNNS 408, pp. 151–158, 2022.
https://doi.org/10.1007/978-3-030-96641-6_18

that the parameters of the dispersed phase of milk and dairy products, which include the size of micelles and ξ-potential, significantly affect the quality of products, which indicates the need for their control in the framework of research and production [14, 18].

Acoustic and electroacoustic spectroscopy is one of the modern research methods that can be applied in the food industry, pharmaceuticals, medicine, nanotechnology, etc. [3, 4, 6, 9, 15, 16]. Acoustic and electroacoustic spectroscopy is based on the method of ultrasonic sounding of the samples placed in a closed chamber. The method allows to characterize opaque and concentrated (without preliminary dilution) media [8].

2 Materials and Methods

The purpose of the research is to study the possibility of using acoustic and electroacoustic spectroscopy in the dairy industry to study the processes occurring at the nanoscale.

Research objectives:

1) to study the possibility of using acoustic and electroacoustic spectroscopy to study the dispersed phase of milk;
2) to study the possibility of using acoustic and electroacoustic spectroscopy to study the process of milk fermentation.

The attenuation spectra, particle size of the dispersed phase of milk and ξ-potential samples were studied by acoustic and electroacoustic spectroscopy at the DT-1202 unit (produced by Dispersion Technology Inc., USA) [2, 7]. The electrical conductivity and active acidity (pH) of the medium were determined using an electrical conductivity meter for aqueous media and a combined LAQYA model 9615 electrode (manufacturer Horiba, Japan), which are included in the DT-1202 spectrophotometer kit.

To study the process of fermenting milk, it was used milk produced by the «DPS» (JSC «Dairy Plant Stavropol») with 5% of fat fraction and starter culture of lactic acid bacteria *Lactococcus lactis, Subsp. lactis* and *Streptococcus thermophiles* («Biokom» LLC, Russia). The dose of introducing the starter was 5% of the mass of milk. Fermentation took place at a temperature of 25 °C [5].

3 Results and Discussion

At first, it was considered to analyze the possibility of using acoustic and electroacoustic spectroscopy to study the dispersed composition of the colloid system of milk. To decide the task there, samples of pasteurized milk with 5% of fat fraction and a 3,2% of protein fraction were used. The data obtained are presented in Figs. 1 and 2.

The distribution of milk dispersed phase particles by size is bimodal. It is shown that the first fraction has a diameter of 65 nm and corresponds to the protein component of milk (casein). The diameter of the second fraction is 300 nm and corresponds to the fat component of milk [10, 13, 20]. It is important to note that in the sample there is a significant predominance of the first fraction over the second.

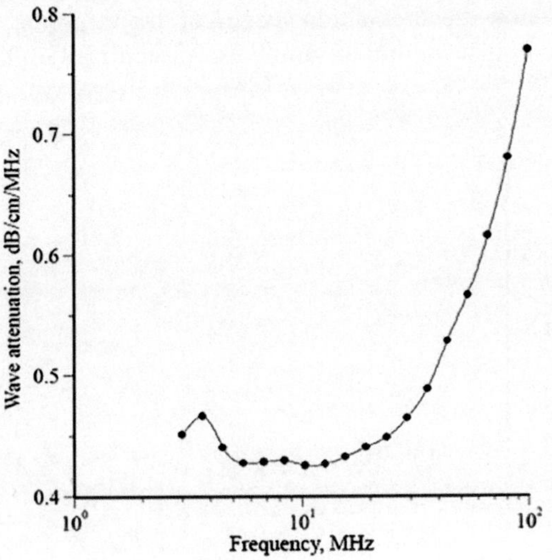

Fig. 1. Spectrum of attenuation of ultrasonic waves in milk (5% fat fraction).

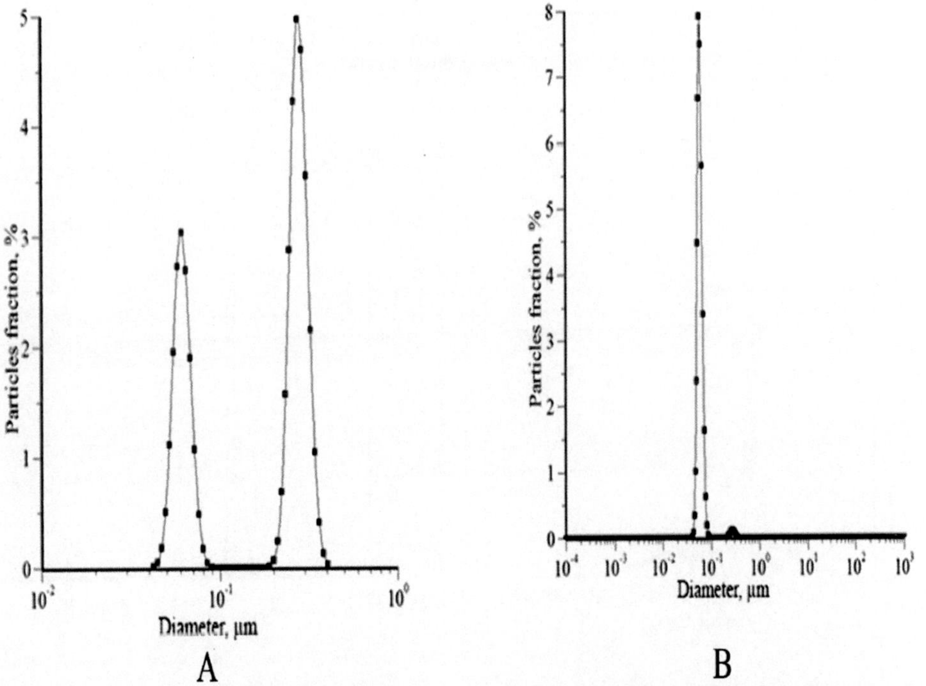

Fig. 2. Histograms of milk dispersed phase particles size distribution: A) distribution by mass, B) distribution by the number of particles.

In Figs. 3 and 4 show the attenuation spectra of the ultrasonic wave and histograms of the particle diameter distribution of samples of pasteurized milk with a fat content of 1.5 to 5.0%.

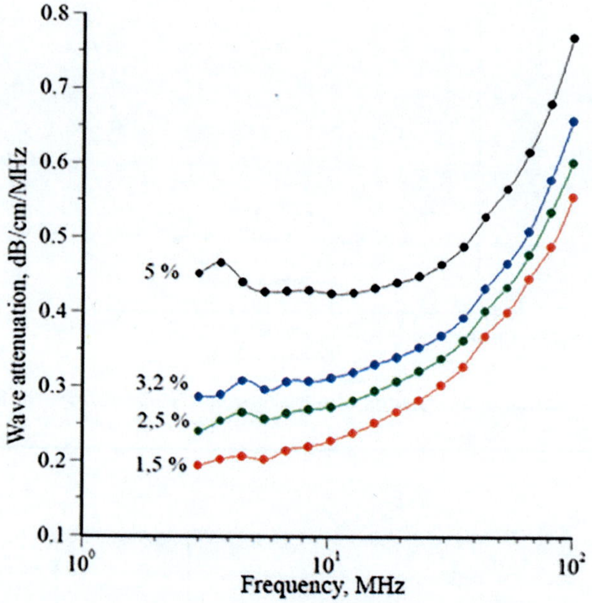

Fig. 3. Attenuation spectra of ultrasonic waves in milk samples.

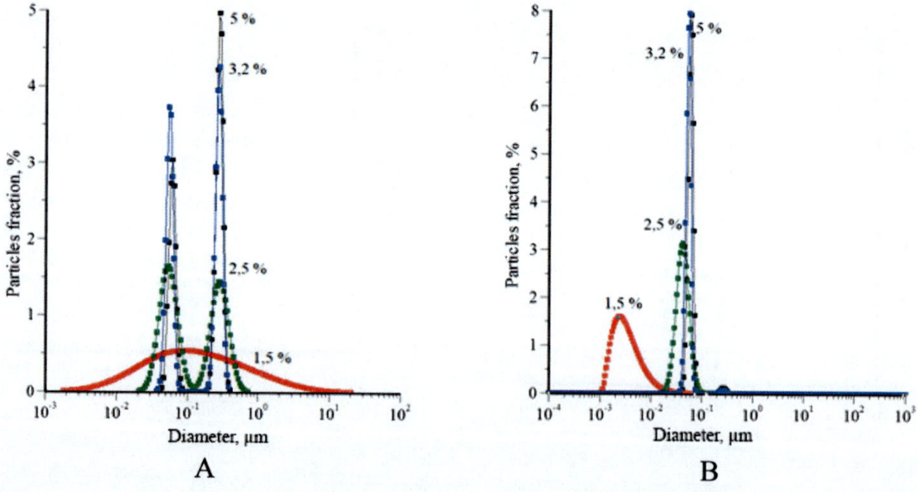

Fig. 4. Histograms of the milk dispersed phase particles distribution by dimensions with different fat content: A) distribution by mass, B) distribution by the number of particles.

Analysis of Figs. 3 and 4 showed that a bimodal distribution of particles is observed in samples with a content of 2.5, 3.2 and 5%. The diameter of the protein fraction is 70 nm, the diameter of the fat fraction is 320 nm. The data obtained are in good agreement with Fig. 1 and Fig. 2. However, in a sample with a fat fraction content of 1.5%, a monomodal distribution of particles with an average diameter of 110 nm and a high polydispersity from 0.1 to 2000 nm is observed, which indicates a significant statistical error that occurs during the study of this sample. The detected statistical error occurs due to the significant prevalence of the number of casein micelles in comparison with fat balls (protein fraction content 3.2%, fat–1.5%). As a result, it can be concluded that when analyzing dairy products by acoustic and electroacoustic spectroscopy, it is necessary to take into account the ratio of protein and fat fractions.

At the next stage, the milk fermentation process was investigated. Figures 5, 6, 7 and 8 shows the dependences of pH, electrical conductivity, ζ-potential and particle size of the fermentation time.

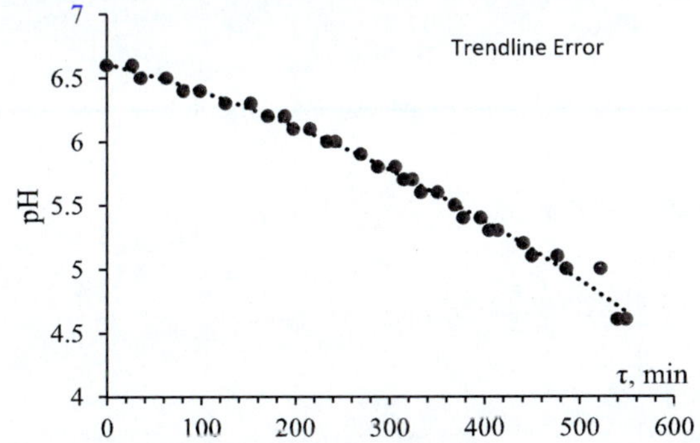

Fig. 5. Dependence of the active acidity of milk at time of fermentation.

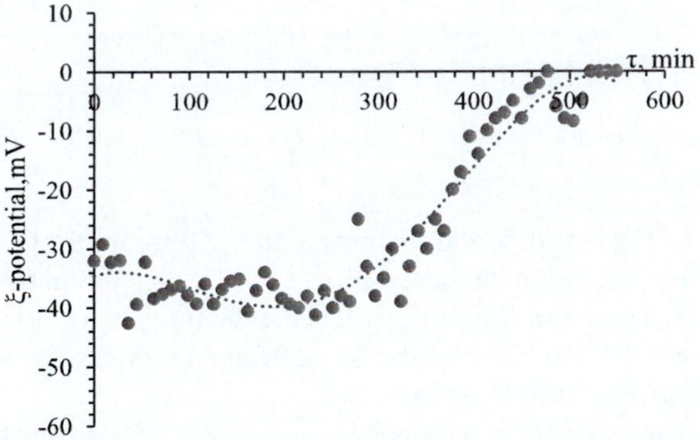

Fig. 6. Dependence of ζ-potential of particles of dispersed phase of milk at time of fermentation.

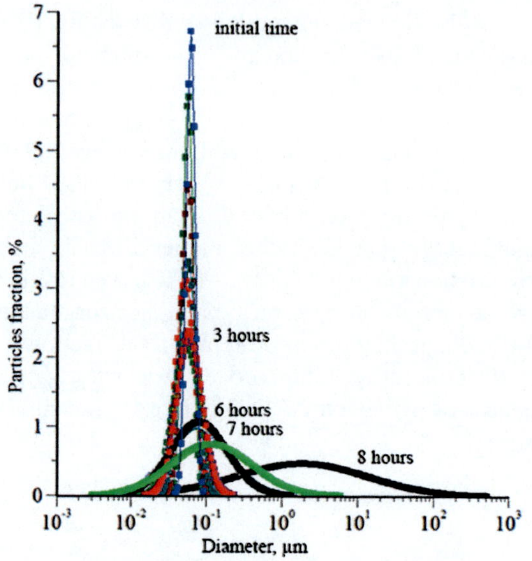

Fig. 7. Histograms of milk dispersed phase particles distribution from time of fermentation.

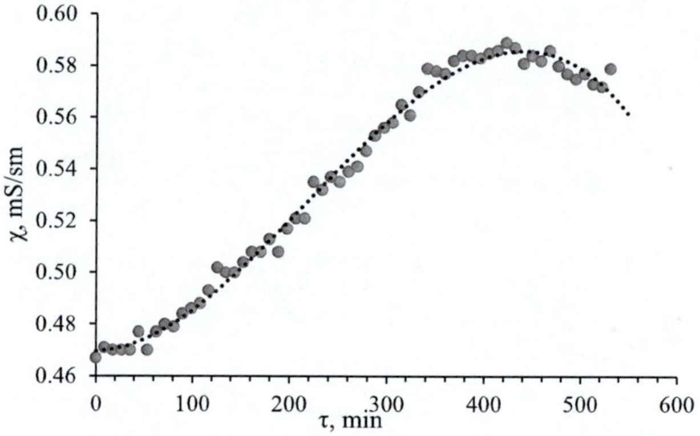

Fig. 8. Dependence of electrical conductivity of milk at time of fermentation.

The analysis of Figs. 5, 6, 7 and 8 showed that 480 min after the start of the milk fermentation process, a significant increase in the size of the protein fraction occurs from 70 to 1500 nm (Fig. 7), an increase in electrical conductivity by 0.11 μS/cm (Fig. 8), the pH of the sample reaches the value of the isoelectric point of casein (Fig. 5), ζ-potential takes values close to zero (Fig. 6) [19].

Discovered changes of milk parameters are directly related to the transformation of the type of colloid system from free (micellar) to structurally-bound (gel) due to the loss of surface charge and coagulation, followed by coalescence of casein micelles to form a gel structure of a protein clot. Described dependencies clearly demonstrate the capabilities of acoustic and electroacoustic spectroscopy methods for the study and control of various processes occurring in milk and dairy products at the nanoscale.

4 Conclusion

As a result of the conducted research, it can be concluded that acoustic and electroacoustic spectroscopy is a promising method for evaluating the parameters of the dispersed phase of milk and dairy products (the size of the protein and fat fraction and ζ-potential) in the food industry. It is important to note that when analyzing dairy products by acoustic and electroacoustic spectroscopy, it is necessary to take into account the ratio of protein and fat fractions, which is associated with the occurrence of a statistical error in the study of samples with low fat fraction content.

Acknowledgments. The work was carried out using the Center for Collective Use of the North Caucasus Federal University (NCFU) with the financial support of the Ministry of Education and Science of the Russian Federation (project № RF - 296.61321X0029, agreement № 075-15-2021-687).

References

1. Agostoni, C., Turck, D.: Is cow's milk harmful to a child's health? J. Pediatr. Gastroenterol. Nutr. **53**(6), 594–600 (2011)
2. Blinov, A.V., Khramtsov, A.G., Snezhkova, J.J.: Detergent-disinfectant agent based on colloidal silver stabilized by quaternary ammonium compounds for the dairy industry. J. Hygienic Eng. Des. **27**, 3–11 (2019)
3. Blinov, A.V., Siddiqui, S.A., Ibrahim, S.A.: Investigation of the influence of Zinc-containing compounds on the components of the colloidal phase of milk. Arab. J. Chem. **14**(7), 103229 (2021)
4. Bonacucina, G., et al.: Acoustic spectroscopy: a powerful analytical method for the pharmaceutical field? Int. J. Pharm. **503**(1–2), 174–195 (2016)
5. Boor, K.: A 100-Year Review: Microbiology and safety of milk handling. J. Dairy Sci. **100**(12), 9933–9951 (2017)
6. Demchenkov, E.L., Nagdalian, A.A., Budkevich, R.O., Oboturova, N.P., Okolelova, A.I.: Usage of atomic force microscopy for detection of the damaging effect of $CdCl_2$ on red blood cells membrane. Ecotoxicol. Environ. Saf. **208**, 111683 (2021)
7. Dukhin, A.S., Goetz, P.J.: Characterization of Liquids, Nano- and Microparticulates, and Porous Bodies Using Ultrasound. Elsevier, New York (2010)
8. Dukhin, A.S., Goetz, P.J., Travers, B.: Use of ultrasound for characterizing dairy products. J. Dairy Sci. **88**(4), 1320–1334 (2005)
9. Guerin, M., Seaman, J.C.: Characterizing clay mineral suspensions using acoustic and electroacoustic spectroscopy-A review. Clays Clay Miner. **52**(2), 145–157 (2004)
10. Huppertz, T., Fox, P.F., Kelly, A.L.: The caseins: structure, stability, and functionality. In: Proteins in Food Processing, pp. 49–92 (2018)
11. Kucheryavskiy, S., Melenteva, A., Bogomolov, A.: Determination of fat and total protein content in milk using conventional digital imaging. Talanta **121**, 144–152 (2014)
12. Luisa, B.G.: Handbook of Milk Composition. Elsevier, New York (1995)

13. McMahon, D.J., Oommen, B.S.: Casein micelle structure, functions, and interactions. In: Advanced Dairy Chemistry, pp. 185–209 (2013)
14. Raikos, V.: Effect of heat treatment on milk protein functionality at emulsion interfaces. A review. Food Hydrocolloids **24**(4), 259–265 (2010)
15. Rzhepakovsky, I., et al.: Anti- arthritic effect of chicken embryo tissue hydrolyzate against adjuvant arthritis in rats (X- ray microtomographic and histopathological analysis). Food Sci. Nutr. **00**, 1–22 (2021). https://doi.org/10.1002/fsn3.2529
16. Simon, L.L., Simone, E., Abbou Oucherif, K.: Crystallization process monitoring and control using process analytical technology. In: Process Systems Engineering for Pharmaceutical Manufacturing, pp. 215–242 (2018)
17. Singhal, S., Baker, R.D., Baker, S.S.: A comparison of the nutritional value of cow's milk and nondairy beverages. J. Pediatr. Gastroenterol. Nutr. **64**(5), 799–805 (2017)
18. Temerbayeva, M., Rebezov, M., Yessimbekov, Z.: Development of yoghurt from combination of goat and cow milk. Ann. Res. Rev. Biol. **23**(6), 1–7 (2018)
19. Urbienė, S., Leskauskaitė, D.: Formation of some organic acids during fermentation of milk. Pol. J. Food Nutr. Sci **15**(56), 277–281 (2006)
20. Zhao, Z., Corredig, M.: Colloidal properties of casein micelles suspensions as a function of pH during concentration by osmotic stressing. Food Hydrocolloids **60**, 445–452 (2016)

Technological and Nutritional Potential of Lentil in the Turkey Cutlets Production

Michael Schetinin[1,2], Marina Vaytanis[1,3](✉), Olga Musina[1], and Zoja Khodyreva[1,3]

[1] Polzunov Altai State Technical University, Barnaul, Russia
[2] Moscow State University of Food Production, Moscow, Russia
[3] Altai State University, Barnaul, Russia

Abstract. The aim of the work is to evaluate the technological and nutrient potential of lentil flour in the technology of turkey meat products and to provide a justification for using this potential. The object of the study is minced turkey with the addition of lentil flour in an amount of 5 to 25% and cutlets made from it. The subject of the study is the water absorption capacity, water binding capacity and oil absorption capacity of minced turkey meat with the addition of lentil flour, as well as organoleptic, nutrient and microbiological properties of cutlets. Standard and special research methods were used. The amount of lentil flour corresponding to the maximum organoleptic assessment of cutlets, which is 20%, was established. The introduction of lentil flour gives minced turkey a plastic consistency, increases the mass of minced meat by 74.0–79.5%, pH by 0.73–0.96 units, increases the water absorption capacity by 12.5–25.7%, water binding capacity by 6.1–21.8% and oil absorption capacity by 3.4–16.8%, which reduces losses during heat treatment. The addition of lentil flour to the composition of minced meat leads to an increase in the content of proteins, fats, carbohydrates and dietary fiber, vitamin A, B_1, PP, potassium, magnesium, calcium, iron, copper, zinc and selenium in cutlets. The best quality of minced meat and turkey cutlets are noted in the sample with 20% lentil flour. The use of lentil flour in the production of cutlets expand the range of food with increased nutrient balance, save raw meat and lead to excellent consumer properties.

Keywords: Chopped meat · Minced meat · Semi-finished products · Turkey meat · Lentils · Legumes · Cutlets

1 Introduction

Production and consumption of meat products in the world is increasing annually [1]. Eating meat and meat products on a regular basis is part of the concept of a healthy diet, meat has increased nutritional and biological value and is recommended for inclusion in the human diet on a regular basis [2]. Meat is the best source of high-quality protein, balanced in the composition of amino acids and having a high degree of digestibility, in addition, meat proteins contain biologically active peptides released as a result of

V. Kurchenko et al. (Eds.): ICAETT 2021, LNNS 408, pp. 159–166, 2022.
https://doi.org/10.1007/978-3-030-96641-6_19

enzymatic hydrolysis, which have a positive effect on human health (antimicrobial, antioxidant, antihypertensive and immunomodulatory activity) [3].

Among all types of animal protein in terms of production and sales, the first place is occupied by poultry meat [4]. Thus, according to the North American Meat Institute, in 2017, the consumption of turkey meat in the United States reached 2.7 billion kilograms, and turkey became one of the most consumed types of processed meat in America, and according to forecasts of the USDA, the consumption of turkey meat per capita by 2025 will reach 7.8 kg per person [5]. In Germany, poultry meat production increased in 2019 by 4.5% compared to the previous year [6]. Chicken and turkey meat is popular among the Russian population [7].

Turkey meat and products based on it are becoming increasingly popular in the global consumer market for a number of reasons, the main of which are as follows. Turkey meat is characterized by high biological value and low in fat and cholesterol, excellent organoleptic indicators, and also meets the public demand for a healthy diet [4, 8–12]. Turkey meat has a high content of complete proteins (19.1–22.9 g/100 g), unsaturated fatty acids (68.3% of the sum of fatty acids), contains vitamins - A, E, D, as well as B vitamins, macronutrients are represented by potassium, calcium, sodium, magnesium and phosphorus, trace elements - iron, zinc and selenium [7, 13–15].

The development of technologies and inevitable changes in the lifestyle of modern man have led to an increase in demand for semi-finished products, and the scientific and applied problem is that these products must be simultaneously high-quality, minimally processed, with a composition understandable to the consumer, easy to prepare (cook), with a minimum amount of food additives and a long shelf life. Recently, the criterion of health benefits of such semi-finished products has come to the fore [6, 16].

The combination of minced meat with raw materials of plant origin makes it possible both to diversify the existing range of chopped products, and to increase the technological properties of minced meat, and the nutritional value of semi-finished products [9, 10, 17, 18]. The presented scientifically based technology allows to solve the above problem of improving the quality of minced products from turkey meat and, accordingly, at the same time saving meat raw materials.

For such a combination with minced turkey meat, lentils are a promising raw material. Lentils (Lens culinaris) is a grain crop belonging to the legume family (Fabaceae), which plays a significant role in the global production of legumes along with beans, peas and chickpeas [19].

Lentils can be considered as an inexpensive source of protein, the content of which reaches 21–31% [20, 21], lentil proteins are rich in lysine, leucine, arginine, aspartic and glutamic acid [22]. At the same time, lentil proteins are limited to sulfur-containing amino acids (methionine and cysteine) and tryptophan [23], which confirms the relevance of combining lentils with other sources of protein, such as poultry meat, for example, turkey. This solution allows to achieve an adequately balanced amino acid profile of food products.

Lentil carbohydrates (62–69%) are represented mainly by starch (35–53%) with a low glycemic index and dietary fiber (5–20%), as well as oligosaccharides [24, 25]. Lentils can also be considered as a source of folic acid, zinc (4.8 mg/100 g) and iron (7.5 mg/100 g). Phenols found in lentils in much higher concentrations than in other legumes have antioxidant, antidiabetic, anticarcinogenic and anti-inflammatory properties [26].

Lentils are popular among consumers and food producers. So, lentil flour has found application in bakery products (bread, cakes, crackers), pasta, extruded snacks, dressings, soups, dairy and meat products. In the latest works of 2021, lentil flour is positioned by scientists as an underestimated promising ingredient for the food industry. In particular, based on the study of food and technological properties of lentil flour (solubility, emulsification, gelatinization, foaming), its high potential for the preparation of high-quality food products (meat products, yogurts, gluten-free bakery products) is shown [26]. Lentil flour can be used to reduce the lipid content of food products and its glycemic index. For example, the partial replacement of pork with lentil flour has a positive effect on the physico-chemical and organoleptic characteristics of low-fat burgers, which were highly appreciated by consumers for both appearance and taste [27]. Due to the fact that its starch, dietary fiber and proteins containing polar amino acids participate in hydrophilic interactions and form hydrogen bonds with water molecules, lentil flour is characterized by a high moisture-retaining ability. And since its proteins contain non-polar amino acids that interact with lipids and retain them, lentil flour also has a high fat-retaining ability. According to the literature, the amount of lentil flour introduced into meat products is usually from 4 to 15% [26].

2 Materials and Methods

The following objects were used in this work: turkey meat of the 1st category according to GOST 31473-2012; whole grain lentil flour according to STO 1172225000287-001-2017 (Altai groats RF); minced turkey with different amounts of plant raw materials; turkey cutlets with lentil flour.

Water absorption capacity was determined using a fat meter for milk, water binding capacity - pressing, pH minced meat system was measured using a pH meter with a preliminary dilution of distilled water in a ratio of 1: 10, oil absorption capacity - using a refractometer. Tasting of turkey cutlets with lentil flour was carried out according to GOST 9959-2015. Microbiological indicators in cooked cutlets were carried out according to GOST R 54354-2011.

3 Results

Solving the above problem, recipes of minced turkey meat with the addition of lentil flour were compiled. To do this, the meat part (turkey meat) was replaced with lentil flour, in order to establish the amount of plant component that provides the best organoleptic and technological indicators of the products. For this purpose, the following minced models were compiled and studied: check sample - minced meat without adding lentil flour; sample No. 1 - minced meat with 5% replacement of meat part with lentil flour; sample No. 2 - minced meat with 10% replacement of meat part with lentil flour; sample No. 3 - minced meat with 15% replacement of meat part with lentil flour; sample No. 4- minced meat with 20% replacement of meat part with lentil flour; sample No. 5 - minced meat with 25% replacement of meat part with lentil flour.

The studied samples of minced meat were evaluated according to a set of indicators in comparison with the control. Replacing the meat part with lentil flour gives the minced meat a soft, fairly plastic and homogeneous consistency, in comparison with minced turkey, without replacing the meat part with lentil flour. Samples of minced meat No. 3 and No. 4 are well formed and retain their shape. Introduction of lentil flour into minced meat in the amount of 25% (sample No. 5) leads to a change in consistency, expressed in dryness and crumbliness of minced meat, which will further complicate the process of forming chopped products from such minced meat and affect the consumer properties of finished products. All samples of minced meat in appearance are characterized as a homogeneous, well-mixed minced meat system, except for sample No. 5, which forms cracks on the surface. With an increase in the dose of flour, there is a slight visible inclusion of plant raw materials in minced meat (sample No. 5). Lentil flour does not affect the smell of the studied samples of minced meat. The best organoleptic indicators of minced turkey meat are observed in samples No. 3 and No. 4 with the addition of 15% and 20% of lentil flour. These samples are characterized by a good ability to mold products from them.

With an increase in the amount of lentil flour introduced into the minced turkey meat system, the mass of minced meat changes by 74.0–79.5% compared to the control. The maximum mass of minced meat is noted in Sample No. 5 and is 79.5%.

Figure 1 presents the results of studies of the functional and technological properties of minced meat samples.

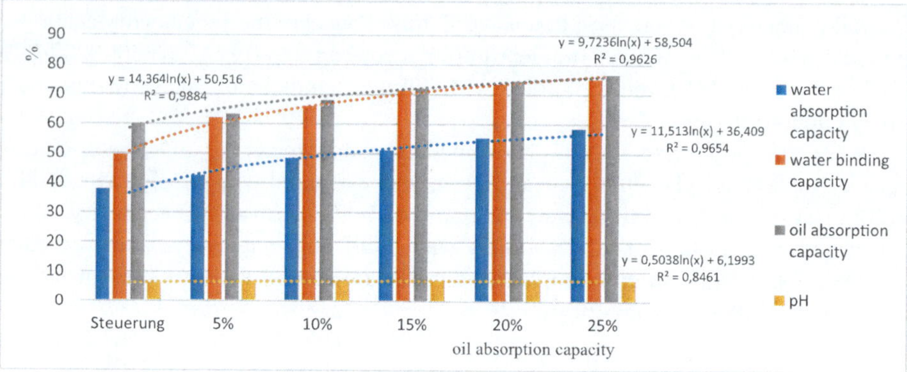

Fig. 1. Functional and technological indicators of minced turkey with lentil flour

The introduction of lentil flour changes the active acidity of minced meat by 0.73–0.96 units in comparison with the control sample (Fig. 1). With this change in the pH of the medium in the alkaline side, changes in the water binding capacity occur by 12.5–25.7%, water absorption capacity by 6.1–21.8%, as well as by 3.4–16.8% in oil absorption capacity. Changes in the indicators of water binding capacity, water absorption capacity and oil absorption capacity in the direction of increase, provides an increase in the juiciness of finished products and reduces technological losses during subsequent heat treatment. High values of these indicators are observed in samples No. 4 and No. 5. However, sample No. 5 was rejected for low organoleptic indicators.

Thus, the optimal amount of lentil flour is 20% (sample No. 4), this sample provides the best quality indicators. In order to confirm this amount of lentil flour, cutlets were produced and a tasting assessment was carried out in comparison with the control. Analysis of the organoleptic profiles of the samples showed that the highest score was noted in sample No. 4 with the introduction of 20% lentil flour. Cutlets prepared from this sample have a homogeneous mass on the cut, while a juicy consistency, delicate taste and characteristic smell are noted.

Microbiological indicators of the quality of turkey meat cutlets were investigated during their storage in a chilled state (at a temperature of plus 4 + 2 °C for 12 h) and in a frozen state (at a temperature of 18 °C for three months). Microbiological studies of cutlets with the introduction of lentil flour in the amount of 20% with various methods and shelf life, indicate that the introduced plant component allows you to get products with good sanitary and hygienic indicators.

The nutritional value of chopped turkey cutlets with the addition of lentil flour in the amount of 20% in comparison with the control sample is presented in Tables 1 and 2.

Table 1. Nutritional value of turkey cutlets with 20% replacement of meat part with lentil flour and cutlets, without replacing the meat part

Nutrient	Content in cutlets, without replacing the meat part by 100 g	Content in cutlets made from turkey with 20% replacement of meat part with lentil flour in 100 g	Percentage of daily value coverage %
Caloric content, kcal	274,0	324,0	14,0
Proteins, g	14,8 ± 0,2	18,6 ± 0,3	28,5
Fats, g	21,9 ± 0,3	22,4 ± 0,3	31,6
Carbohydrates, g	4,7 ± 0,1	12,6 ± 0,2	3,5
Fibre, g	0,59 ± 0,02	2,3 ± 0,05	11,5

The nutritional value of turkey cutlets was determined by the calculation method, taking into account the losses during heat treatment of products. Analysis of the chemical composition of the presented samples (Table 1), showed an increase in the content of proteins, fats, carbohydrates and a significant increase in dietary fiber. It is worth noting that turkey cutlets with lentil flour in the amount of 20% fill the daily need for proteins by 28.5%, fats by 31.6%. Cutlets contain dietary fiber, in 100 g their amount is 2.3 g, which helps to meet 11.5% of the daily needs of the body in this nutrient.

According to the results presented in Tables 1 and 2, it can be noted that the combination of turkey meat with lentil flour leads to the enrichment of products (cutlets) with dietary fiber, an increase in the content of vitamins: retinol, thiamine, nicotinic acid in comparison with the sample, without replacing the meat part with lentil flour. In addition, it leads to an increase in the content of macronutrients - sodium, potassium, magnesium, calcium and trace elements - iron, copper, zinc and selenium. Turkey cutlets

Table 2. Vitamin-mineral composition in turkey cutlets with the addition of lentil flour in comparison with the control sample

Product name	Minerals, mg/100 g								Vitamins, mg/100 g			
	Na	K	Ca	Mg	P	Fe	Cu	Zn	A	B1	B2	PP
Turkey cutlets	108,9	181,3	27,7	18,5	164,8	1,7	105,6	2,2	0,03	0,1	0,23	6,1
Daily requirement coverage, %	8,3	7,2	2,8	4,6	20,6	9,4	10,6	18,3	0,33	6,7	12,8	38,1
Turkey cutlets with lentil flour	135,8	263,2	41,0	31,2	223,3	3,7	222,1	2,6	0,74	0,2	0,25	7,0
Daily requirement coverage, %	10,4	10,5	4,1	7,8	27,9	20,6	22,2	21,7	8,2	13,3	13,9	43,8

with 20% replacement of the meat part with lentil flour in the amount of 100 g provide a daily need for vitamin PP by 43.8%, in macroelements: phosphorus by 27.9%; in trace elements: iron by 20.6%, copper by 22.2% and zinc by 21.7%.

4 Conclusion

Formulations of minced meat based on turkey with lentil flour have been developed and the quality of minced compositions in comparison with the control has been evaluated. Replacement of the meat part with lentil flour in the amount of 20% ensures the best organoleptic parameters of the minced meat system. Based on the data obtained, recipes for chopped products (cutlets) from turkey with lentil flour were developed. From the resulting minced meat, chopped products (cutlets) were developed and their tasting evaluation was carried out using the expert method. Microbiological indicators of turkey cutlets with 20% replacement of meat part with lentil flour in comparison with control are determined.

The nutritional value of the studied samples is calculated. It was found that cutlets with the addition of 20% lentil flour in the amount of 100 g replenish the daily rate by 28.5% in proteins, by 31.6% in fats, by 11.5% in dietary fiber, by 43.8% in vitamin PP, in macroelements: by 27.9% phosphorus; in trace elements: 20.6% iron, copper 22.2% and zinc 21.7%. The addition of lentil flour in the production of chopped semi-finished products is quite a promising direction and allows you to expand the range of combined meat semi-finished products and enrich them with biologically valuable substances.

Acknowledgments. The research was performed within the project 075-00316-20-01 (FZMMM-2020-0013, mnemocode 0611-2020-013) from the Ministry of Science and Higher Education of the Russian Federation.

References

1. González, N., Marquès, M., Nadal, M., Domingo, J.: Meat consumption: Which are the current global risks? A review of recent (2010–2020) evidences. Food Res. Int. **137**, 109620 (2020)
2. Palman, Y., Leo, R., Pulvirenti, A., Green, S.J., Hayouka, Z.: Antimicrobial peptide cocktail activity in minced turkey meat. Food Microbiol. **92**, 103580 (2020)
3. Martini, S., Conte, A., Tagliazucchi, D.: Comparative peptidomic profile and bioactivities of cooked beef, pork, chicken and turkey meat after in vitro gastro-intestinal digestion. J. Proteomics **208**, 103500 (2019)
4. Cavalcanti, É.N.F., et al.: Breast meat quality of turkey breeder hens at disposal age affected by deep pectoral myopathy. Poult. Sci. **100**(8), 101259 (2021)
5. Carvalho, L.T., et al.: Quality of turkeys breast meat affected by white striping myopathy. Poult. Sci. **100**(4), 101022 (2021)
6. Kluth, I., Teuteberg, V., Ploetz, M., Krischek, C.: Effects of freezing temperatures and storage times on the quality and safety of raw turkey meat and sausage products. Poult. Sci. **100**(9), 101305 (2021)
7. Moiseeva, N.S., Chekryga, G.P., Motovilov, O.K., Golub, O.V.: Fluorescence method in measuring the degree of brine penetration into boneless whole-muscle turkey meat. Food Proces. Tech. Technol. **51**(2), 209–219 (2021)
8. Khodyreva, Z.R., Schetinin, M.P., Musina, O.N., Schetinina, E.M., Vaytanis, M.A.: Development of a daily diet for children with celiac disease in municipal pre-school educational institutions. Voprosy pitanija [Nutrition Issues] **90**(2), 110–116 (2021)
9. Vaytanis, M.A., Khodyreva, Z.R.: The use of hemp flour in the production of minced meat semi-finished products. Bull. KSAU **1**, 126–133 (2021)
10. Barkova, V.G., Velichko, N.A., Ivanova, O.V.: The development of the compounding of meat and cereal pate from turkey meat. Bull. KSAU **5**, 167–173 (2019)
11. Gálvez, F., Domínguez, R., Pateiro, M., Carballo, J., Tomasevic, I., Lorenzo, J.M.: Effect of gender on breast and thigh turkey meat quality. Br. Poult. Sci. **59**(4), 408–415 (2018)
12. Zampiga, M., Tavaniello, S., Soglia, F., Petracci, M., Mazzoni, M., Maiorano, G.: Comparison of 2 commercial turkey hybrids: productivity, occurrence of breast myopathies, and meat quality properties. Poult. Sci. **98**(5), 2305–2315 (2019)
13. Arakcheeva, E.N., Zabashta, N.N., Golovko, E.N., Lisovitskaya, E.P.: Research on turkey meat production for baby food products. Sci. J. Kuban State Agrarian Univ. **162**, 247–259 (2020)
14. Barone, A.M., Banovic, M., Asioli, D., Wallace, E., Ruiz-Capillas, C., Grasso, S.: The usual suspect: how to co-create healthier meat products. Food Res. Int. **143**, 110304 (2021)
15. Meretukova, F.N., Abregova, N.V.: Research of quality indicators of semi-finished turkey meat products prepared using sous-vide technology. New Technol. **17**(2), 48–55 (2021)
16. Bıyıklı, M., Akoğlu, A., Kurhan, Ş., Akoğlu, İ.T.: Effect of different Sous vide cooking temperature-time combinations on the physicochemical, microbiological, and sensory properties of turkey cutlet. Int. J. Gastron. Food Sci. **20**, 100204 (2020)
17. Chernenkov, E.N., Chernenkova, A.A., Kaluzhina, O.: Quality assessment of kupat from turkey meat with wheat talcane. Vsyo o myase [All about meat] **5**, 395–398 (2020)
18. Musina, O., Putnik, P., Koubaa, M., Barba, F.J., Greiner, R., Roohinejad, S.: Application of modern computer algebra systems in food formulations and development: a case study. Trends Food Sci. Technol. **64**, 48–59 (2017)
19. Kumar, S., Pandey, G.: Biofortification of pulses and legumes to enhance nutrition. Heliyon **6**(3), e03682 (2020)
20. Carcea, M.: Nutritional value of grain-based foods. Foods 9(4), 504–509 (2020)

21. Saricaoglu, F.T.: Application of high-pressure homogenization (HPH) to modify functional, structural and rheological properties of lentil (*Lens culinaris*) proteins. Int. J. Biol. Macromol. **144**, 760–769 (2020)
22. Monnet, A., Laleg, K., Michon, C., Micard, V.: Legume enriched cereal products: a generic approach derived from material science to predict their structuring by the process and their final properties. Trends Food Sci. Technol. **86**, 131–143 (2019)
23. Semba, R.D., Ramsing, R., Rahman, N., Kraemer, K., Bloem, M.W.: Legumes as a sustainable source of protein in human diets. Glob. Food Secur. **28**, 100520 (2021)
24. Revilla, I., Lastras, C., González-Martín, M.I., Vivar-Quintana, A.M., Morales-Corts, R., Gómez-Sánchez, M.A.: Predicting the physicochemical properties and geographical origin of lentils using near infrared spectroscopy. J. Food Compos. Anal, **77**, 84–90 (2019)
25. Graf, D., Monk, J.M., Wu, W., Wellings, H.R., Robinson, L.E., Power, K.A.: Red lentil supplementation reduces the severity of dextran sodium sulfate-induced colitis in C57BL/6 male mice. J. Funct. Foods **64**, 103625 (2020)
26. Romano, A., Gallo, V., Ferranti, P., Masi, P.: Lentil flour: nutritional and technological properties, in vitro digestibility and perspectives for use in the food industry. Curr. Opin. Food Sci. **40**, 157–167 (2021)
27. Argel, N.S., Ranalli, N., Califano, A.N., Andrés, S.C.: Influence of partial pork meat replacement by pulse flour on physicochemical and sensory characteristics of low-fat burgers. J. Sci. Food Agric. **100**(10), 3932–3941 (2020)

Application of Chitosan for Fermented Dairy Products with *Lactobacillus delbrueckii* subsp. *Bulgaricus* Manufacturing

Vladimir Kurchenko[1], Aleksei Lodygin[2(✉)], Tatsiana Halavach[1], Ivan Evdokimov[2], and Maria Shramko[2]

[1] Belarussian State University, Minsk, Republic of Belarus
kurchenko@tut.by
[2] North Caucasus Federal University, Stavropol, Russia
ievdokimov@ncfu.ru

Abstract. The influence of chitosan with molecular weight of 350 kDa different concentrations on changes in active and titratable acidity, lactic acid and lactose content, the count of lactic acid microorganisms and relative abundance of aroma-forming compounds during fermentation of L. delbrueckii subsp. bulgaricus was studied. It is shown that on the 17th day of storage of the product in an experimental sample containing chitosan at a concentration of 0.01%, the titratable acidity was 138 °T, and in the control sample 203 °T. The count of lactic acid microorganisms in experimental sample was $6.34 * 10^6$ CFU per cm^3, and in the control sample $3.69 * 10^3$ CFU per cm^3. The interaction of chitosan with lactobacilli does not lead to significant change in the composition and content of aroma substances. The use of chitosan at concentration of 0.01% allows for a 3-fold increase in the shelf life of the fermented dairy product.

Keywords: Chitosan · Lactic acid microorganisms · Fermentation · Aroma-forming and flavour compounds · Fermented dairy products · Shelf life

1 Introduction

One of the most common groups of functional food is fermented dairy intended for dietetic and clinical nutrition [3]. To obtain these products, starter cultures based on pure cultures of lactic acid microorganisms of the subspecies Lactobacillus delbrueckii subsp. bulgaricus are widely used [12]. The production fermented dairy products is characterized by a tendency to improve traditional technologies by enriching products with functional ingredients such as oligosaccharides, essential macro-and micronutrients, dietary fibres [2, 5]. These compounds affect the intestinal microflora, showing prebiotic properties [5, 15]. The most well-known prebiotics that affect the human body, selectively enhancing the growth and activity of bacteria of the gastrointestinal tract, are inulin, oligosaccharides such as xylooligosaccharide, galactooligosaccharide, isomaltooligosaccharide and chitosan, which selectively stimulate the growth of Bifidobacterium [2, 5, 15].

V. Kurchenko et al. (Eds.): ICAETT 2021, LNNS 408, pp. 167–175, 2022.
https://doi.org/10.1007/978-3-030-96641-6_20

Chitosan is a biogenic heteropolymer of N-acetylglucosamine and glucosamine [16]. Previous studies have shown that chitosan has a wide range of biological activity: it reduces cholesterol, exhibits anti-cancer and fungicidal properties [1, 6, 9]. Chitosan characterized by a large molecular weight, high viscosity. The amino groups of the polymer are in a protonated form at a pH below 6.3 and it has a negative charge [1, 6].

There are a limited number of publications devoted to the effect of chitosan on bifidobacteria and lactobacilli [4, 14]. In this regard, it seemed appropriate to investigate the mechanism of the effect of chitosan on the process of lactic acid fermentation. In gram-positive bacteria Lactobacillus delbrueckii subsp. bulgaricus cells, the main target for interaction with chitosan may be teichoic acids, which are negatively charged by numerous phosphorus acid residues [4]. Teichoic acids in the composition of cell walls are also associated with positively charged proteins – autolysins, which play an important role in the degradation of murein. This peptidoglycan is necessary for the growth and division of bacterial cells. In addition to the components of the bacterial cell wall, the cytoplasmic membrane is a target for chitosan. When interacting with it, chitosan disrupts the permeability of the plasmalemma, which can lead to a drop in the membrane potential. The change in the conformation of membrane proteins caused by the action of chitosan is important. They can affect the permeability of the cytoplasmic membrane and the activity of enzymes involved in energy metabolism [13].

The main purpose of this research was to study the effect of high-molecular chitosan on the lactic acid fermentation process induces by Lactobacillus delbrueckii subsp. bulgaricus.

2 Materials and Methods

The following object were use in the framework of this study:

- dry skim milk that meets the requirements of GOST 33629-2015 "Preserved milk products. Dry milk", (manufacturer: JSC "Stavropolskij Dairy Plant", Russia);
- chitosan with a molecular weight of 350 kDa and a degree of deacetylation of 95%, meets the requirements of TU 9289-046-04689375-96 (manufacturer: OOO "Bioprogress", Russia);
- pure starter culture of Lactobacillus delbrueckii subsp. bulgaricus (manufacturer: JSC "Stavropolskij Dairy Plant", Russia), meeting the requirements of SanPiN 2.3.2.1078-01.

Dry skim milk was reconstituted in distilled water at a temperature of 30–35 °C to a mass concentration of total solids of 10.0%, which was controlled refractometrically according to GOST R 54668-2011 "Milk and milk processing products. Methods for determining of moisture and dry matter mass concentration". Reconstituted skim milk contained 0.15% of fat, 3.2% of protein, 5% of lactose. Chitosan solution with mass concentration of 1% and molecular weight of 350 kDa in 10% lactic acid solution was added to the experimental samples of reconstituted skim milk to a final concentration of 0.0025, 0.005, 0.0075 and 0.01%.

The starter culture of Lactobacillus delbrueckii subsp. bulgaricus was introduced into the experimental and control samples of skim milk after pasteurization and chilling at

43–45 °C in an amount of 3% of the total volume of samples. The end of the fermentation process was determined by the coagulum density, as well as by the titratable and active acidity. After that, the samples were hermetically sealed and stored for 17 days at (4 ± 2) °C, controlling the temperature by the thermometric method according to GOST 31981-2013 "Yoghurts. General technical conditions". Sampling for research during the storage of the product was carried out under aseptic conditions.

Following parameters of control and experimental samples were determined in three repeats during storage: active acidity (pH) potentiometrically according to GOST 32892-2014 "Method of active acidity measuring", titratable acidity titrometrically according to GOST 3624-92 "Titrimetric methods for acidity determining" and the count of lactic acid microorganisms (CFU per cm^3) according to GOST 10444.11-2013 "Methods for lactic acid microorganisms determining". The content (molar concentration, mole per litre) of lactic acid and lactose was determined by calculation formulas according to the previously described method [13].

Studies of the composition of secondary metabolites that determine fermented dairy product organoleptic properties were carried out by the GC-MS method. Biologically active substances were extracted from freeze-dried samples with ethyl alcohol solution of 75% concentration, the analysis of which was carried out on an Agilent 6850 gas chromatograph equipped with an Agilent 5975B mass detector. The percentage composition of secondary metabolites was calculated from the peak areas without using correction coefficients. The qualitative analysis is based on a comparison of the mass spectra of the components corresponding to the data of the NIST0.5a mass spectrum library.

3 Results

The influence of various concentrations of high-molecular chitosan on the physical-chemical properties of fermented dairy product during the cultivation of Lactobacillus delbrueckii subsp. bulgaricus and long-term storage was studied.

3.1 The Effect of High-Molecular Chitosan on the Cultivation of Lactobacillus delbrueckii subsp. bulgaricus

The accumulation of lactic acid was observed in the control sample during the cultivation of L. delbrueckii subsp. bulgaricus starter culture for 4 h at a temperature of 43–45 °C. Because of this process, the active acidity decreased from pH 6.0 to 5.05, as shown in Table 1.

Caseins are at their isoelectric point at pH level below 5.0. As a result, their hydration decreases and aggregation of hydrophobic particles is observed, resulting in the formation of a coagulum [8]. The formation of a coagulum occurred after 20 h of fermentation in experimental samples containing chitosan at concentrations of 0.0025, 0.005, 0.0075 and 0.01%. With an increase in the chitosan concentration, a decrease in titratable acidity and the accumulation of lactic acid was observed, while the active acidity in the samples reached to 4.9–5.0 (Table 1).

High-molecular chitosan slows down the process of lactic acid fermentation and the formation of a coagulum during the process of Lactobacillus delbrueckii subsp. bulgaricus cultivation.

Table 1. The effect of chitosan with a molecular weight of 350.0 kDa different concentrations on pH, titratable acidity, lactic acid and lactose concentrations during the cultivation of Lactobacillus delbrueckii subsp. bulgaricus

Parameters	Duration of fermentation, hours	Control samples	Experimental samples with addition of chitosan, per cent			
			0.0025	0.005	0.0075	0.01
pH	0	6.00 ± 0.2	5.97 ± 0.1	5.94 ± 0.2	5.94 ± 0.2	5.98 ± 0.1
	24	4.93 ± 0.1	4.95 ± 0.2	4.99 ± 0.1	4.98 ± 0.1	5.01 ± 0.2
Titratable acidity, °T	0	35 ± 1	41 ± 2	43 ± 1	43 ± 2	41 ± 2
	24	90 ± 3	89 ± 2	82 ± 2	83 ± 1	75 ± 1
Lactic acid concentration, mole per l	0	0.035	0.041	0.043	0.043	0.041
	24	0.090	0.089	0.082	0.083	0.075
Lactose concentration, mole per l	0	0.1373	0.1358	0.1353	0.1353	0.1358
	24	0.1235	0.1238	0.1255	0.1253	0.1273

3.2 The Influence of High-Molecular Chitosan on Fermented Dairy Product Shelf Life

The maximum titratable acidity of fermented dairy product 140 °T according to the requirements of the "Technical Regulations of the Customs Union 033/2013" in the control sample was reached on 5th day of storage. On the 17th day of storage, this value was exceeded by 145%. In the experimental samples with an increase in the chitosan content of 0.0025, 0.005, 0.0075 and 0.01%, the best indicators corresponding to the requirements for fermented dairy products were achieved on the 7th, 9th, 13th and 17th days of storage, respectively, as shown in Table 2.

The optimal concentration of chitosan with a molecular weight of 350 kDa, which allows reducing the titratable acidity for 17 days of storage from 203 to 138 °T, is 0.01%, The shelf life of fermented dairy product at this concentration increases by more than 3 times compared to the control sample, meeting the requirements of the technical regulations.

As indicated in Table 2, an increase in the chitosan concentration in the samples leads to a decrease in the intensity of lactic acid fermentation.

The most pronounced process of lactose mechanism by homofermentative lactic acid fermentation mechanism slowing down occurred in a sample containing chitosan at a concentration of 0.01%.

The concentration of lactic acid in this sample decreased by 32% compared to the control sample during 17 days of storage. A decrease in the intensity of lactose assimilation by L. delbrueckii subsp. bulgaricus cells may be determined by the interaction of chitosan with the plasma membrane of bacteria, which leads to a violation of its permeability to the enzyme β-galactosidase, which catalyses the hydrolysis of lactose to glucose and galactose [8, 11].

Table 2. The effect of chitosan with a molecular weight of 350 kDa different concentrations on titratable acidity of control and experimental samples of fermented dairy products during storage

Duration of storage, days	Titratable acidity, °T, of control samples	Titratable acidity, °T, of experimental samples with addition of chitosan, per cent			
		0.0025	0.005	0.0075	0.01
1	90 ± 3	89 ± 2	82 ± 2	83 ± 1	75 ± 1
3	117 ± 2	110 ± 3	106 ± 1	104 ± 3	100 ± 4
5	**140 ± 4**	132 ± 3	118 ± 3	110 ± 2	103 ± 2
7	157 ± 3	**145 ± 2**	132 ± 3	116 ± 4	114 ± 1
9	180 ± 2	164 ± 1	**148 ± 3**	128 ± 4	126 ± 4
11	185 ± 4	170 ± 3	155 ± 2	138 ± 2	129 ± 4
13	192 ± 5	179 ± 2	161 ± 4	**142 ± 5**	131 ± 3
15	198 ± 4	185 ± 4	164 ± 2	145 ± 3	132 ± 4
17	203 ± 2	191 ± 5	186 ± 2	155 ± 3	**138 ± 4**

Along with this, structural changes in the cytoplasmic membrane can cause a slowdown in the active transport of lactose hydrolysis products into lactobacillus cells [7, 10, 11]. This is evidenced by the preservation of optimal active acidity in experimental samples containing high-molecular chitosan, slowing the formation of lactic acid and reducing the consumption of lactose, as indicated in Table 3.

Slowing down of lactic acid formation and maintaining of lactose high concentration in samples containing chitosan at concentrations of 0.0075, 0.01%, leads to maintenance of a high level of lactic acid bacteria count during storage [13]. This is confirmed by the data of lactic acid microorganisms in the control and experimental samples for the 17th day of storage quantitative accounting (Fig. 1).

The count of Lactobacillus delbrueckii subsp. bulgaricus in a sample of a fermented dairy product containing 0.01% of chitosan exceeded more than 1000 times relative to the control sample. The lactic acid fermentation process was slowed down at this concentration of chitosan, which allowed increasing the shelf life of fermented dairy product while maintaining a high content of lactobacilli, low concentration of lactic acid and good organoleptic properties.

3.3 The Influence of High-Molecular Chitosan on the Content of Aroma-Forming and Flavour Compounds in a Fermented Dairy Product

High-molecular chitosan, interacting with the membranes of lactobacilli, can significantly affect the metabolic processes occurring in them. This can lead to a change in the composition of aromatic and flavouring compounds that determine the organoleptic properties of fermented dairy product. The composition of secondary metabolites in the

Table 3. The effect of chitosan with a molecular weight of 350.0 kDa different concentrations on tpH, concentration of lactic acid and lactose after 17 days of storage of the fermented dairy product

Parameters	Duration of storage, days	Control samples	Experimental samples with addition of chitosan, per cent			
			0.0025	0.005	0.0075	0.01
pH	1	4.93 ± 0,1	4.95 ± 0,2	4.99 ± 0,1	4.98 ± 0,1	5.01 ± 0,2
	17	3.82 ± 0,1	3.85 ± 0,1	3.87 ± 0,1	3.96 ± 0,1	4.04 ± 0,1
Lactic acid concentration, mole per l	1	0.090	0.089	0.082	0.083	0.075
	17	0.203	0.191	0.186	0.155	0.138
Lactose concentration, mole per l	1	0.1235	0.1238	0.1255	0.1253	0.1273
	17	0.0953	0.0983	0.0995	0.1073	0.1115

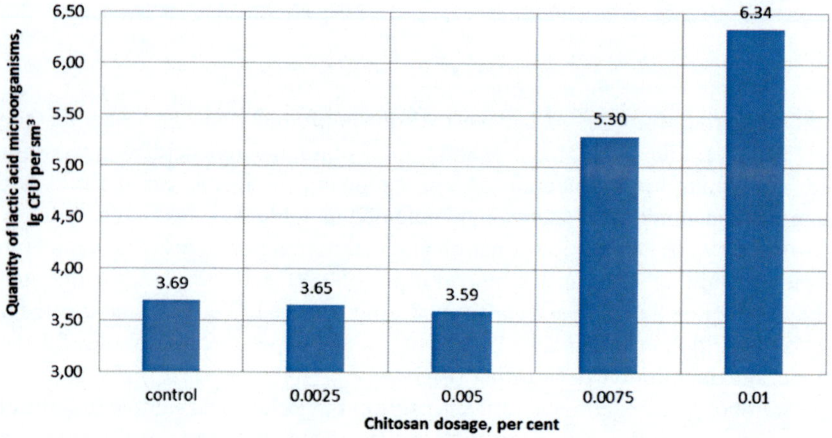

Fig. 1. The count of lactic acid microorganisms in the fermented dairy product on the 17th day of storage, depending on different concentrations of chitosan with a molecular weight of 350 kDa

fermented dairy product on the 17th day of storage, which is presented in Table 4, was studied using GC-MS.

It was established that control and in the experimental samples containing 0.01% of chitosan characterised by a presence of wide range of secondary metabolites such as alcohols (isopropanol); aldehydes (phenylacetaldehyde); acids (methylbutanoic, benzoic, pentadecanoic, myristic, palmitic, stearic). Interaction of chitosan with L. delbrueckii subsp. bulgaricus cells leads to a slight change in the composition and content of a number of secondary metabolites that determine the organoleptic properties of fermented milk products.

Table 4. The relative abundance of aroma-forming and flavour compounds in fermented dairy product on the 17th day of storage

No	Name of substance CAS	Chemical formula	Relative abundance, per cent	
			control sample	chitosan 0.01 per cent
		Alcohols		
1	2-Propanol 000067-63-0	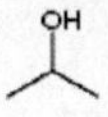	5.17	
2	2-Furanmethanol 000098-00-0	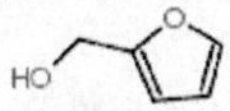		2.05
		Aldehydes		
3	Benzeneacetaldehyde 000122-78-1	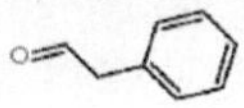	2.12	
4	4H-Pyran-4-one, 3-hydroxy-2-methyl 000118-71-8	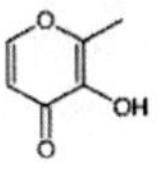	1.37	8.85
5	2,3-Dihydro-3,5-dihydroxy-6-methyl-4H-pyran-4-one 028564-83-2	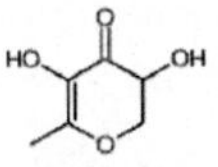	34.71	20.70
		Acids		
6	Butanoicacid, 2-hydroxy-3-methyl-004026-18-0	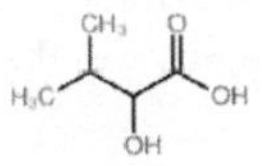	2.03	4.60
7	Benzoicacid 000065-85-0	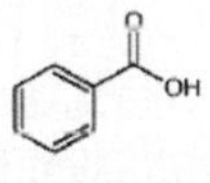	9.94	9.62
8	Pentadecanoicacid 001002-84-2	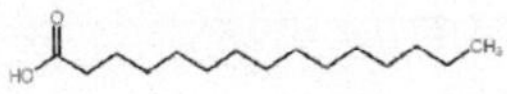	0.78	1.28
9	Tetradecanoicacid (Myristic acid) 000544-63-8	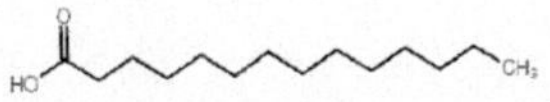	2.04	2.73
10	n-Hexadecanoicacid 000057-10-3	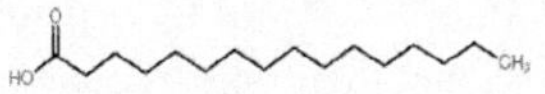	6.05	7.91
11	cis-Vaccenicacid 000506-17-2			21.58
12	Octadecanoicacid 000057-11-4	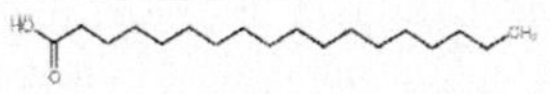	1.55	2.93

Thus, the effective interaction of L. delbrueckii subsp. bulgaricus cells with chitosan occurs due to the presence in its structure of about 1850 amino groups, which in an acidic media carry out multipoint ionic interaction with negatively charged molecules of teichoic acid of lactobacilli membranes. With an increase in the concentration of chitosan, processes of lactose assimilation by Lactobacillus delbrueckii subsp. bulgaricus and lactic acid accumulation slows down. As a result of this processes, the shelf life of fermented dairy product is significantly increased, while maintaining a high level of lactic acid bacteria count and improved organoleptic properties of target product.

4 Discussion and Conclusion

The formation of a coagulum occurs in the control sample after 4 h of skim milk fermentation by starter culture of L. delbrueckii subsp. bulgaricus. The formation of a coagulum slows down and ends after 20 h of lactic acid microorganisms cultivation in experimental samples containing chitosan in concentrations from 0.0025 to 0.01%.

When storing a lactic acid product for 17 days, The maximum titratable acidity exceeded the norm of 140 °T at 7, 9, 13 and 17 days during storage of fermented dairy products containing chitosan in concentrations of 0.0025, 0.005, 0.0075 and 0.01% respectively.

Slowing the formation of lactic acid and maintaining of lactose high concentrations in samples containing chitosan stimulated the growth of lactic acid bacteria during storage to $6.34 * 10^6$ CFU per cm^3.

Interaction of high-molecular chitosan with membranes of lactobacilli L. delbrueckii subsp. bulgaricus does not lead to a significant change in the composition and content of some secondary metabolites that determine the organoleptic properties of fermented dairy products.

The use of chitosan with a molecular weight of 350 kDa at a concentration of 0.01% allows to slow down the process of lactic acid fermentation induced by L. delbrueckii subsp. bulgaricus, which significantly increases the shelf life of the fermented dairy product with the preservation of a low concentration of lactic acid, a high content of lactobacilli and good organoleptic properties.

References

1. Aam, B.B., Heggset, E.B., Norberg, A.L., Sørlie, M., Vårum, K.M., Eijsink, V.G.H.: Production of chitooligosaccharides and their potential application in medicine. Mar. Drugs **8**, 1482–1517 (2010)
2. Ashaolu, T.J.: Safety and quality of bacterially fermented functional foods and beverages: a mini review. Food Qual. Safety **4**, 123–127 (2021)
3. Granato, D., Barba, F.J., Bursać Kovačević, D., Lorenzo, J.M., Cruz, A.G., Putnik, P.: Functional foods: product development, technological trends, efficacy testing, and safety. Annu. Rev. Food Sci. Technol. **11**, 93–118 (2020)
4. Hatti-Kaul, R., Chen, L., Dishisha, T., El Enshasy, H.: Lactic acid bacteria: from starter cultures to producers of chemicals. FEMS Microbiol. Lett. **365**, fny213 (2018). https://doi.org/10.1093/femsle/fny213

5. Lee, H.-W., Park, Y.-S., Jung, J.-S., Shin, W.-S.: Chitosan oligosaccharides, dp 2–8, have prebiotic effect on the Bifidobacterium bifidium and Lactobacillus sp. Anaerobe **8**(6), 319–324 (2003)
6. Kean, T., Thanou, M.: Biodegradation, biodistribution and toxicity of chitosan. Adv. Drug Deliv. Rev. **62**, 3–11 (2010)
7. Kong, M., Chen, X.G., Xing, K., Park, H.J.: Antimicrobial properties of chitosan and mode of action: a state of the art review. Int. J. Food Microbiol. **144**, 51–63 (2010)
8. Kowalewska-Piontas, I., Bednaski, W., Bieleska, M.: Comparison of the hydrolytic and trans-galactolisation properties exhibited by β -galactosidase of different origin. Pol. J. Nat. Sci. **11**, 7–16 (2002)
9. Kulikov, S., et al.: Molecular weight and pH aspects of the efficacy of oligochitosan against methicillin-resistant *Staphylococcus aureus* (MRSA). Carbohydrate Polymers **87**, 545–550 (2012)
10. Riaz Rajoka, M.S., et al.: Chitin/chitosan derivatives and their interactions with microorganisms: a comprehensive review and future perspectives. Crit. Rev. Biotechnol. **40**, 365–379 (2020)
11. Ristić, T., Lasič, S., Kosalec, I., Bračič, M., Fras-Zemljič, L.: The effect of chitosan nanoparticles onto Lactobacillus cells. React. Funct. Polym. **97**, 56–62 (2015)
12. Rivera-Espinoza, Y., Gallardo-Navarro, Y.: Non-dairy probiotic products. Food Microbiol. **37**, 1–11 (2010)
13. Shramko, M.I., Lodygin, A.D., Evdokimov, I.A., Sushinskaya, N.V., Kurchenko, V.P.: Influence of oligochitosans and highly molecular chitosan on *Lactobacillus Bulgaricus* cultivation. IOP Conf. Series: Earth Environ. Sci. **548**, 082066 (2020)
14. Soccol, C.R., Machado Prado, M.R., Blandon Garcia, L.M., Rodrigues, C., Bianchi Pedroni Medeiros, A., Thomaz, S.V.: Current developments in probiotics. J. Microb. Biochem. Technol. **7**, 1 (2014). https://doi.org/10.4172/1948-5948.1000175
15. Zaitoun, M., Ghanem, M., Harphoush, S.: Sugars: types and their functional properties in food and human health. Int. J. Public Health Res. **6**(4), 93–99 (2018)
16. Zou, P., et al.: Advances in characterization and biological activities of chitosan and chitosan oligosaccharides. Food Chem. **190**, 1174–1181 (2016)

The Study of the Quantitative Content of Flavonoids and Biological Activity of the Herba Thlaspi Arvense L.

Kira Pupykina[1], Rashit Farkhutdinov[2(✉)], Albina Fedorova[2(✉)], Ekaterina Koroleva[1], and Pavel Akhmadiev[2]

[1] Bashkir State Medical University, Ufa, Russia
[2] Bashkir State University, Ufa, Russia

Abstract. A method has been developed for quantifying the amount of flavonoids in herba *Thlaspi arvense* L. by differential spectrophotometry. Optimal conditions for the extraction of flavonoids from herba *Thlaspi arvense* L. were selected and the quantitative content of the sum of flavonoids in terms of luteolin-7-glucoside was determined, which was 1,12 ± 0,024%. An assessment of the biological activity of the herba *Thlaspi arvense* L. was carried out, which showed that the thickness of the spermatogenic epithelium in rats of the experimental group against the background of intragastric administration of an aqueous extract of a plant object increased by 100% compared to the control group, and the diameter of the cross-section of the convoluted seminal tubules increased by 15% ($p < 0.05$). The index of spermatogenesis, reflecting the number of generations of spermatogenic cells in the wall of convoluted seminal tubules, increased in the testes of experimental rats compared with the control group. In male rats, after 21 days of application of an aqueous extract of *Thlaspi arvense* L., receptive and perceptive sexual motivations increased in sexual behavior. A reduction in the latent period of the first "emotional" approach to an intact female rat, an increase in the number of such "emotional" approaches and an increase in the duration of the "courtship" period for females were noted. In receptive sexual behavior in males, an increase in the number of female coverings was observed by 70.48% .

Keywords: *Thlaspi arvense* L. · Flavonoids · Spermatogenesis · Sexual motivations

1 Introduction

Currently, the study of drugs based on medicinal plant raw materials is still quite widespread and promising direction. This is due to the content of a large amount of biologically active substances (BAS) in plant raw materials, which provide a wide range of pharmacological effects. Along with this, phytopreparations have fewer side effects, since they interact with the human body in a complex way. Thus, the search for new sources of biologically active substances is an urgent task of modern pharmacy [5].

V. Kurchenko et al. (Eds.): ICAETT 2021, LNNS 408, pp. 176–183, 2022.
https://doi.org/10.1007/978-3-030-96641-6_21

One of the little-studied plant species is the *Thlaspi arvense* L. of the cabbage family (*Brassicaceae*), which is a low (10–50 cm) annual herbaceous plant with erect, branched stems. The leaves are simple, the lower ones are petiolate, oblong-oval, the stem ones are sessile, swept, the edge of the leaf blade is whole or rarely toothed. The flowers consist of 4 sepals and 4 petals, white, small, six stamens, one pistil, collected in a brush. The fruits are round-shaped pods, flattened, about 15 mm in diameter, the seeds are small, furrowed. Blooms from May to August [1, 7, 8].

In folk medicine, *Thlaspi arvense* is widely used as an antibacterial agent, anthelmintic, diuretic, potency-enhancing, hemostatic, astringent, etc. However, in scientific works, the *Thlaspi arvense* is often considered as a weed-field plant [3, 13].

A review of the literature sources suggests that the chemical composition of the *Thlaspi arvense* has not been studied enough. Flavonoids, saponins, glycoside (sinigrin), ascorbic acid, as well as higher fatty acids (linolenic, linoleic, stearic, etc.) were found in the herba *Thlaspi arvense* L. [4, 7, 12].

Thus, a detailed comprehensive study of the chemical composition of herba *Thlaspi arvense* L. and its biological activity is an urgent area of research.

The aim of the study was to study the quantitative content of flavonoids in the herba *Thlaspi arvense* L. and to determine the biological activity of the aqueous extract on model animals.

Research objectives:

- to develop a methodology for the quantitative determination of flavonoids in herba *Thlaspi arvense* L.;
- to evaluate with the help of histological studies the effect of course intake of the aqueous extract of herba *Thlaspi arvense* L. on the size of the seminal tubules in male rats and the presence of spermatozoa in the lumen of the seminal tubules;
- to determine the effect of course intake of water extract of herba *Thlaspi arvense* L. on receptive and perceptual sexual motivations of male rats in a behavioral test.

2 Materials and Methods

As an object of research, we used herba *Thlaspi arvense* L. harvested in the Ufa district of the Republic of Bashkortostan in 2020. The raw materials were dried by the air-shadow method. The studies were carried out with dried material. The raw materials were packed and stored at room temperature, in a dry, well-ventilated room, without direct sunlight.

Studies of quantitative determination of the amount of flavonoids in herba *Thlaspi arvense* L. were carried out using the method of differential spectrophotometry at the Department of Pharmacognosy with a course in botany and the basics of phytotherapy of BSMU [4].

The method of quantitative determination of flavonoids: about 2.0 g (exact weight) of raw materials crushed to the size of particles passing through a 2 mm sieve was placed in a flask with a 250 mL slot, 100 mL of 70% ethyl alcohol was added and weighed with an error of ±0.01 g. The flask was attached to a reverse refrigerator and heated in a boiling water bath for 60 min, periodically shaking to flush the raw material particles from the walls. The flask with the contents was cooled to room temperature, weighed and, if necessary, brought to the initial mass with 70% ethyl alcohol. Extraction was filtered into a 100 mL volumetric flask through a paper filter previously moistened with 70% alcohol (*solution A*).

2.0 mL of solution A was placed in a measuring flask with a capacity of 25 mL, 2 mL of a 2% alcohol solution of aluminum chloride in 95% alcohol was added and the volume of the solution was brought to the mark with 95% alcohol (*solution B*). Comparison solution: 2.0 mL of solution A, 1 mL of 3% acetic acid was added and 70% alcohol was brought to the mark in a measuring flask with a capacity of 25 mL.

After 30 min, the optical density of the investigated solution with aluminum chloride was measured on a spectrophotometer in a cuvette with a layer thickness of 10 mm at a wavelength of 395 nm using a comparison solution.

In parallel, under the same conditions, the optical density of a solution complex of a standard sample (SS) of luteolin-7-glucoside with aluminum chloride was measured: 2.0 mL of a solution of luteolin-7-glucoside was placed in two measuring flasks with a capacity of 25 mL, 2 mL of a 2% alcohol solution of aluminum chloride in 95% alcohol was added to one flask, and 1 mL of 3% acetic acid was added to the other and brought to the mark with the appropriate alcohol, stirred and the optical density was measured after 30 min.

Preparation SS luteolin-7-glucoside: 0.05 g of a standard sample of luteolin-7-glucoside (exact suspension) pre-dried at a temperature of 130–135 °C for 3 h, dissolved in 85 mL of 95% ethyl alcohol when heated in a water bath with periodic stirring. The solution is cooled, quantitatively transferred to another 100 mL volumetric flask, the volume of the solution was brought to the mark with the same alcohol and mixed.

Preparation of a 2% aluminum chloride solution: 2.0 g of aluminum chloride (GOST 3759–75) was dissolved in 40 mL of 95% ethyl alcohol in a 100 mL volumetric flask and the volume of the 95% alcohol solution was brought to the mark.

The content of the sum of flavonoids in absolutely dry raw materials in terms of luteolin-7-glucoside was calculated by the formula:

$$x = \frac{\text{A} \times \text{a0} \times 100 \times 2 \times 25 \times 100 \times 100}{\text{Ao} \times \text{a} \times 2 \times 25 \times 100 \times 100 - \text{W})} \tag{1}$$

A - the optical density of the test solution
A_0 - the optical density of CO luteolin-7-glucoside
a_0 - the mass of CO luteolin-7-glucoside, gr
a - the weight of the raw material suspension, gr
W - the mass loss during drying, %

Pharmacological studies: the experiments were conducted on male rats of the *Wistar* line at the age of 1.5 years, which were kept in standard vivarium conditions of the Department of Physiology and General Biology of BASHGU. 20 sexually mature rats were selected for the experiment, the average weight of which is 250-300g. The rats of the experimental and control groups were kept at 12-h daylight hours on a standard diet with briquetted feed, with free access to water and food. Estrating female rats of the *Wistar* line at the age of 6 months were used in conducting behavioral tests. For 21 days, the control group was injected with a dose of 1 mL/kg of 0.9% aqueous sodium chloride solution (NaCl), the test group was injected with an aqueous extract of herba *Thlaspi arvense* L., which was prepared as follows: the exact suspension of vegetable raw materials is poured with the necessary amount of water in a ratio of 1:10, stirred, covered with a lid and insisted on a boiling water bath with frequent stirring for 15 min, then the extraction is cooled for 30 min and filtered. An aqueous extract in the volume of 1 mL/kg was injected through a probe into the stomach of experimental animals for 21 days.

All experiments were conducted in strict accordance with the existing generally accepted international principles of humane treatment of laboratory animals (Directive of the Council of the European Union of 24.11.1986, 86/609/EEC).

Histological studies: rat testis samples were fixed in neutral 10% Lilly formalin, dehydrated in alcohols of ascending concentration and poured into paraffin. A series of frontal sections with a thickness of 10–12 microns was prepared. Microscopy of the obtained histological preparations was carried out using a Micmed-5 light-optical microscope (LOMO, Russia). Photographing was carried out using the Levenhuk C510 camera (5M pixels). Studies of the structural changes of the testes were carried out in the field of view of the microscope Micmed-5 (LOMO, Russia) at magnification of 100 and 400 times. To obtain the results, the following parameters were analyzed: the thickness of the spermatogenic epithelium (microns), the diameter of the cross-section of the convoluted seminal tubules (microns), the index of spermatogenesis (number).

Behavioral test: sexual behavior (paired with intact estrating females) of male rats was studied in the "open field" installation, modified for the "zoo social preferences area" (PZP), which was a circular arena with sides divided by opaque plastic partitions into 4 compartments connecting in the center. In the first, second and third compartments were placed, respectively: a drinking bowl, a feeder with grain, a plastic house. The fourth compartment of the installation was free, the test male was placed in it 20 min before the start of the test. Later, the female rat was placed in the central part. Observations of the male after the female was hooked were conducted for 60 min. It was noted: proceptive sexual motivations (the latent period of the first "emotional" approach of the male to the female, the time from the female's planting to the first manifestation by the male of the elements of "courtship", such as licking, sniffing and grooming); the number of "emotional" approaches of the male to the female; the duration of sexual activity (the total period of time spent by the male on "courting" the female and covering her) and receptive sexual motivations - the number of coverings.

Statistical data processing was carried out using a licensed package of application programs "Statistica Trial" (Stat Soft Inc., USA).

3 Results

Flavonoids of medicinal plants have a wide range of pharmacological activity and are essential in the formation of a therapeutic effect [2, 10, 11], therefore, studies have been conducted to study the quantitative content of this group in herba *Thlaspi arvense* L. At the first stage, spectral characteristics of alcohol extracts of herba *Thlaspi arvense* L. were studied in comparison with flavonoid witness substances and spectral characteristics of extracts with the addition of a complexing additive - aluminum chloride

(III) solution, with which flavonoids form a complex stable in an acidic environment and a bathochromic shift of the absorption band of flavonoids from 330–350 nm to 390–410 nm was observed [6, 9]. When adding a solution of aluminum chloride to the water-alcohol extracts of herba *Thlaspi arvense* L., an absorption maximum was observed in the region of 395 nm, which coincided with the maximum absorption of a solution of luteolin-7-glucoside with aluminum chloride, therefore, this wavelength was chosen as an analytical one for carrying out a quantitative determination technique (Fig. 1). As a control, the studied extraction from herba *Thlaspi arvense* L. without the addition of aluminum chloride was used, which made it possible to exclude the influence of concomitant substances on the determination results. In addition, the parameters of flavonoid extraction from the herba *Thlaspi arvense* L.were studied and optimal conditions were selected: the extractant is ethyl alcohol 70%, the degree of grinding of raw materials is 2 mm, the ratio of raw materials and extractant is 1:100, the extraction time is 60 min with a single extraction. Based on the results obtained, a method for quantifying the amount of flavonoids in the herba *Thlaspi arvense* L. is proposed.

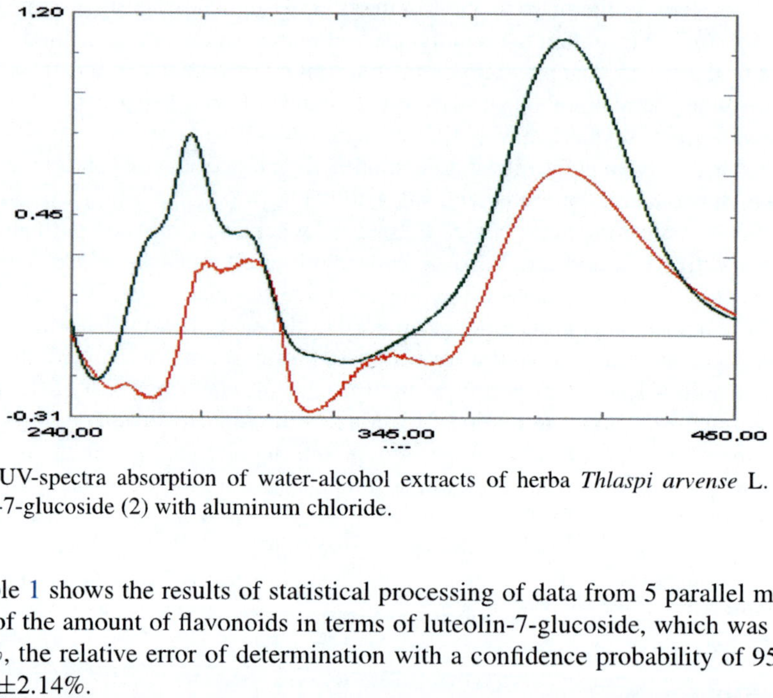

Fig. 1. UV-spectra absorption of water-alcohol extracts of herba *Thlaspi arvense* L. (1) and luteolin-7-glucoside (2) with aluminum chloride.

Table 1 shows the results of statistical processing of data from 5 parallel measurements of the amount of flavonoids in terms of luteolin-7-glucoside, which was 1.12 ± 0.024%, the relative error of determination with a confidence probability of 95% was within ±2.14%.

Histological examination of the structure of cross sections of the seminal tubules in the experimental and experimental groups showed that normal and mature spermatozoa were observed in the testes of the control group in the lumen of the seminal tubules. The space between them was filled with a small volume of interstitial tissue and Leydig cells. In the experimental group of rats, the cells of the spermatogenic epithelium, compared with the control group, have a significantly more pronounced epithelial thickness and diameter of the seminal tubules. Sertoli cells and all cell lines of spermatogenesis, such as

Table 1. Indicators of flavonoid content in herba *Thlaspi arvense* L.

№ п/п	The content of the sum of flavonoids in terms of luteolin-7-glucoside,%	Metrological characteristics
1	1.12	Xcp. = 1.12 $S\hat{y} = \sqrt{\Sigma}(X - X_{cp.})^2/n(n-1)$ = 0.00866 $\Delta X = t \cdot S\hat{y} = 0.024$ $\varepsilon_{OTH.} = \Delta X \cdot 100/X_{cp.} = 2.14\%$
2	1.09	
3	1.14	
4	1.11	
5	1.13	

spermatogonia, primary spermatocytes and spermatids, were visible, and a large number of spermatozoa exist in the lumen of these tubules [2].

Table 2 presents the results of determining the potential generative activity of the testis of male rats. Such indicators as the thickness of the spermatogenic epithelium (microns), the diameter of the cross-section of the convoluted seminal tubules (microns) were determined, the index of spermatogenesis was calculated.

Table 2. Quantitative characteristics of changes in the generative activity of the testes of intact rats after a course intake of an aqueous extract of *Thlaspi arvense* L. (M ± m)

The name of the indicator	Control	Experience
Thickness of spermatogenic epithelium (microns)	41.79 ± 3.36	83.76 ± 8.60*
Diameter of the cross-section of the convoluted seminal tubules (microns)	149.51 ± 6.49	171.97 ± 6.50*
Spermatogenesis index (number)	3.25 ± 0.07	4.69 ± 0.06

* *Significant differences in* $p < 0.05$.

The thickness of the spermatogenic epithelium in rats of the experimental group against the background of intragastric administration of an aqueous extract of the *Thlaspi arvense* L. increased by 100% compared to the control group ($p < 0.05$), and the diameter of the cross-section of the convoluted seminal tubules increased by 15% ($p < 0.05$). The spermatogenesis index, reflecting the number of generations of spermatogenic cells in the wall of the convoluted seminal tubules, is the most important quantitative indicator characterizing the generative activity of the testis. In the testes of experimental rats (Table 2) compared with the control group, the index of spermatogenesis increases, these differences are significant.

The results obtained are most likely due to the fact that the extract of *Thlaspi arvense* L. due to the content of a complex of biologically active substances has a positive effect on metabolic processes, has antioxidant properties, which leads to improved blood circulation and affects the physiological functions of the spermatogenic epithelium [10].

From the results of the studies presented in Table 3, it can be seen that rats of the *Wistar* line of the experimental group (after 21 days of application of the *Thlaspi arvense*

L. extract), relative to the control group, an increase in receptive and perceptive sexual motivations was noted in the sexual behavior of male rats. In the perceptive sexual behavior of experienced males, a reduction in the latency period of the first "emotional" approach to an intact female rat was noted by 26.4%, an increase in the number of such "emotional" approaches by 17.8% and an increase in the duration of the "courtship" period for females by 29.9%. At the same time, in the receptive sexual behavior of the experimental group of males, an increase in the number of female coverings was observed by 70.48%.

Table 3. Indicators of sexual motivation in male rats of the *Wistar* line during the test "site of zoosocial preferences", during 60 min of observations, paired with intact female rats (M ± m)

Group	The latent period of the first approach to the female, c	The number of "emotional" approaches to the female, pc	Total courtship time, c	Number of coatings, pc
Control	6.6 ± 1.8	24.2 ± 1.0	716.4 ± 90.3	0.8 ± 0.1
Experience	4.8 ± 1.3	29.4 ± 1.3	1022.6 ± 174.4*	2.7 ± 0.2

* *Significant differences in* $p < 0.05$.

4 Discussion and Conclusion

At this stage of research, one of the groups of biologically active substances of herba *Thlaspi arvense* L. flavonoids has been studied. It was found that the optimal parameters for the extraction of flavonoids from herba *Thlaspi arvense* L. are the use of 70% ethyl alcohol as an extractant, the grinding of raw materials is 2 mm, the extraction time is 60 min with a single extraction, the optical density of the studied solutions was measured at a wavelength of 395 nm and the quantitative content of the sum of flavonoids in terms of luteolin-7-glucoside was 1.12 ± 0.024%. As a result of the study of the biological activity of the aqueous extract of herba *Thlaspi arvense* L.on experimental animals, a positive dynamics of the effect of the studied object on experimental animals was noted, namely, it was found that after its course application, in rats of the experimental group, compared with the control group, the thickness of the spermatogenic epithelium, the diameter of the cross-section of the convoluted seminal tubules increased, the index of spermatogenesis increased, and receptive and perceptive sexual motivations in the sexual behavior of male rats increased.

Thus, it can be noted that *Thlaspi arvense* L. is a promising medicinal plant and further more detailed study of its chemical composition and biological activity will allow to substantiate from a scientific point of view its possibilities of application in medical practice.

References

1. Altendorfa, K., Isbellb, T., Wysea, D.L., Andersona, J.A.: Significant variation for seed oil content, fatty acid profile, and seed weight in natural populations of field pennycress (*Thlaspi arvense* L.). Ind. Crops Prod. **129**, 261–268 (2019)
2. Andriyanenkov, A.V.: Study of pharmacological efficacy of thick extracts of Thlaspi arvense L. and sandy esparcet on a model of benign prostatic hyperplasia in rats. Sci. J. Sciencerise **10**(4(15)), 46–51 (2015). https://doi.org/10.15587/2313-8416.2015.52263
3. Chopra, R., Nicole Folstad, M., Marks, D.: Combined genotype and fatty-acid analysis of single small field pennycress (*Thlaspi arvense*) seeds increases the throughput for functional genomics and mutant line selection. Indust. Crops Prod. **156**, 112823 (2020). https://doi.org/10.1016/j.indcrop.2020.112823
4. Claver, A., Rey, R., López, M.V., Picorel, R., Alfonso, M.: Identification of target genes and processes involved in erucic acid accumulation during seed development in the biodiesel feedstock Pennycress (Thlaspi arvense L.). J. Plant. Physiol. **208**, 7–16 (2016)
5. Claver, A., et al.: Functional analysis of β-ketoacyl-CoA synthase from biofuel feedstock Thlaspi arvense reveals differences in the triacylglycerol biosynthetic pathway among Brassicaceae. Plant Mol. Biol. **104**(3), 283–296 (2020). https://doi.org/10.1007/s11103-020-01042-7
6. Esfahanian, M., et al.: Generating pennycress (*Thlaspi arvense*) seed Triacylglycerols and Acetyl-Triacylglycerols containing medium-chain fatty acids. Orig. Res. **10**, 1–14 (2021)
7. Sleptsov, I.V., Zhuravskaya, A.N.: Dynamics of flavonoid accumulation in leaves of amaranthus Retroflexus, agastache rugosa and Thlaspi arvense L. collected in central Yakutia. Chem. Plant. Raw. Mater. **3**, 67–72 (2016)
8. Sleptsova, I.V., Shashurina, M.M., Zhuravskaya, A.N.: Short-term excitation of a permanent magnetic field on the physiological, morphological and biochemical characteristics of seedlings of Amaranthus retroflex, Agastache rugosa and Thlaspi arvense. Plant. Physiol. **66**(1), 66–72 (2019)
9. Tsogtbaatar, E., Cocuron, J.-C., Sonera, M.C., Alonso, A.P.: Metabolite fingerprinting of pennycress (Thlaspi arvense L.) embryos to assess active pathways during oil synthesis. J. Exp. Botany. **66**(14), 4267–4277 (2015)
10. Utomo, B., Daningtia, N.R., Yuliani, G.A., Yuniarti, W.M.: Effects of a standardized 40% ellagic acid pomegranate (Punica granatum L.) extract on seminiferous tubule histopathology, diameter, and epithelium thickness in albino Wistar rats after heat exposure. Veterinary World. **12**(8), 1261–1265 (2019)
11. Waltera, N.S., Gorkia, V., Singhb, R., Salunkeb, D.B., Kaura, S.: Exploring the antiplasmodal efficacy of erucic acid and its derivative isolated from *Thlaspi arvense* D. C. (Brassicaceae). S. Afr. J. Bot. **139**, 158–166 (2021)
12. Zhao, R., et al.: One-pot process for simultaneously obtaining oil and sinigrin from field pennycress (*Thlaspi arvense*) seeds using microwave-assisted biphasic extraction. Indust. Crops Prod. **166**, 113483 (2021). https://doi.org/10.1016/j.indcrop.2021.113483
13. Zhao, R., et al.: Biodiesel preparation from Thlaspi arvense L. seed oil utilizing a novel ionic liquid core-shell magnetic catalyst. Indust. Crops. Prod. **162**, 113316 (2021)

The Obtaining an Extract from the Herba of Monarda and Determining the Indicators of Its Quality

Kira Pupykina, Ekaterina Krasyuk(✉), and Galiya Shaydullina

Bashkir State Medical University, Ufa, Russia
arrabal@bk.ru

Abstract. The article presents the results of studying the conditions for obtaining a liquid extract from the herba *Monarda*. At the initial stage, studies were carried out to establish the optimal extraction parameters for obtaining a liquid extract from the herba *Monarda*. It was found that the highest yield of extractive substances was observed when 40% ethyl alcohol was used as an extractant and the degree of grinding of the raw material was 2 mm. The liquid extract was obtained by the method of percolation in the ratio of raw material and extractant (1:1). In order to further use the extract, it was necessary to determine its quality indicators: density, which was 1.15 ± 0.02 g/cm^3, dry residue - $27.05 \pm 0.69\%$, pH of the investigated extract - 6.5 ± 0.03. The content of biologically active substances in the liquid extract was studied, which showed the presence of a sufficient amount for the manifestation of the therapeutic effect. The quantitative content of ascorbic acid and tannins in the extract was determined by titrimetric methods of redox titration and it was found that it was $0.043 \pm 0.013\%$ and $0.871 \pm 0.032\%$, respectively. The content of flavonoids was determined by differential spectrophotometry in terms of luteolin, which was $0.223 \pm 0.006\%$. On the basis of the studies carried out, indicators of the quality of the liquid extract are proposed, which can later be used for its standardization.

Keywords: *Monarda* · Extract · Percolation · Flavonoids · Tannins · Ascorbic acid · Luteolin

1 Introduction

Recently, one of the promising areas of scientific research is the search and identification of new types of essential oil plants as sources of valuable biologically active substances, which makes it possible to expand the assortment of medicinal herbal remedies used for the prevention and treatment of many diseases. The advantage of medicinal plants is their softness of action, the possibility of long-term use in the treatment of chronic diseases, the minimum number of side effects and the possibility of a rational combination of medicinal plants both among themselves and with synthetic drugs [9, 10, 16, 17, 23]. Since a limited number of essential oil plants grow in natural conditions in the temperate zone of Russia, including in the Republic of Bashkortostan, the issues of introducing

V. Kurchenko et al. (Eds.): ICAETT 2021, LNNS 408, pp. 184–190, 2022.
https://doi.org/10.1007/978-3-030-96641-6_22

plants of different taxa to provide the domestic raw material base, replacing imported plant materials valuable for practical use and creating medicinal products on their basis are being resolved [6, 9]. In addition, extensive research is being conducted on the study of essential oil and aromatic plants, which are used in the medical and food industries, and have economic value [11, 22]. In this regard, plants of the genus *Monarda* L. are interesting for study, which are characterized by strong polymorphism, and according to a number of important morphological characters are classified as promising [5, 8, 12, 15, 21]. The herba *Monarda* contains a rich set of biologically active substances, primarily essential oils, flavonoids, tannins, as well as carotenoids, organic acids, polysaccharides, which cause bactericidal, antiviral, antifungal, antibiotic, anthelmintic activity, exhibit immunomodulatory, anti-inflammatory, sedative effects, restore disturbed redox processes in the body, stimulate the regeneration of damaged areas of the skin and mucous membranes [1, 3, 7, 13, 14, 18, 23]. Therefore, an in-depth study of the chemical composition and biological activity of plants of the genus *Monarda* L. is an urgent task.

The aim of the study is to study the parameters of the extraction of biologically active substances from the herba *Monarda* to obtain an extract based on it.

Research objectives:

- to select the optimal conditions for the output of extractive substances;
- to determine the quality indicators of the liquid extract from the herba *Monarda*;
- to carry out a quantitative analysis of the content of active substances in a liquid extract from the herba *Monarda*.

2 Materials and Methods

The object of the study was *Monarda* herba harvested in the South Ural Botanical Garden-Institute of the Ufa Federal Research Center of the Russian Academy of Sciences in the flowering phase and used to obtain the extract. Liquid extract from herba monarda was obtained by percolation, which is based on continuous filtration of plant raw materials at room temperature by sequentially circulating portions of the extractant and collecting the resulting fractions. The extract was obtained in several stages. The first portion of the raw material intended for loading was pre-soaked with half the amount of extractant relative to the mass of the raw material and left for 4–6 h to swell. Then the raw materials were placed in a percolator and infused for 24 h, with a double volume of extractant relative to the mass of the raw material. After a specified period of time, sequential percolation was carried out until the vegetable raw materials were completely depleted with the separation of extracts. The first portion in the amount of 85% in relation to the mass of raw materials was collected in a separate container, then another container was used and percolation was carried out until the raw materials were completely depleted. At the same time, weaker extracts were obtained, which were evaporated under vacuum at a temperature of 50–60 °C to 15% relative to the mass of the raw material swirled into the percolator. After cooling, the condensed residue was dissolved in the first portion of extraction and the finished product was obtained [2, 4, 19, 20, 25].

The content of extractive substances was determined by the gravimetric method after a single extraction of plant materials with a solvent, drying and weighing the dry residue. Determination of the quantitative content of biologically active substances was carried out by chemical and spectral methods. The determination of the content of ascorbic acid in the extract was carried out by redox titration using a solution of 2,6-dichlorophenolindophenolate natrii (0.001 mol/L) as a titrant, titration was carried out in an acidic medium until the test solution became pink. Tannins were determined by redox titration, using a solution of indigosulfonic acid as an indicator, and a solution of permanganate kalii (0.02 mol/L) as a titrant, titration was carried out until the test solution became yellow, recalculated for tannin. To determine the content of flavonoids, the method of differential spectrophotometry was used with the addition of a complexing additive – 1% alcoholic solution of aluminum chloride. The optical density of the colored solutions was measured on a *Shimadzu UV*-1800 spectrophotometer at a wavelength of 398 nm, the cuvette layer thickness was 10 mm. The recalculation was carried out for the dominant substance in the sum of flavonoids - luteolin.

For the obtained liquid extract from the herba *Monarda*, quality indicators were determined. Density was determined using a pycnometer. Determination of the dry residue was carried out by evaporating the liquid extract in a water bath to a dry residue, followed by bringing the residue to constant weight in an oven at a temperature of 105 °C and weighing on an analytical balance. The determination of the pH of the liquid extract was carried out by the potentiometric method on an Ionomer pH-150 MI device.

Statistical processing of the obtained results was carried out by standard methods of variation statistics (programs “Excel 7.0”, “Statistica 5.0”, “Statistica 6.0”).

3 Results

At the initial stage of obtaining a liquid extract from the herba *Monarda*, it was necessary to choose an extractant at which the highest yield of extractive substances was observed. For this, ethyl alcohol of various concentrations and water were used (Table 1). Based on the data obtained, it was found that the optimal extractant, in which the highest yield of extractive substances is observed, is ethyl alcohol at a concentration of 40%.

Table 1. Indicators of the content of extractive substances in the herba *Monarda*, depending on the extractant, %

№ п/п	Extractant name	f	x	S_x	P, %	t(P,*f*)	$E\alpha$	ε, %
1	Ethyl alcohol 95%	4	23.40	0.3072	95	2.78	0.854	3.65
2	Ethyl alcohol 70%	4	21.83	0.1803	95	2.78	0.501	2.33
3	Ethyl alcohol 40%	4	51.00	0.7990	95	2.78	1.974	3.87
4	Ethyl alcohol 30%	4	39.60	0.4102	95	2.78	1.144	2.88
5	Water	4	47.47	0.339	95	2.78	1.110	2.34

Further, the yield of extractive substances was studied depending on the degree of grinding of the raw material, using sieves with different hole diameters: 0.5 mm; 1 mm,

2 mm; 3 mm and 5 mm (Table 2). The results of the study showed that the maximum yield of extractives from herba *Monarda* is achieved when using raw materials crushed to a particle size passing through a sieve with a hole diameter of 2 mm.

Table 2. Indicators of the content of extractive substances depending on the degree of grinding of raw materials, %

№ п/п	Grinding degree, mm	f	x	Sx	P, %	t(P,*f*)	*Eα*	ε,%
1	0.5	4	31.54	0.377	95	2.78	1.050	3.33
2	1.0	4	45.23	0.561	95	2.78	1.560	3.45
3	2.0	4	48.25	0.503	95	2.78	1.399	2.90
4	3.0	4	32.54	0.307	95	2.78	0.855	2.63
5	5.0	4	27.36	0.197	95	2.78	0.550	2.01

A liquid extract from the herba *Monarda* was obtained by the percolation method with the following extraction parameters: extractant - ethyl alcohol 40%, the degree of grinding of the raw material −2 mm, the ratio of raw material and extractant 1:1. The resulting extract was a dark brown liquid with a characteristic thymol-carvacrol odor and a bitter taste.

For the obtained extract, quality indicators were determined: density, pH of the extract, dry residue and quantitative content of biologically active substances (Tables 3 and 4).

Based on the results of determining the quality indicators of the liquid extract from the herba *Monarda*, it was found that the density of the extract was 1.15 ± 0.02 g/cm^3, the dry residue was 27.05 ± 0.69%, the pH of the investigated extract was 6.5 ± 0.03 (Table 3).

Table 3. Indicators of the quality of liquid extract from herba *Monarda*

№ п/п	f	x	Sx	P. %	t(P.*f*)	*Eα*	ε,%
Density, g/cm^3							
1	4	1.15	0.0055	95	2.78	0,02	1.34
Dry residue, %							
2	4	27.05	0.2481	95	2.78	0.62	2.55
pH extract							
3	4	6.50	0.0129	95	2.78	0.03	0.53

In the liquid extract, the main groups of biologically active substances (BAS) were determined, which together determine the pharmacological effect of the extract: ascorbic acid, tannins and flavonoids. Comparative spectral characteristics of a liquid extract from the herba *Monarda* and a solution of a standard sample of a flavonoid - luteolin with the addition of a complexing additive - an alcohol solution of aluminum chloride, shown in Fig. 1, confirm the coincidence of their absorption maxima at a wavelength of 398 nm.

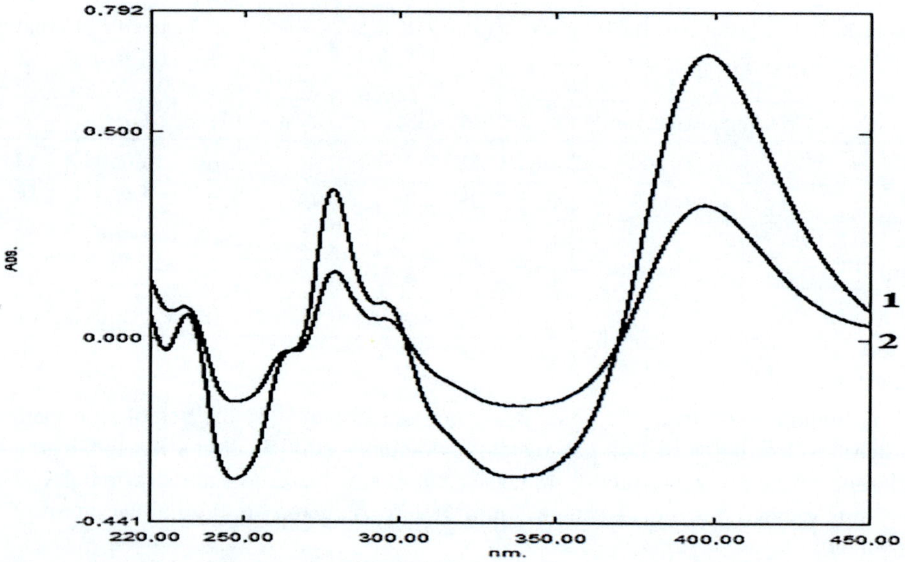

Fig. 1. Differential UV-spectra of a standard solution of luteolin (1), liquid extract of *Monarda* (2).

Based on the results of the quantitative determination of biologically active substances given in Table 4, it can be noted that the content of ascorbic acid in the liquid extract was 0.043 ± 0.013%, tannins - 0.871 ± 0.032%, the amount of flavonoids in terms of luteolin - 0.223 ± 0.006%.

Table 4. Indicators of the quantitative content of biologically active substances in the liquid extract, %

№ п/п	Name of the BAS group	f	x	Sx	P. %	t(P.*f*)	*Eα*	ε,%
1	Ascorbic acid	4	0.043	0.048	95	2.78	0.013	3.14
2	Tannins	4	0.871	0.010	95	2.78	0.032	3.47
3	Flavonoids, calculated as luteolin	4	0.223	0.022	95	2.78	0.006	2.85

4 Discussion and Conclusion

Thus, the optimal conditions for obtaining a liquid extract from the herba *Monarda* were selected, namely, the extractant was 40% ethyl alcohol, the degree of grinding of the raw material was 2 mm, which made it possible to extract the maximum amount of biologically active substances. The indicators of the quality of the liquid extract were determined: density - 1.15 ± 0.02 g/cm^3, dry residue - 27.05 ± 0.69%, pH - 6.5 ± 0.03 and the content of the main groups of biologically active substances: ascorbic acid - 0.043 ± 0.013%, tannins - .871 ± 0.032%, the sum of flavonoids in terms of luteolin - 0.223 ± 0.006%.

Summarizing the results obtained, it should be noted that plants of the genus *Monarda* L. are promising objects of research, since they have a complex and varied chemical composition, a wide range of pharmacological activity and can be sources of valuable biologically active compounds and drugs.

References

1. Contaldo, N., et al.: Phytochemical effects of phytoplasma infections on essential oil of Monarda fistulosa L. Bull. Insectol. **64**(Suppl. 1), S177–S178 (2011)
2. Davituliani, V.V., Korotkov, V.P., Nikolaev, A.M.: The physicochemical properties of a liquid extract of Licorice root. Pharm. Chem. J. **3**(4), 233–234 (1969)
3. Di Vito, M., et al.: Monarda citriodora hydrolate vs essential oil comparison in several anti-microbial applications. Ind. Crops Prod. **128**, 206–212 (2019)
4. Dzharov, V.V., Mishra, A.P., Shariati, M.A., Atanassova, M.S., Plygun S.: Phytochemical contents in solid-liquid extraction of aqueous alcoholic extract of chicory (*Cichorium intybus l.*) leaves. Foods Raw Mater. **4**(2), 32–37 (2016)
5. Ghosh, M.L., Damron, D.S., Schepetkin, I.A., Quinn, M.T.: Sential oils from Monarda fistulosa: chemical composition and activation of transient receptor potential A1 (TRPA1) channels. Molecules **25**(21), 4873 (2020)
6. HayashI, T., HeinS, R.D., CameroN, A.C., Carlson, W.H.: Ethephon influences flowering, height, and branching of several herbaceous perennials. Sci. Hortic. **90**(3), 305–324 (2001)
7. Kurkin, V.A, Lapina A.S., Daeva E.D., Kadentsev V.I.: Flavonoids from herb of *Monarda fistulosa*. Chem. Nat. Compound. **56**(2), 242–245 (2020)
8. Kurkin, V.A, Tsibina A.S.: The development of the new approaches to the standardization of the *Monarda fistulosa L.* herbs. Res. J. Pharm. Technol. **14**(7), 3916–3920 (2021)
9. Kruglov, D.S., Kruglova, M.J., Olennikov, D.N.: Correlation between the microelement profile and essential oil composition of plants from the Filipendula genus. Russ. J. Bioorg. Chem. **46**(7), 1378–1384 (2020)
10. Lovkova, M.Y., Kliment'eva, N.I., Buzuk, G.N., Sokolova, S.M.: Chemical features of medicinal plants (review). Appl. Biochem. Microbiol. **37**(3), 229–237 (2001)
11. Marchese, A., Orhan, E., Daglia, M., Barbieri, R.: Antibacterial and antifungal activities of thymol: a brief review of the literature. Food Chem. **01**, 402–414 (2016)
12. Malankina, E.L., Kuzmenko A.N., Evgrafov A.A, Zaitchik B.T.: Content and composition of wild bergamot (*Monarda fistulosa L.*) essential oil at different phenological phases. Moscow Univ. Chem. Bull. **75**(6), 391–394 (2020)
13. Mattarelli, P., Modesto, M., Barbanti, L., Bellardi, M.G.: Chemical composition and antimicrobial activity of essential oils from aerial parts of Monarda didyma and Monarda fistulosa cultivated in Italy. J. Essent. Oilbear. Plants **20**(1), 76–86 (2017)
14. Nikitina, A.S.: Component composition of the essential oil of the herb *Monarda fistulosa L.* from the collection of the Nikitsky Botanical Garden .Chem. Veg. Raw Mater. **2**, 55–62 (2018)

15. Oparin, R.V., Pokrovsky, L.M., Vysochina, G.I., Tkachev, A.V.: Investigation of the chemical composition of the essential oil of Monarda fistulosa L. and Monarda didyma L., cultivated in the conditions of western Siberia. Chem. Veg. Raw Mater. **3**, 19–24 (2020)
16. Rozzi, N.L., Phippen, W., Simon, J.E., Singh, R.K.: Supercritical fluid extraction of essential oil components from lemon-scented botanicals. LWT Food Sci. Technol. **35**(4), 19–324 (2002)
17. Nekratova, N.A., Shurupova, M.N.: Resources of medicinal plants in the Kuznetsky alatau. Int. J. Environ. Stud. **71**(5), 656–666 (2014)
18. Shanaida, M., Hudz, N., Jasicka-Misiak, I., Wieczorek P.P.: Polyphenols and pharmacological screening of a *Monarda fistulosa* L. dry extract based on a hydrodistilled residue byproduct. Front. Parmacol. **12**, 563 (2021)
19. Smolyakova, I.M., Andreeva, V.Y., Kalinkina, G.I., Shchetinin, P.P., Avdeenko, S.N.: Development of extraction techniques and standardization methods for a common lady's Mantle (Alchemilla vulgaris) extract. Pharm. Chem. J. **45**(11), 675–678 (2012)
20. Smolyakova, I.M., Andreeva, V.Y., Kalinkina, G.I., Avdeenko, S.N., Shchetinin, P.P.: Development of technology and methodology for standardization of the extract of the common cuff. Chem. Pharma. J. **45**(11), 27–30 (2011)
21. Tabanca, N., Ali, A., Wang, M., Khan, S.I., Khan, I.A.: Bioassay-guided investigation of two monarda essential oils as repellents of yellow fever mosquito aedes aegypti. J. Agric. Food Chem. **61**(36), 8573–8580 (2013)
22. Tsuro, M., Inoue, M., Kameoka, H.: Variation in essential oil components in regenerated lavender (Lavandula vera) plants. Sci. Hortic. **88**(4), 309–317 (2001)
23. Zhilyakova, E.T., Novikov, O.O., Naumenko, E.N.: Study of Monarda fistulosa essential oil as a prospective antiseborrheic agent. Bull. Exp. Biol. Med. **148**(4), 612 (2009)
24. Zheljazkov, V.D., Astatkie, T.: Effect of distillation waste and plant hormones on spearmint growth and composition. J. Sci. Food Agric. **91**(6), 1135–1141 (2011)
25. Ziyamukhamedova, M.M., Nazarova, Z.A., Faizullaeva, N.S.: Obtaining handeliae trichophyllae liquid extract. Pharm. Chem. J. **40**(10), 569–571 (2006)

Using the Flax Seeds and the Flax Oil in the Production of Chopped Semi-finished Chicken Meat Products in Order to Enrich Them with Polyunsaturated Fatty Acids

Isabella Stefanova[1] and Veronika Borisova[2](✉)

[1] All-Russian Scientific Research Institute of Poultry Processing Industry – Branch of FSC ARRTPI RAS (ARSRIPPI), P/o Rzhavki, Russia
dp@vniipp.ru

[2] FSBEI HE Smolensk State Agricultural Academy, Smolensk, Russia

Abstract. The research results of the 'Uralsky' flax cultivation in the Central Russia and obtaining the seeds stock and its seeds for the purpose of enriching the chopped semi-finished poultry products (cutlets) are presented in the article. Obtaining the plant-based functional substances from the flax seeds, as well as identifying the feasibility of using the locally updated plant raw material for producing the enriched semi-finished products with certain effects has been the purpose of the study. With this end in view, the research has been made to develop the chopped semi-finished poultry products enriched with the flax seeds of 'Uralsky' oil flax variety, as well as with the locally-updated oil flax seeds, and with the flax oil. The fatty acids composition of the samples enriched with the flax oil have been identified. The research techniques are as follows: evaluation of the organoleptic properties for the developed semi-finished products and for the reference sample, fatty acids identification for the semi-finished products enriched with the flax seeds and the flax oil on the 'Kristall 5000.2' gas chromatograph, amino acid protein composition identification for the developed semi-finished products using the amino acid detector YL 9100 HPLS System. The study revealed that using the locally-updated flax seeds enriches the organoleptic properties of the semi-finished products. The evaluation of the fatty acids composition identified a lower content of the polyunsaturated fatty acids in the developed samples of semi-finished products supplemented with the locally-updated flax seeds. This study has never before been conducted in the Smolensk Region and has a scientific novelty.

Keywords: Oil flax · 'Uralsky' variety · Locally-updated flax · Flax seeds · Flax oil · Fatty acids composition · Organoleptic properties

1 Introduction

At the present time, nutritionists and scientists develop rations enriched with functional ingredients of certain effect, as well as high protein foods [1–3]. The polyunsaturated fatty acids are important nutritional ingredients. The polyunsaturated fatty acids

V. Kurchenko et al. (Eds.): ICAETT 2021, LNNS 408, pp. 191–199, 2022.
https://doi.org/10.1007/978-3-030-96641-6_23

Omega-3, Omega-6 are essential for humans and play important roles in their organisms. The Omega-3 has an important role in decreasing the risk of cardiovascular disease, atherosclerotic damages etc. It has been revealed, that deficiency of the Omega-3 in the human ration is the principal factor contributing to the rapid growth of non-infection diseases both in frequency and in severity. Particular importances for human organism have linoleic and linolenic polyunsaturated acids which are the building blocks of the cell membranes. These acids ensure the healthy development and adaptation of the human organism to the environmental risk factors. The ratio of these acids in the human organism is also important. The optimal ratio is as follows: Omega-6:Omega-3 – 2:1 [4–8].

The fish fat can be the source of the polyunsaturated fatty acids. However, the fish population decreases dramatically. It should be also noted, that water pollution results in fish contamination with the heavy metal, which impacts negatively the quality of the harvested fats. This has led to the necessity of considering alternative sources of polyunsaturated fatty acids [9, 10].

The scientific literature cites plant-based sources of polyunsaturated fatty acids. The study [11] considers sesame seeds as the source of the polyunsaturated fatty acid. The study [12] cites seaweed as a prospective source of the polyunsaturated fatty acids. The flax seeds are also high in polyunsaturated fatty acids [13, 14].

The enrichment with Omega-3 and Omega-6 can be done during lifetime by adding food supplements with high content of these fatty acids [15, 16].

Thus, the study on enriching the meat products with polyunsaturated fatty acids is relevant nowadays.

2 Materials and Methods

The purpose of this research is to study the prospective of cultivating the ‘Uralsky’ variety oil flax in the Smolensk Region for seeds both for further cultivation and for biologically active food supplements to porridge, yogurt, and other fermented milk products. These seeds can also be used as enriching raw material for chopped semi-finished poultry products (cutlets).

The research tasks are as follows:

- cultivating the test samples of the ‘Uralsky’ oil flax variety in the Smolensk Region;
- obtaining the fully ripe seeds of the oil flax for further use;
- comparing the organoleptic properties of the locally-updated flax and the original seeds of the ‘Uralsky’ flax variety;
- developing the technology of the enriched semi-finished products supplemented with the 5% of the grounded flax seeds and of the semi-finished products supplemented with the 5% of the flax oil;
- identifying the organoleptic properties of the developed semi-finished products; comparing them with the reference sample (general-purpose semi-finished product);
- identifying the fatty acids composition of the developed semi-finished products.

The study applies organoleptic evaluation methods based on GOST 31470-2012 to examine the developed semi-finished products. The four-point gauge of organoleptic properties has been developed to determine how well the sample meets them and to identify the best indicators. The best indicator corresponds to four points, the worst – to one point. The fatty acids composition of the semi-finished products was assessed with the 'Kristall 5000.2' gas chromatograph. The chromatograms were processed by marking the peaks and calculating their areas with the 'Chromatech Analytic 2.6' software. Data collection for this study was conducted based on the publicly available information in the domestic and foreign press presented by oilseeds researchers, oil flax in particular. The study on cultivation of the locally-updated oil flax and material collection was made on the trial field of the Smolensk State Agrarian Academy.

3 Results

The flax was sown in the beginning of April. The weeding and the herbicide treatment were done as required. In June, the good shoots bloomed. The Fig. 1 illustrates the blooming flax field. Apparently, the seeds sprouted well, and the flax rose straight.

Fig. 1. The blooming field of the locally-updated oil flax.

The flax was harvested in August as far as the seed boxes were ripe. One square meter yielded 677 plants. After harvesting the plants, the seeds were extracted, analyzed and compared with the original seeds of the oil flax. The Fig. 2 shows the flax seeds of the 'Uralsky' variety and the seeds of the oil flax harvested in the trial field. Apparently, the seeds harvested in the trial field indistinguishable from the original seeds.

The organoleptic examination of the seeds (taste, aroma) revealed no differences.

Fig. 2. Photo: a) Oil flax seeds of the 'Uralsky' variety; b) locally-updated oil flax seeds.

The harvested seeds have been used for producing the enriched semi-finished products. The enriched semi-finished products were supplemented with the 5% of the grounded flax seeds and with the 5% of the flax oil. The examination results of the organoleptic properties are listed in the Table 1. The regular indicators have been identified given the positive impact on the organoleptic samples profile.

Table 1. The organoleptic properties of the baked semi-finished products

Indicators	Reference sample	5% of 'Uralsky' variety flax seeds	5% of locally-updated flax seeds	5% of flax oil
Exterior	The surface is well-breaded, without cracks, smooth. The shape is regular 4			
Consistence	Loose 3	Not very dense 4	Not very dense 4	Loose 4
Cut appearance	Minced meat is homogeneous, highly porous 3	Homogeneous with incorporation of uniformly distributed grounded flax seeds 4	Homogeneous with incorporation of uniformly distributed grounded flax seeds 4	Minced meat is homogeneous, porous 4
Odor	Nice, peculiar to the thermally processed poultry 4			The smell of the flax oil is present 3

(*continued*)

Table 1. *(continued)*

Indicators	Reference sample	5% of 'Uralsky' variety flax seeds	5% of locally-updated flax seeds	5% of flax oil
Colour	The crust is straw-yellow, the internal part is homogeneous grey 4	The crust is straw-yellow, the internal part is homogeneous grey with incorporation of uniformly distributed grounded flax seeds 4	The crust is straw-yellow, the internal part is homogeneous grey with incorporation of uniformly distributed grounded flax seeds 4	The crust is straw-yellow, the internal part is homogeneous grey 4
Taste	Nice, peculiar to the thermally processed poultry 3	Nice, peculiar to the thermally processed poultry, the flax flavour is tasted when chewing, the aftertaste is pleasant 4	Nice, peculiar to the thermally processed poultry, the strong flax flavor is tasted when chewing, the aftertaste is pleasant 4	After the thermal processing, the bitter aftertaste is available when chewing 3

The obtained results are represented graphically on thc Fig. 3.

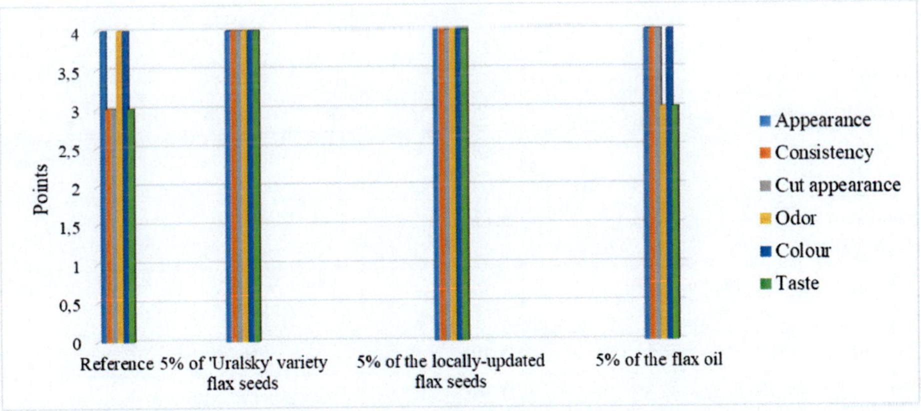

Fig. 3. The graphical representation of the organoleptic indicators of the semi-finished product samples.

The fatty acids composition of the semi-finished products has been identified. The results are represented in the Table 2.

Table 2. The fatty acids composition of the semi-finished product samples

The fatty acid name	The portion of the fatty acid, % of the total of the fatty acids		
	+5% flax oil	+5% 'Uralsky' variety oil flax	+5% locally-updated oil flax
Butyric C 4:0	0.02	0.02	0.28
Caproic C 6:0	-	-	0.03
Caprylic C 8:0	0.004	0.006	0.09
Capric C 10:0	0.03	0.04	0.02
Decenic C 10:1	0.005	0.005	0.3
Undecanoic C 11:0	-	-	0,07
Lauric C 12:0	0.05	0.06	0.15
Tridecanoic C 13:0	-	0.005	0.14
Myristic C 14:0	0.70	0.71	0.49
Myristoleic C 14:1	0.07	0.07	0.53
Pentadecanoic C 15:0	0.05	0.05	0.75
Pentadecenic C 15:1 (c10)	-	-	-
Palmitic C 16:0	20.63	20.78	22.38
Palmitoleic C 16:1	2.68	2.69	3.04
Heptadecanoic (margaric) C 17:0	0.19	0.19	0.86
Margarolic C 17:1	0,15	0,13	0.59
Stearic C 18:0	8.16	7.90	10.20
Oleic C 18:1 (c9)	32.50	32.21	30.23
Elaidic C 18:1 (t9)	1.84	1.77	1.88
Linoleic C 18:2 n6	23.57	28.56	22.98
Alpha Linoleic C 18:2 n3	0.11	0.13	0.77
Gamma Linoleic C 18:3 n6	0.08	0.11	0.47
Alpha Linoleic C 18:3 n3	7.98	3.06	1.63
Arachidic C 20:0	0.12	0.09	0.34
Eicosenic (gondoic) C 20:1 (c11)	0.35	0.36	0.55
Eicosaenoic C 21:0	0.32	0.40	0.65

(continued)

Table 2. (*continued*)

The fatty acid name	The portion of the fatty acid, % of the total of the fatty acids		
	+5% flax oil	+5% 'Uralsky' variety oil flax	+5% locally-updated oil flax
Eicosatrienoic C 20:3 (c8,11,14) n6	0.05	0.09	-
Eicosatrienoic C 20:3 (c11,14.17) n3	0.29	0.46	0.52
Arachidonic C 20:4	0.03	0.10	
Behenic C 22:0	0.03	-	0.35
Saturated	30.30	30.25	36.15
Monounsaturated	37.60	37.24	36.85
Polyunsaturated, incl.:	32.11	32.51	27.02
Omega-6	23.73	28.86	24.62
Omega-3	8.38	3.65	2.40

4 Conclusion

The study has achieved the set out goal and solved the research tasks. The test sample seeds of the locally-updated oil flax have been grown and harvested. It was revealed that the number of plants harvested from 1 m^2 amounts to 677. The seeds have a good germination. Taking into account the high enough sowing density (the optimal seeds number for sowing is 500–700), no plant lodging was observed. Therefore, it can be noted, that the agro-technical soil properties and the climate are good enough for cultivating the locally-updated flax.

The appearance analysis of the original flax seeds of the 'Oleiferous' variety and the seeds derived from the locally-updated samples revealed no differences. The organoleptic evaluation also showed no differences.

The prototypes of the semi-finished poultry products supplemented with the 5% of the flax seeds of the 'Uralsky' variety, with the 5% of the locally-updated flax seeds, and with the 5% of the flax oil have been developed. The organoleptic assessment has not detected differences in the samples with the flax seeds. The non-enriched reference sample has been developed for comparisons. It has been found out that the developed seeds-supplemented semi-finished products have higher average scores in consistency, appearance, and taste. These semi-finished products have a more fine texture, while being more succulent, than the others. Their cut surface had a lesser looseness. The taste of the samples supplemented with the flax seeds was more delicious and vigorous. The addition of the flax oil to the samples caused deterioration in both taste and smell.

The fatty acids composition analysis of the semi-finished products has shown that the samples supplemented with the seeds of the locally-updated flax have more saturated acids (36.15), than the other samples. However, the content of the polyunsaturated fatty acids in these samples was less than in the samples supplemented with the flax oil or with the oil flax – 27.02; 32.51 and 32.11 respectively. The content of the polyunsaturated fatty acids in the samples supplemented with the seeds of the 'Uralsky' flax variety and with the flax oil was approximately the same. The samples supplemented with the seeds of the 'Uralsky' flax variety have more Omega-6 (28.86), than the samples supplemented with the seeds of the locally-updated flax (24.62) or the samples supplemented with the flax oil (8.36). The samples supplemented with the flax oil scored the highest in Omega-3 content (8.36), while the samples supplemented with the seeds of the locally-updated flax scored the lowest (2.4) in this category.

As a result, it can be stated that supplementing the semi-finished products with the flax seeds enriches their organoleptic properties. The results obtained in this study revealed the possibility of using the locally-updated oil flax for producing the enriched semi-finished products. The study involving the fertilizer application in order to obtain the locally-updated flax with a more balanced fatty acids composition is planned to continue.

References

1. Abuajah, C.I., Ogbonna, A.C., Osuji, C.M.: Functional components and medicinal properties of food: a review. J. Food Sci. Technol. **52**(5), 2522–2529 (2014). https://doi.org/10.1007/s13197-014-1396-5
2. Parkhurst, C.R., Mountney, C.J.: Poultry Meat and Egg Production, Springer, Boston (1988)
3. Borisova, V.L., Terentyev, S.E., Stefanova, I.L., Sazonova, E.A., Kramlikh, O.Y.: Effective use of meat of meat-and-egg chicken and eggs for the production of specialized semi-finished products. In: IOP Conference Series: Earth and Environmental Science (2021). https://doi.org/10.1088/1755-1315/699/1/012022
4. Hegde, M.V., Zanwar, A.A., Ghorpade, P.B.: Flax bio-village concept. In: Hegde, M.V., Zanwar, A.A., Adekar, S.P. (eds.) Omega-3 Fatty Acids Keys to Nutritional Health, pp. 11–20. Springer, Cham (2016). https://doi.org/10.1007/978-3-319-40458-5_2
5. Tur, J.A., Bibiloni, M.M., Sureda, A., Pons, A.: Dietary sources of omega 3 fatty acids: public health risks and benefits. Br. J. Nutr. **107**(2), S23-52 (2012)
6. Simopoulos, A.P.: The importance of the ratio of omega-6/omega-3 essential fatty acids. Biomed. Pharmacother. **56**(8), 365–379 (2002)
7. Calder, P.C.: n-3 polyunsaturated fatty acids, inflammation, and inflammatory diseases. Am. J. Clin. Nutr. **83**(6), 1505S-1519S (2006)
8. Pan, A., et al.: α-Linolenic acid and risk of cardiovascular disease: a systematic review and meta-analysis. Am. J. Clin. Nutr. **96**(6), 1262–1273 (2012)
9. Strobel, C., Jahreis, G., Kuhnt, K.: Survey of n-3 and n-6 polyunsaturated fatty acids in fish and fish products. Lipids Health Dis. **30**(11), 144 (2012)
10. Bosch, A.C., O'Neill, B., Sigge, G.O., Kerwath, S.E., Hoffman, L.C.: Heavy metals in marine fish meat and consumer health: a review. J. Sci. Food Agric. **96**(1), 32–48 (2016)
11. Borisova, V.L., Stefanova, I.L., Terentyev, S.E., Sazonova, E.A., Kulikova, M.G.: The use of sesame seeds in the enrichment of semi-finished products from poultry meat. AIP Conf. Proc. **2419,** 030002 (2021). https://doi.org/10.1063/5.0069105
12. Adarme-Vega, T.C., Lim, D.K.Y., Timmins, M., Vernen, F., Li, Y., Schenk, P.M.: Microalgal biofactories: a promising approach towards sustainable omega-3 fatty acid production. Microb. Cell. Fact. **11**, 96 (2012)
13. Singh, K.K., Mridula, D., Rehal, J., Barnwal, P.: Flaxseed: a potential source of food, feed and fiber. Crit. Rev. Food Sci. Nutr. **51**(3), 210–222 (2011)

14. Rodriguez-Leyva, D., Bassett, C.M.C., McCullough, R., Pierce, G.N.: The cardiovascular effects of flaxseed and its omega-3 fatty acid, alpha-linolenic acid. Can. J. Cardiol. **26**(9), 489–496 (2010)
15. Bhalerao, S.S., et al.: Studies in production of omega-3 chicken meat I. Ind. J. Poult. Sci. **45**(3), 273–279 (2012)
16. Bhalerao, S.S., Hegde, M.V., Katyare, S.S., Kadam, S.S.: Promotion of omega-3 chicken meat production: an Indian perspective. World's Poult. Sci. J. **70**, 365–373 (2014)

Investigation of the Possibility of Using Chondroitin Sulfate in Milk Raw Materials by the Organoleptic Method

Alla Novokshanova(✉) and Dmitriy Nikityuk

Federal Research Centre of Nutrition, Biotechnology and Food Safety, Moscow, Russia

Abstract. This work aims to establish the possibility of using chondroitin sulfate (CS) in milk raw materials by an organoleptic method. The mass fractions of protein, fat, lactose, dry substances, and active and titratable acidity were determined in milk raw materials. Sensory assessment was performed by a group of qualified experts on a five-point scale. To assess the reliability, the ranking method was used to determine the weight coefficients of organoleptic indicators (taste, smell, colour and consistency). In experimental samples of dairy raw materials (skim milk, buttermilk, and curd whey), CS was added with constant stirring from 0.05 to 0.50% with a step of 0.05%. Control samples of dairy raw materials did not contain CS. All samples were pasteurized at a temperature of (90 ± 2) °C, cooled to a temperature of (20 ± 2) °C, and organoleptic analysis was performed. The quality of sensory assessment was monitored by calculating Kendall's concordance ratios. Concordance coefficients in the assessment of skim milk, buttermilk and curd whey equal to 0.840, 0.936 and 0.904, respectively which showed high consistency in the opinions of the tasters' opinions. All samples with CS, regardless of its amount, had good organoleptic characteristics. The data suggest that the creation of functional products with a chondroprotective profile based on dairy raw materials is promising.

Keywords: Skim milk · Buttermilk · Curd whey · Chondroitin sulfate · Organoleptic characteristics

1 Introduction

WHO statistics show that approximately 80% of the population suffers from various diseases of the musculoskeletal system. Diseases of the musculoskeletal system are diagnosed in both elderly and young people, in most cases in the working age range of 30–50 years, determine the global need for rehabilitation services, and often lead to disability of the population. Despite different approaches to the treatment and regeneration of joint tissue, there are currently no effective solutions to this problem [6].

The adverse consequences of diseases of the musculoskeletal system for society are enormous, in terms of both direct health care costs, and indirect costs (absenteeism, reduced productivity, etc.) [14].

V. Kurchenko et al. (Eds.): ICAETT 2021, LNNS 408, pp. 200–206, 2022.
https://doi.org/10.1007/978-3-030-96641-6_24

The factors that contribut to the development of diseases of the musculoskeletal system, are physical exertion, overweight, and lack of traditional sources of chondroprotective compounds in the diet.

For example, in a study of the actual diets of people with increased physical activity using the questionnaire-weight method, most of the subjects did not have sufficient traditional nutritional sources of chondroprotective substances to satisfy their physiological needs [7].

Scientific, technical and patent literatures, show that the number of food products of this type is very limited. Usually, products to prevent joint diseases are enriched with collagen or its hydrolysates [3–5]. However, in such cases, the protein part of the product is somewhat depleted, since collagens do not contain all essential amino acids. Another option for chondroprotective products is tablet and encapsulated forms, which are dietary supplements to food, instead of traditional food products.

A successful example of the introduction of chondroprotectors into food products is described in the development of diets and nutrition programs for athletes in artistic gymnastics of Russian national teams [9]. In this study, a combination of CS and glucosamine sulfate was used as a chondroprotective additive. The rations developed in this case can be used in special center for athletes, where there is a full-time dietitian and conditions for preparing food. For a wide range of athletes, people involved in physical culture, or those whose activities are associated with prolonged static and mechanical overload of the joints, specialized food products with chondroprotective properties are not available.

The duration of drug intake is also of great importance in the prevention and maintenance therapy of diseases of the musculoskeletal system. Therefore, food products with chondroprotective action should be acceptable to the majority of the population. Undoubtedly, such products include dairy products, which are present in the diets of the population of all age groups.

In addition to substantiating the relevance of using functional food ingredients, the aspect of their safety in finished products is of great importance. The possible interaction of chondroprotectors with the food matrix, the impact of various technological operations, or both of these factors can simultaneously significantly affect the final qualitative and quantitative composition of the ingredients. In a study by Uzzan N., Nechrebeki J., Labuza T. P. [13], the kinetics of degradation of various nutrient ingredients were determined using high performance liquid chromatography.

The results showed very good stability of CS in a dairy environment, while glucosamine proved to be of limited stability during any processing or storage. Glucosamine also destabilizes the milk protein system at or above the boiling point, causing precipitation to precipitate in heat exchangers [13].

Based on the foregoing, the study used various raw milk and CS.

2 Materials and Methods

The subjects of the study were industrially obtained samples of skim milk, buttermilk and milk whey. This raw material was selected due to its low fat content, and to further maximize its use for food purposes. Experimental samples contained different amounts of CS in milk raw materials. The corresponding combinations without CS served as control samples. All studies were performed in triplicate.

2.1 A Subsection Sample

The purpose of the work is to establish the possibility of using CS in milk raw materials by the organoleptic method.

The research objectives included:

- preparation of prototypes with different amounts of CS;
- performance of a sensory evaluation of the control and prototypes;
- analysis of experimental data.

The amount of CS introduced was calculated considering the standards of uniform sanitary-epidemiological and hygienic requirements that are in force in the customs territory of the Eurasian Economic Union. This document regulates the maximum daily consumption of CS 1.2 g and an adequate level of consumption –0.6 g [12]. If the content of a certain ingredient in a product is at least 15% of the adequate consumption level, with regular use, that product will exhibit functional properties for that ingredient. Therefore, in the case of CS, this amount is 0.09 g.

Thus, at this stage, we analyse samples of dairy raw materials – skim milk, buttermilk and curd whey with a mass fraction of CS from 0.05% to 0.50% and a step of 0.05%.

The mass fractions of protein, fat, lactose, and solids in milk raw materials were determined by the instrumental expression method using a MilkoScan FT 120 infrared analyser.

The active acidity was determined by the potentiometric method on a pH metre. The titratable acidity was determined by the indicator titration method based on the determination of the equivalence point in the presence of the phenolphthalein indicator. The unit of measurement of titratable acidity was Turner's degree (°T).

CS was added to the dairy raw materials with constant stirring. Then the samples were pasteurized at a temperature of (90 ± 2) °C and cooled to a temperature of (20 ± 2) °C, and organoleptic analysis was performed.

For the sensory analysis of the samples, a commission of five qualified experts was formed. Each expert analysed the samples independently of the other experts.

To assess the reliability, the ranking method was used to determine the weight coefficients, otherwise referred to as significance indicators, which were assessed by experts depending on their importance. The least important indicator received 1 point. The next, least important indicator is 2 points, etc.

The results of the examination of the encrypted samples were recorded for taste, smell, colour and consistency, and then summarized in the form of a final score in points in the tasting card. In this case, the quality of the indicator was assessed at 5 points if there were no deviations from the predetermined requirements for organoleptic properties. With minimal deviations, the indicator was estimated at 4 points. In the case of noticeable deviations, the result was assessed from 3 to 2 points, in the case of very significant deviations, it was 1 point or less.

To assess the consistency of expert opinions, Kendall's concordance coefficients were calculated:

$$W = \frac{12 \cdot S}{m^2(n^3 - n)}, \tag{1}$$

where W is the coefficient of concordance; S is the sum of squares of deviations (d2) from the mean (d); n is the number of analysed indicators; m is the number of tasters.

3 Results

The quality and physicochemical parameters of skim milk, buttermilk and milk whey satisfied the requirements for this raw milk in Russia (Table 1).

Table 1. Physical and chemical indicators of milk byproducts

Dairy raw materials	Mass fraction, %			
	Fat	Protein	Lactose	Dry matter
Skimmed milk	0.05 ± 0.01	3.36 ± 0.02	4.73 ± 0.12	9.08 ± 0.52
Buttermilk	0.42 ± 0.04	3.09 ± 0.01	4.31 ± 0.26	8.09 ± 0.21
Curd whey	0.05 ± 0.02	0.46 ± 0.04	4.10 ± 0.02	5.72 ± 0.04

The CS addition 0.0 to 0.5% of the mixture volume did not affect the acidity indices of the studied milk raw materials. The titratable acidity of both control and experimental skim milk samples was 15.0–17.0 °T, that of buttermilk was 14.0–16.0 °T, and that of whey was 64.0–67.0 °T. The active acidity of both control and experimental samples of skim milk and buttermilk was 6.31–6.39 °T, and that of whey was 5.81–5.87 °T.

By analysing the significance of the factors, we found that the distribution of organoleptic indicators and weight coefficients for all types of milk raw materials with CS as shown in Table 2 taste was the most significant, since its weight coefficients are the greatest (Table 2). The least significant indicator was colour.

Since there is an element of subjectivity in the organoleptic examination [14], to assess the consistency of tasters' opinions, the concordance coefficient was calculated, which can vary in the range of 0.0–1.0; 0.0 corresponds to complete inconsistency, and 1.0 – corresponds to full agreement of the experts. The quality of the assessment is considered satisfactory if the value of the concordance coefficient exceeds 0.4–0.5 and high if the value is 0.7–0.8. The calculations showed good agreement of tasters' opinions in evaluating skim milk, buttermilk and curd whey, since the concordance coefficients were 0.840, 0.936 and 0.904, respectively

The final assessment of the samples in points was calculated considering the contribution of each indicator, as the sum of the product of the assessment of the quality indicator by the corresponding weight coefficient. Considering the coefficients of the significance of individual criteria, the estimates of raw material samples depending on the CS amount and type of raw material, are presented in Table 3.

Table 2. The location of factors in terms of significance in the assessment of samples with CS

Indicators	Rango amount			Weight coefficients (g_i)		
	1	2	3	1	2	3
Color	6	5	5	012	0.10	0.10
Consistency	9	10	10	0.18	0.20	0.20
Smell	17	16	17	0.34	0.32	0.34
Taste	18	19	18	0.36	0.38	0.36

1 – Skimmed milk, 2 – Buttermilk, 3 – Curd whey

Table 3. Summary table of organoleptic indicators of samples with CS, taking into account the coefficients of significance and the boundaries of the absolute error, points ($\pm\Delta$)

Mass fraction CS, %	Skimmed milk	Buttermilk	Curd whey
0.05	4.98 ± 0.02	4.99 ± 0.01	4.99 ± 0.01
0.10	4.97 ± 0.02	4.97 ± 0.02	4.98 ± 0.02
0.15	4.90 ± 0.05	4.96 ± 0.04	4.97 ± 0.03
0.20	4.86 ± 0.06	4.90 ± 0.04	4.88 ± 0.04
0.25	4.79 ± 0.06	4.85 ± 0.06	4.79 ± 0.04
0.30	4.72 ± 0.05	4.81 ± 0.08	4.72 ± 0.05
0.35	4.63 ± 0.05	4.74 ± 0.07	4.62 ± 0.05
0.40	4.54 ± 0.06	4.68 ± 0.08	4.55 ± 0.05
0.45	4.47 ± 0.05	4.58 ± 0.09	4.49 ± 0.08
0.50	4.35 ± 0.14	4.46 ± 0.10	4.40 ± 0.10

All samples with CS, regardless of its amount, had good organoleptic characteristics. The colour and consistency were identical to those in the control samples. A decrease in severity of milk taste, and the appearance of a low-intensity mineral taste, odour and taste of medicines were observed only in some samples containing more than 0.3% CS.

4 Discussion and Conclusion

The medical and biological substantiation of the composition of products with chondroprotective properties is based on studies of the tissue of the articular cartilage. Cartilage tissue is formed by chondrocytes and an extracellular matrix, which consists of collagen and proteoglycans: CS and hyaluronic acid [1, 2, 8, 15, 16].

Although CS is a high-molecular-weight compound, the intestinal barrier is not absolutely insurmountable for it and many biopolymers, such as chicken egg albumin or milk immunoglobulins. More than 70% of cholesterol is absorbed in the digestive tract. The positive effect of chondroprotectors on the functional activity of the musculoskeletal system has been proven in clinical studies. Therefore, the use of CS is allowed in the territory of the countries of the EurAsEC Customs Union, and as part of specialized food products for athletes, the recommended level of consumption of CS for adults is 600 mg/day [12].

When creating new food products, especially those containing specific ingredients, it is first advisable, to proceed from organoleptic characteristics.

The sensory evaluation data of samples with CS in Table 3 indicate the possibility of using this ingredient in milk raw materials even in an amount exceeding the adequate level of consumption, and suggest that the creation of functional products of a chondroprotective profile based on milk raw materials has great prospects.

Acknowledgments. The research was conducted with the financial support of the Ministry of Science and Higher Education of the Russian Federation (research project No. 0410-2020-0017).

The author expresses gratitude to Anna Aleksandrovna Ababkova, Candidate of Technical Sciences, for her assistance in the preparation of prototypes.

References

1. Antiushko, D.P., Nezdoliy, A.O.: The biological activity of the composition with glucosamine in the diet of rats in the modeling of cartilage and connective tissue damage. Voprosy pitaniia [Probl. Nutr.] **86**(1), 72–75 (2017)
2. Arianova, E.A., Bogachuk, M.N., Perederyaev, O.I.: Determination of chondroitin sulfate in food supplements by capillary zone electrophoresis. Voprosy pitaniia [Probl. Nutr.] **82**(3), 67–71 (2013)
3. Flechner, K., Alf, D.: Results of a postmarketing surveillance study of collagen hydrolysate CH-Alpha. Ortop. Prax. **41**(2), 486–494 (2005)
4. Fernandez, J.L., Peres, O.M.: Effects of gelatine hydrolysates in the prevention of athletic injuries. Arch. Med. Deporte **15**, 277–282 (1998)
5. Hercberg, S.: Development of a nutritional policy designed to meet public health challenges. Sante Publique **26**(3), 281–286 (2014)
6. Moura, C., et al.: A brief review on processes for cartilage repair. Appl. Mech. Mater. **890**, 229–236 (2019)
7. Novokshanova, A.L.: About the classification of specialized sports nutrition products. IOP Conf. Ser. Earth. Environ. Sci. **677**, 032054 (2021)
8. Salazar, J., et al.: Glucosamine for osteoarthritis: biological effects, clinical efficacy, and safety on glucose metabolism. Artritis. **2014**, 432–463 (2014)
9. Tokayev, E.S., et al.: Development of diets and nutrition programs for athletes. Meat. Technol. **6**, 6–9 (2010). https://www.elibrary.ru/download/elibrary_26596455_49219491.pdf
10. Towhed, T.E., et al.: Glucosamine therapy for treating osteoarthritis. Cocchrane. Database. Syst. Rev. **2**, CD002946 (2005)
11. Towhed, T.E., Anastassiades, T.: Glucosamine therapy for osteoarthritis: an update. J. Rheumatol. **34**(9), 1787–1790 (2007)
12. Uniform sanitary and epidemiological and hygienic requirements for products (goods) subject to sanitary and epidemiological supervision (control). Approved by the decision of the Customs Union Commission No. 299 dated May 28, 2010. (as amended on November 10, 2015)
13. Uzan, N., Nechrebeki, J., Labuza, T.P.: Thermal and storage stability of nutraceuticals in milk deverage dietary supplement. J. Food. Sci. **72**, E109–E114 (2007)

14. Yakuba, Y., Khalaphyan, A.A., Temerdashev, Z.A., Bessonov, V.V., Malinkin, A.D.: Flavouring estimation of quality of grape wines with use of methods of mathematical statistics. Voprosy pitaniia [Probl. Nutr.] **85**(5), 93–99 (2016)
15. Yakupov, R.R., Karimova, L.K., Nafikov, R.G.: Dystrophic changes in the locomotor system of manual laborers. Hyg. Sanit. **4**, 55–58 (2010)
16. Yarema, N.M.: ω-3 polyunsaturated fatt yacids use for optimization of children inflammatory joints diseases treatment. Voprosy pitaniia [Probl. Nutr.] **86**(2), 84–90 (2017)

Properties of Cyclodextrins Nanocomplexes with Peptides and Fat-Soluble Vitamins

Tatsiana Halavach[1], Natalia Sushiskaya[1], LudmilaAlieva[2], Svetlana Lodygina[2], and Aleksey Lodygin[2](✉)

[1] Belarussian State University, Minsk, Republic of Belarus
[2] North Caucasus Federal University, Stavropol, Russia

Abstract. Beta-cyclodextrin (β-CD) nanocomplexes with fat-soluble vitamins D_3, A and peptides of whey protein hydrolysate have been developed for functional foods design. Peptides of whey protein hydrolysate with molecular weights of 150–6000 Da (WPHP) were obtained. They have hypoallergenic, antioxidant, genoprotective and other functional properties. They were incorporated into β-CD using the coprecipitation method to reduce their bitter taste. The taste of peptides decreased to moderately bitter in the resulting complex of WPHP:β-CD. Fat-soluble vitamins inclusion complexes D_3:β-CD and A:β-CD allowed them to be converted from a solution in olive oil into a powdered form. Multicomponent composition (MCC) on theirbase was developed, 100 g of which contained 30 g of WPHP, 1.06 mg of vitamin D_3 (42500 IU), 86.0 mg of vitamin A (250,000 IU) and 4 g of olive oil. It is shown that the antioxidant activity of the obtained samples of inclusion complexes decreases in the series: D_3:β-CD > WPHP:β-CD > MCC > β-CD. It was shown that WPHP:β-CD and MCC exhibit an antimutagenic effect, preventing base pair replacement mutations by 20% and a reading frame shift by 15% in the Ames test on Salmonella typhimurium TA 98 and Salmonella typhimurium TA 100 strains on the model of induced mutagenesis. The resulting powdered forms of fat-soluble vitamins and peptides are easily dosed and can be used in the development of various functional foods.

Keywords: Cyclodextrins · Vitamin A · Vitamin D_3 · Peptides · Inclusion complexes · Functional foods

1 Introduction

The current state of human health is largely determined by the nature and structure of nutrition. Violation of nutrition structure is the main factor causing irreparable damage to our health [1, 4, 21, 25]. The WHO Regional Committee for Europe has proposed an action plan in the field of food to expand the production of specially created food, or, in other words, wider use of functional foods [24].

In this regard, functional foods (FF) are food products that, through the addition or elimination of certain food ingredients, are changed in such a way that they have

V. Kurchenko et al. (Eds.): ICAETT 2021, LNNS 408, pp. 207–220, 2022.
https://doi.org/10.1007/978-3-030-96641-6_25

a regulating effect on physiological functions, biochemical reactions of a person, contribute to reducing the risk of any disease and have a superior effect on human health in comparison with traditional foods [1, 18, 24].

The issues of food manufacturing for children, athletes and the elderly people occupy an important place among the FF. This is due to the fact that up to 10% of young children suffer from food allergies (FA). It is caused by the development of sensitization of the patient's body to food allergens [2, 9, 10]. The clinical manifestation of food allergy is most often atopic dermatitis, a chronic allergic inflammation of the skin. The main cause of the disease is impaired functioning of the immune system [9]. Allergen-specific IgG and IgE were found in young children with clinical signs of FA not only to cow's milk proteins, but also to the most common food antigens of animal and vegetable origin: poultry and fish meat, fruits, fermented milks and gluten-containing products [2, 10].

Ingredients of natural products are modified for reduction their allergenicity in such a way that they begin to show hypoallergenic physiological activity. Whey proteins are widely used in baby food.

The composition of whey proteins includes β-lactoglobulin (β-lg), α-lactoalbumin (α-la) and bovine serum albumin (BSA) which accounts for 53.5%, 21.1%, 6.2% respectively. All these proteins are allergens [9, 10]. The greatest allergenic potential has β-lh, which is not hydrolyzed by pepsin. Its structure contains 7 antigenic determinants that cause the synthesis of allergen-specific IgE (see Fig. 1). As can be seen from Fig. 1, β-lh contains linear antigenic determinants. It is necessary to carry out enzymatic hydrolysis to reduce its allergenic potential, which will result in peptides that do not contain amino acid sequences, form antigenic determinants and can cause the formation of allergen-specific IgG and IgE. Hydrolysates are divided into partial and deep depending on the depth of proteolysis.

Partial hydrolysates with an intermediate degree of hydrolysis contain peptides of various lengths and a minimum amount of free amino acids. Deep hydrolysates are represented by short-chain peptides with a molecular weight of 3–5 kDa or less. An increase in the degree of protein hydrolysis leads to a decrease in their allergenic properties [8, 9]. It should be noted that the products of whey protein hydrolysis have antioxidant, antimicrobial and antimutagenic effects [8, 9, 20].

Low molecular weight peptides obtained by deep enzymatic hydrolysis of whey proteins have a bitter taste, which makes it difficult to use them for creating for baby food formula development [9]. It is required to design multicomponent compositions, which contain both hydrophilic and hydrophobic compounds in the framework of FF development. One of the ways to solve this problem is the application of cyclodextrin inclusion complexes with peptides of whey protein hydrolysate, fat-soluble vitamins and other ingredients [10, 16, 27].

Natural products intended for child nutrition are additionally enriched with any functional ingredients, in particular fat-soluble vitamins. However, their practical use for creating multicomponent foods development is limited due to their high hydrophobicity and poor solubility in water [6, 15, 23, 26].

Cyclodextrins (CD) are obtained by enzymatic hydrolysis of starch. Depending on the type of CD, they contain from 6 to 8 glucose residues, which form the tor [3, 7, 12, 17, 22, 26]. All OH groups in the CD are located on the outer hydrophilic surface of

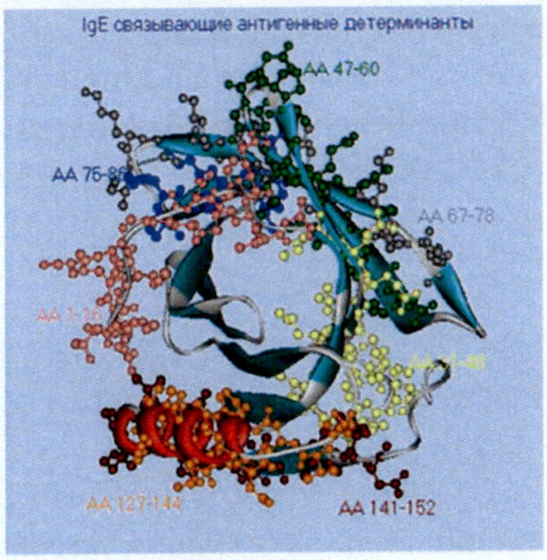

A

MKCLLLALALTCGAQALIVTQTMKGLDIQKVAGTWYSLAMAASDISLL
DAQSAPLRVYVEELKPTPEGDLEILLQKWENGECAQKKIIAEKTKIPAVF
KIDALNENKVLVLDTDYKKYLLFCMENSA
EPEQSLACQCLVRTPEVDDEALEKFDKALKALPMHIRLSFNPTQLEEQC
HI

B

Fig. 1. Tertiary (A) and primary (B) structure of β-lg with indication of antigenic determinants [9]

the molecule, and the inner cavity is hydrophobic. Hydrophobic substances are able to integrate into this cavity in aqueous solutions, forming inclusion complexes – clathrates. Such molecular containers are able to hold molecules of nonpolar substances in the inner cavity, and due to the hydrophilic outer surface, they give them greater solubility, stability, change their taste, color and smell [7, 22, 26]. CD belong to the 5th class of toxicity and are "non-toxic compounds". They are recommended as a food additive (E459) [19]. The stability and bioavailability of CD clathrates with peptides and fat-soluble vitamins plays an important role in the manufacturing of specialized foods for child nutrition.

Fat-soluble vitamins and other hydrophobic substances are able to integrate into CD internal cavity, forming powdery forms [12]. Due to the complexation of CD with fat-soluble vitamins and other hydrophobic substances, they acquire greater solubility, become stable in storage processes [7, 22, 26]. WPHP, which have hypoallergenic properties, are widely used in the dietetics of baby food [2, 8, 9, 13]. At the same time, the peptides included in the protein hydrolysate have a bitter taste. They can be included in the CD structure to reduce this taste. Peptide clathrates are included in formulas that also contain various vitamins and minerals [12, 15, 17, 26]. Vitamins D_3 and A must be converted into powdered form by complexation with CD for inclusion in MCC. MMC suitable for specialized foods can be designed on the basis of powdered complexes of inclusion of peptides, fat-soluble vitamins and hydrophilic macro- and micronutrients in CD.

The aim of the study was to develop a method for obtaining of hypoallergenic peptides from enzymatic whey proteins hydrolysates and their inclusion in beta-CD, as well as obtaining of beta-CD nanocomplexes with fat-soluble vitamins D_3 and A.

2 Materials and Methods

The objects of the study were beta-CD nanocomplexes with a fraction of whey proteins enzymatic hydrolysate peptides, fat-soluble vitamins D_3, A and their multicomponent mixtures.

Vitamin D_3 containing 0.425 mg (corresponds to 17,000 IU of cholecalciferol) in 1 ml of vegetable oil (Farmland, Belarus); vitamin A containing 34.4 mg (corresponds to 100,000 IU of retinol acetate) in 1 ml of vegetable oil (Technologist, Ukraine); beta-cyclodextrin ("Sigma", USA, CAS 7585-39-9, E459); whey protein concentrate (JSC "Shchuchinsky cheese and butter making plant", Belarus); alkalase (protease from *Bacillus licheniformis*, activity 2.64 E per g, "Sigma", USA).

Preparation of whey proteins enzymatic hydrolysate and determination of the composition of peptides. Enzymatic hydrolysis of whey proteins was carried out by alkalase at a protein concentration of 2%, temperature of 50 °C and pH 8.0 during 2 h. Whey protein hydrolysate was filtered using Spin X UF Concentrator 20 filters (Corning, England) with a throughput of 5 kDa to obtain a low molecular weight fraction of peptides [9].

The composition of WPHP was analysed using the Agilent 1290 chromatography-mass spectrometric system (Agilent, USA) with a high-resolution Q-TOF 6550 mass spectrometric detector in the mode of positive electrospray ionization (ESI+). HPLC analysis was performed using a Hypersil Gold column (100 × 2.1 mm, 1.9 μm, Agilent, USA). The column was balanced with 0.1% aqueous formic acid solution. The separation of the samples was carried out using a linear gradient of acetonitrile 5–95% for 55 min at a temperature of 45 °C; the flow rate of the mobile phase was 200 μl per min; the sample volume was 15 μl; detection was carried out at 230 and 280 nm. Parameters of the ionization source: gas temperature –290 °C; gas flow –12 l per min; shell gas temperature –325 °C; shell gas flow –9 l per min. The voltage on the fragmenter was set to 150 V; the spectrum recording range was 100–3200 m/z (mass-to-charge ratio).

Preparation of β-cyclodextrin inclusion complexes with vitamins D_3, A and a fraction of peptides of milk whey protein hydrolysate. The method of coprecipitation was used to obtain complexes of inclusion of vitamins D_3, A and WPHP with β-CD. Beta-CD manufactured by Roquette (France) was used to obtain clathrates with WPHP. Solutions containing β-CD and hydrolysates were prepared in a mass ratio of 2:1 (based on the dry matter content). The obtained solutions of cyclic oligosaccharide and peptides were incubated for 4 h at a temperature of 50 °C under constant stirring conditions (200 rpm). A similar experiment was carried out at an incubation temperature of 22 °C. Clathrate samples were freeze-dried at a temperature of minus 53 °C, a pressure of 0.1 atm for 24–48 h aor subsequent analysis.

The method of coprecipitation was used to obtain β-CD complexes with vitamins D_3 and A. Solution of β-CD was prepared, which was mixed with a solution of vitamins D_3 or A in olive oil in a ratio of 2:1. The resulting mixture was incubated at 25 °C with stirring and subsequent cooling to 4 °C. The resulting clathrates were separated by centrifugation for 10 min at 3000 rpm and dried by lyophilisation [8, 11].

Analytical Methods

Evaluation of organoleptic properties of liquid samples of WPHP:β-CD was performed according to the procedure described in [11].

The vitamin content in the obtained complexes D_3-CD and A-CD was determined using the gravimetric method. Powder of complexes weighing 2 g was washed with 4 ml of hexane for the elution of vitamins and fatty acids included in the clathrate. The supernatant was separated by centrifugation. The extraction procedure was repeated 3 times. The β-CD precipitate was dried and its mass was determined. The difference in the mass of the obtained β-CD and D_3-CD or A-CD complexes was used to determine the content of vitamins in the inclusion complex. The obtained β-CD complex with hexane was used as a control [11, 22].

Thermogravimetric method was used to analyse the formation of D_3-CD, A-CD and WPHP-CD complexes and their stability during thermal decomposition. The measurements were carried out using the thermal analysis system TA-4000 "Mettler Toledo" Switzerland. The weight of the test sample was 20.0 mg. Temperature programming was used in the range of 25–550 °C, the rate of temperature rise was 5 °C per min. The analysis time is 110 min [20]. The calculation of the effective activation energy (Ea) was carried out according to the Broido method.

Determination of Antioxidant Activity of Inclusion Complexes

The ORAC (Oxygen Radical Absorbance Capacity) fluorimetric method was used to assess the antioxidant activity (AOA) of samples: D_3:CD, A:β-CD and WPHP:β-CD [23]. The method is based on measuring of the decrease in fluorescence intensity of fluorescein (FL) during its interaction with oxygen radicals. Antioxidants in the reaction medium, interacting with oxygen-containing radicals, slow down the free radical oxidation of FL. The AOA of D_3:CD, A:β-CD and WPHP:β-CD complexes was determined by their ability to bind free radicals formed in the Fenton system. Fluorescence measurements were carried out on an RF-5301 PC fluorimeter (Shimadzu, Japan). The fluorescence intensity was recorded at a wavelength of 514 nm. The excitation of wavelength was equals to 490 nm.

The calculation of AOA values was carried out according to the degree of fluorescence intensity (A, %), calculated by the formula:

$$A = \frac{FI}{FI_0} \times 100$$

where Fl_0 is the fluorescence intensity of the control sample of FL (a solution of FL without Fe^{2+}, EDTA, hydrolysate and H_2O_2), Fl is the fluorescence intensity of the solution after the addition of an antioxidant.

Graphs of the dependence of the fluorescence intensity (A, %) on the dry matter content in the analysed samples of D_3:CD, A:β-CD and WPHP:β-CD were plotted. According to the obtained equation, the concentration of the IC_{50} sample corresponding to 50% fluorescence inhibition was calculated. Plotting and mathematical processing of research results were carried out using the computer program "Microsoft Office Excel"

(Microsoft Corporation, USA). The results of independent experiments are presented as the arithmetic mean ± confidence interval. The reliability of the differences between the data samples was determined by the method of confidence intervals.

Determination of Antimutagenic Properties of Inclusion Complexes

Histidine auxotrophic strains of Salmonella typhimurium TA100 and TA98 were used as test objects for determining the antimutagenic action of complexes: D_3:CD, A:β-CD and WPHP:β-CD. The presence of the antimutagenic effect of the studied samples was taken into account by reducing the frequency of induced reverse mutations by standard mutagens, according to the previously described method [30]. Sodium azide at a concentration of 10 mcg per cup was used as standard mutagen for S. typhimurium TA 100 strain, and 2–nitrofluorene at a concentration of 10 mcg per cup was used for S. typhimurium TA 98 strain. Various concentrations of inclusion complexes D_3:CD, A:β-CD and WPHP:β-CD were introduced into the work solution. Visual recording of the results was carried out after 72–120 h, registering the number of positive wells [5].

Statistical processing of the results was carried out using generally accepted methods of variational statistics of the Data Analysis package of the Microsoft Office Excel program (Microsoft Corporation).

3 Results

Whey protein hydrolysate intended for functional foods was obtained, from which a fraction of low molecular weight peptides WPHP was isolated by filtration. WPHP was used to obtain a nanocomplex with β-CD. Beta-CD nanocomplexes with fat-soluble vitamins A and D_3 were also prepred for inclusion in composites.

3.1 Characteristics of Whey Proteins Hydrolysate

Enzymatic hydrolysis of whey proteins was carried out using alkalase [9–11]. The protein-peptide composition of WPHP was studied using HPLC-MS.

According to the data of HPLC-MS profiles, which are shown in Fig. 2, WPHP are eluted from the chromatographic column from 1 to 4 min. The retention time of native whey proteins is from 20 to 24 min. According to mass spectroscopy data (see Fig. 3) peptides with a molecular weight of 150–1500 Da have been identified in WPHP.

High molecular weight peptides contain 13–14 amino acid residues. A high signal level is set for single-charge ions with m per z values of 380–900 Da, which is proportional to peptides with a length of 6–8 amino acid residues. Thus, the resulting WPHP will have hypoallergenic properties, since it does not contain peptides, which may include antigenic determinants.

The obtained WPHP was lyophilically dried and used to obtain complexes with β-CD, as well as to analyse their biological activity.

WPHP showed antioxidant, antimicrobial and antimutagenic properties (see Table 1), which are extremely important for the design of functional foods. At the same time, it had a bitter taste, which makes it difficult to use WPHP due to organoleptic attribues. Technology of cyclodextrin nanocomplexes with low molecular weight peptides production has been developed for the WPHP bitter taste reducing.

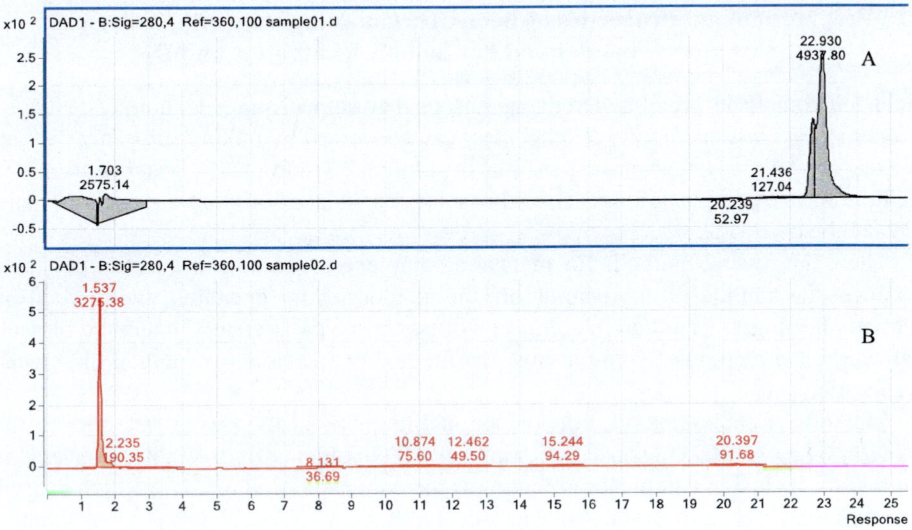

Fig. 2. HPLC-MS profile of whey proteins (A) and hydrolysed whey ultrafiltrate (B)

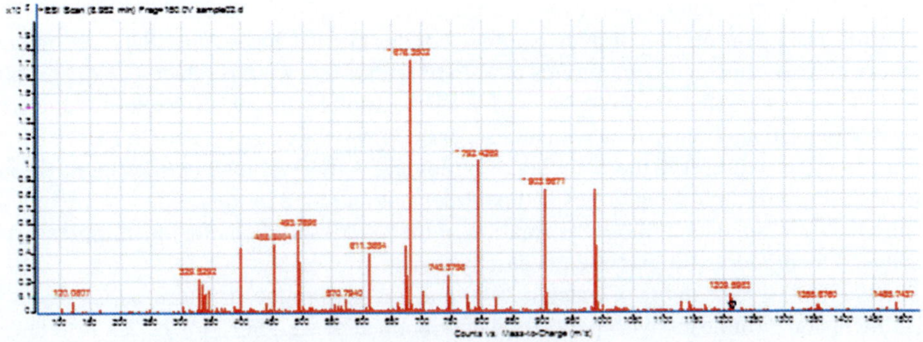

Fig. 3. Mass spectrum of whey protein hydrolysate filtrate

Table 1. Characteristics of deep hydrolysates of milk whey proteins (WPHP) biological activity

Parameters	Value for WPHP
Predominant peptide fraction, Da	680–900
AOA, IC_{50}, μg of protein per ml	14,5 ± 0,3
Antimutagenic activity, *S. typhimurium* TA 98/TA 100, %	19/14
Antimicrobial properties against E. coli ATCC 8739/S. aureus ATCC 6538, %	16,4/7,2

3.2 Preparation and Properties of Beta-CD Nanocomplexes with Biologically Active Substances: Peptides and Fat-Soluble Vitamins A and D_3

Inclusion complexes are obtained using various thermomechanical techniques. Nanostructures of inclusion complexes (cavitates) can be formed by mixing substances either in a dry crystalline state or in suspensions or solutions. Traditionally, water is used as a medium for the complexation reaction, but mixtures with various non-aqueous solvents can also be used [11].

There are several methods for inclusion complexes obtaining: the method of coalescence, the method of coprecipitation, the method of dry kneading, the method of hermetic heating, the method of complex formation in a pasty state, the method of neutralization, the method of spray drying, the method of freezing, the method of solvent evaporation [11, 27].

Methods of coalescence, co-deposition and hermetic heating were tested to obtain WPHP:β-CD, A:β-CD, D_3:β-CD complexes. Technologically, the most convenient method for inclusion complexes obtaining is the coprecipitation method, which is used in this study. Unlike α-cyclodextrin and γ-cyclodextrin β-CD at 25 °C has a low solubility in water, which equals to 18.5 mg per ml. Its solubility increases significantly with an increase in the temperature of the solution and at 50 °C equals to 46.2 mg per ml. The introduction of a dissolved complexant into the system, incubation and subsequent temperature lowering lead to precipitation of the inclusion complex. Lowering the temperature to negative values ensures maximum extraction of the formed complex from the solution. The incorporation of hydrophobic peptides into β-CD structure depends on the temperature. In this regard, clathrates of WPHP with β-CD were produced at 25 °C and 50 °C. Considering that not all peptides of WPHP exhibit hydrophobic properties and are capable to form clathrates, they were not separated from the complexes of WPHP:β-CD. Thus, in order to preserve the complex of biological activity of the peptides included in the WPHP, a mixture of clathrates and peptides not included in the β-CD was obtained using the developed technology. Formation of the WPHP:β-CD inclusion complex was analysed by organoleptic parameters and by thermogravimetry.

WPHP is highly soluble in water at 25 °C and has a pronounced bitterness. Cyclodextrin inclusion complexes with peptides were obtained at weight ratios of 2:1 to reduce the bitterness of WPHP. The incorporation of peptides into the hydrophobic area of cyclodextrinm molecule led to a significant improvement in the organoleptic properties of the resulting clathrate and a decrease in bitterness to a moderately bitter taste (see Fig. 4).

According to organoleptic indicators, the fraction of peptides WPHP has a pronounced bitterness (10 points). Complexes of inclusion WPHP:β-CD provided a significant decrease in the bitterness of peptides (3–4 points) compared with the control sample, depending on the temperature of complexation.

Recently it was shown that the bitter taste of whey protein hydrolysates is given by dipeptides: Phe-Ala, Pro-Pro, Pro-Leu, Leu-Val, Arg-Val and tripeptides: Arg-His-Gly, Ser-Leu-Ala, Leu-Leu-Pro and Try-Try-Gln. Volumetric hydrophobic amino acids in any position in the structure of di- or tripeptides provide the manifestation of bitterness. The bitter taste of Pro-containing peptides is determined by their interaction with the bitter taste receptor [9, 18]. Peptides containing amino acids sequence Leu-Tyr-Phe in

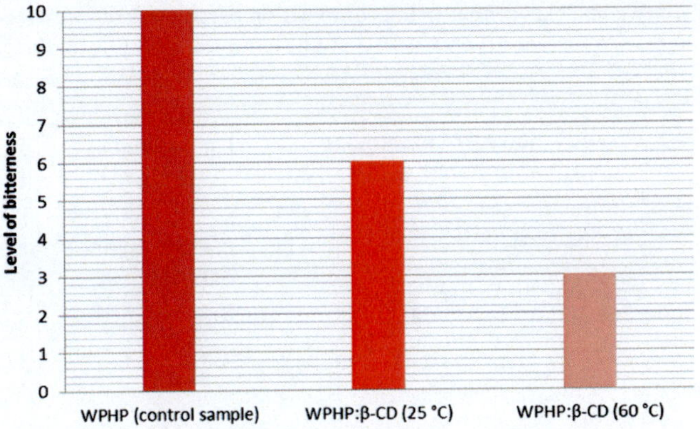

Fig. 4. Organoleptic properties of WPHP filtrate, clathrate WPHP:β-CD, obtained at 25 °C and clathrate WPHP:β-CD, obtained at 60 °C

their structure can also cause the bitterness of whey protein hydrolysates. These peptides included in WPHP can be effectively integrated into the CD structure. As a result obtained clathrates do not interact with bitter taste receptors. Peptides with a higher molecular weight may not form clathrates and retain a weak bitter taste. The proposed technology for WPHP and WPHP:β-CD inclusion complexes production allows to reduce the bitterness of the latter to 3 points, which is relevant to design functional foods. According to the results of gravimetric analysis, 100 g of the resulting complex contains 33.0 g of peptides.

A decrease in the bitterness of peptides indicates their inclusion in β-CD structure. Thermogravimetric analysis was used in order to confirm the formation of WPHP:β-CD clathrates. It is based on fixing of the change in the mass of the test sample during its thermal decomposition in the temperature range of 20–600 °C.

The stages of samples thermal decomposition under conditions of programmable heating at a rate of 5 °C per min are established. Table 2 presents a comparative characteristic of thermal decomposition parameters of WPHP, mechanical mixture of WPHP and β-CD, WPHP:β-CD complex according to the data of DTG/TG profiles. The beta CD sample is characterized by a peak mass loss at 301.8 °C with a maximum thermal degradation rate reaching 0.43 mg per °C. In the case of WPHP, decomposition peaks with maximum of the mass loss rate at 159.6, 203.9, 268.3 and 541.3 °C were detected (0.006, 0.014, 0.29 and 0.40 mg per °C, respectively).

Clathrates of fat-soluble vitamins with CD obtaining have some peculiarities. Due to the fact that fat-soluble vitamins D_3 and A do not differ in increased thermal stability, their complexes with β-CD were obtained at 25 °C. Vitamins D_3 and A dissolved in vegetable oil samples were used in the study for inclusion in clathrates. Completeness of vitamins and fatty acids inclusion after precipitation of D_3:β-CD and A:β-CD complexes by centrifugation was evaluated by their presence in the supernatant using electron spectroscopy, which showed the absence of starting substances in it.

Number of D_3:β-CD and A:β-CD clathrates aqueous solutions with different concentrations were prepared after their freeze drying. Maximum solubility of D_3:β-CD and

Table 2. Comparative analysis of thermal decomposition parameters of WPHP control samples, mechanical mixture of WPHP and β-CD, WPHP:β-CD complex according to the data of DTG/TG profiles

Name of the sample, conditions of the complex formation	Temperature of the maximum destruction rate (TV_{max}), °C	Maximum destruction rate (V_{max}), mg per °C	The amount of sample in the system at TV_{max}, % of the initial content	Activation energy (Ea), kJ per mol
WPHP	268.3	0.029	80.9	76
Mechanical mixture of WPHP and β-CD	297.5	0.29	68.8	118
WPHP:β-CD complex	305.1	0.15	55.6	105

A:β-CD clathrates at 25 °C did not exceed 1.6 mg per ml and 1.4 mg per ml, respectively. The content of fatty acids and vitamins D_3, A in D_3:β-CD and A:β-CD clathrates was determined using the gravimetric method. It is shown that D_3:β-CD and A:β-CD complexes contain 1.06 mg of vitamin D_3 and 86.0 mg of vitamin A in 5 g of the complex.

Thermogravimetric analysis was used to confirm the formation of beta-CD clathrates with fat-soluble vitamins. The stages of thermal decomposition under conditions of programmable heating from 20 °C to 600 °C at a speed of 5 °C per min for freeze-dried preparations D_3:β-CD and A:β-CD are established. The temperature of the maximum rate of oxidative degradation of fat-soluble vitamin clathrates equals to 308.26 °C and corresponds to 1.77 mg per min. Formation of D_3:β-CD and A:β-CD inclusion complexes leads to a change in their physico-chemical properties, which affects the change in the parameters of their thermal decomposition. Comparative analysis of D_3:β-CD complex thermal degradation shows that vitamin D_3 is characterized by a stage of thermal decomposition at a temperature of 220.35–273.44 °C. At the same time, the maximum rate of thermal degradation of β-CD is shifted from 320 °C to 295 °C. These results indicate the formation of an inclusion complex with vitamins and fatty acids of olive oil.

Organoleptic, antioxidant, antimutagenic properties of cyclodextrin nanocomplexes with fat-soluble vitamins and peptides of whey protein hydrolysate were studied in the framework of functional foods development. Also their toxic and hygienic assessment was carried out.

Results of studies of WPHP:β-CD, D_3:β-CD and A:β-CD complexes organoleptic properties, antioxidant and antimutagenic activity are presented in Table 3. Data obtained allows to develop a balanced composition of the MCC from powdered forms of WPHP:β-CD, D_3:β-CD and A:β-CD inclusion complexes. The composition of MCC includes: WPHP: β-CD – 90 g, D_3:CD – 5 g and A:CD – 5 g. The resulting powdered form contains 30 g of whey protein hydrolysate, 1.06 mg of vitamin D_3 (42,500 IU), 86.0 mg of vitamin A (250,000 IU) and 4 g of olive oil. Cyclodextrin nanocomplexes with fat-soluble vitamins D_3 and A are characterised by the following organoleptic attributes: white powder, odorless and tasteless.

Table 3. Organoleptic properties and biological activity of nanocomplexes: WPHP:β-CD, D_3:β-CD and A:β-CD and multicomponent composite (MCC) in a ratio of 90: 5: 5.

Parameters	WPHP:β-CD	D_3:β-CD	A:β-CD	MCC
Bitterness, points	3	0	0	2
AOA, IC_{50}, μg of dry matter per ml	65.7 ± 1.5	80.0 ± 2.2	12.5 ± 0.6	15.1 ± 0.9
Antimutagenic activity, *S. typhimurium* TA 98/TA 100), %	18/12	0/0	0/0	20/15

Analysis of the data presented in Table 3 shows that inclusion in cyclodextrin of whey proteins enzymatic hydrolysis products led to a significant decrease in the bitter taste of peptides to 3 points compared to the control, which has a pronounced bitterness (10 points). Powdered MCC had a white color, odorless, with a weak bitter taste up to 2 points.

Determination of antioxidant activity of WPHP:β-CD, D_3:β-CD and A:β-CD inclusion complexes and multicomponent composite was carried out by the ORAC method. The concentration of the studied samples was calculated, causing 50% inhibition of the formation of reactive oxygen species (IC50). Results presented in Table 3 allows for the conclusion that obtained nanocomplexes are capable of binding free oxygen-containing radicals. Comparative analysis of inclusion complexes samples showed that the anti-radical activity decreases in the series of studied samples: D_3:β-CD > WPHP:β-CD > MCC > A:β-CD.

Strain of S. typhimurium TA 98, which allows identification of DNA damage causing a shift in the reading frame, and strain of S. typhimurium TA 100, with which base pair replacement mutations are identified, were used to analyse the antimutagenic activity of inclusion complexes and MCC in the Ames test. S. typhimurium TA 98 and S. typhimurium TA 100 cells were subjected to induced mutagenesis, leading to an increase in revertants in order to assess the antimutagenic effect of the studied samples. The revealed differences in the number of revertants in the control and experimental samples were statistically significant for the concentration of WPHP:β-CD 0.5 mg per cup.

Experimental data analysis showed that WPHP:β-CD and MCC samples exhibit antimutagenic effects, preventing base pair replacement mutations in S. typhimurium TA 100 strain and reading frame shift in S. typhimurium TA 98.

4 Discussion and Conclusion

Whey proteins enzymatic hydrolysates were produced for design of functional foods. Proteolysis products contain peptides that have hypoallergenic, antioxidant, antimutagenic properties and have a bitter taste. Peptides were included in β-cyclodextrin nanocomplexes to reduce their bitter taste. Such clathrates are not able to interact with receptors, since Pro-containing peptides and some other peptides that cause bitterness enter the hydrophobic cavity of cyclodextrin. As a result, they do not interact with bitter taste receptors.

Nanocomplexes of fat-soluble vitamins A and D_3 were obtained in the framework of multicomponent composite intended for functional foods design. Complexes of inclusion formed by fat-soluble vitamins D_3 and A with β-cyclodextrin led to a change in their physical-chemical properties. They were converted from a liquid state in olive oil to a powdered form. These clathrates had increased thermal stability and solubility in water.

Antioxidant properties and antimutagenic activity of WPHP:β-CD, D_3:β-CD and A:β-CD inclusion complexes and their multicomponent composite (MCC) were studied. Composition-optimised MCC was developed, which includes: WPHP:β-CD – 90 g, D_3:CD – 5 g and A:CD – 5 g. The resulting powdered form contains 30 g of whey protein hydrolysate, 1.06 mg of vitamin D_3 (42,500 IU), 86.0 mg of vitamin A (250,000 IU) and 4 g of olive oil. Comparative analysis of inclusion complexes samples and their multicomponent composite antioxidant activity showed that it was decreasing in the series: D_3:β-CD > WPHP:β-CD > MCC > A:β-CD. Antimutagenic activity of inclusion complexes and MCC was studied Using the model of induced mutagenesis in the Ames test on S. strains. typhimurium TA 98 and S. typhimurium TA 100. It was established that WPHP:β-CD and MCC samples exhibit antimutagenic effects, preventing base pair replacement mutations in S. typhimurium TA100 strain and reading frame shift in S. typhimurium TA98.

Thus, the results of study allow developing multicomponent composite intended for functional foods design. The resulting powdered forms of fat-soluble vitamins and peptides are easily dosed and can be used in the development of various functional foods.

Acknowledgments. The authors express their gratitude to E.I. Taran and N.V. Dudchik for technical assistance in carrying out the research.

References

1. Asgary, S., Rastqar, A., Keshvari, M.: Functional food and cardiovascular disease prevention and treatment: a review. J. Am. Coll. Nutr. **37**(5), 429–455 (2018). https://doi.org/10.1080/07315724.2017.1410867
2. Borovik, T.E., Makarova, S.G., Darchiya, S.N., Gamaleyeva, A.V.: The role of compounds based on hydrolyzed protein in prophylaxis and diet treatment of alimentary allergy in infants. Voprosy sovremennoj pediatrii **9**(1), 150–156 (2010). (In Russian)
3. Das, S.K.: Cyclodextrins – the molecular container. Res. J. Pharm. Biol. Chem. Sci. **4**(2), 1694–1720 (2013)
4. Domínguez, D.L., Fernández-Ruiz, V., Cámara, M.: The frontier between nutrition and pharma: the international regulatory framework of functional foods, food supplements and nutraceuticals. Crit. Rev. Food Sci. Nutr. **60**(10), 1738–1746 (2020). https://doi.org/10.1080/10408398.2019.1592107
5. Dudchik, N.V.: Quantitative evaluation of antimutagenic activity of plant composition in short-term tests. Zdorov'e i okruzhayushhaya sreda (Health Environ.) **24**(1), 218–221 (2014). (In Russian)

6. Edwards, P.J.B., Jameson, G.B., Palmano, K.P., Creamer, L.K.: Heat-resistant structural features of bovine β-lactoglobulin A revealed by NMR H/D exchange observations. Int. Dairy J. **12**, 331–344 (2002)
7. Fernández, M.A., et al.: Complex systems that incorporate cyclodextrins to get materials for some specific applications. Carbohyd. Res. **480**, 12–34 (2019)
8. Görgüç, A., Gençdağ, E., Yilmaz, M.: Bioactive peptides derived from plant origin by-products: biological activities and techno-functional utilizations in food developments – a review. Food Res. Int. **136**(109504), 1–45 (2020). https://doi.org/10.1016/j.foodres.2020.109504
9. Halavach, T.M.: Biologically active properties of hydrolysed and fermented milk proteins. J. Microbiol. Biotechnol. Food Sci. **9**(4), 714–720 (2020)
10. Halavach, T.M., Kurchenko, V.P.: Allergenicity of milk proteins and ways to reduce it. Trudy BGU **5**(1), 9–55 (2010). (In Russian)
11. Halavach T.M., et al.: Antimutagenic and antibacterial activity of β-cyclodextrin clathrates with extensive hydrolysates of colostrum and whey. Biointerface Res. Appl. Chem. **11**(2), 8626–8638. https://doi.org/10.33263/BRIAC112.86268638
12. Jansook, P., Ogawa, N., Loftsson, T.: Cyclodextrins: structure, physicochemical properties and pharmaceutical applications. Int. J. Pharm. **535**(1–2), 272–284 (2018)
13. Kim, S.B., et al.: Enzymatic hydrolysis of heated whey: iron-binding ability of peptides and antigenic protein fractions. J. Dairy Sci. **90**, 4033–4042 (2007)
14. Kim, S.B., et al.: Peptic and tryptic hydrolysis of native and heated whey protein to reduce its antigenicity. J. Dairy Sci. **90**, 4043–4050 (2007)
15. Loftsson, T., Brewster, M.E.: Cyclodextrins as functional excipients: methods to enhance complexation efficiency. J. Pharm. Sci. **101**(9), 3019–3032 (2012)
16. Manzoor, M., Ahmad, M., Bandral, J.D., Gani, A., Singh, J., Shams, R.: Food hydrocolloids: functional, nutraceutical and novel applications for delivery of bioactive compounds. Int. J. Biol. Macromol. **165**(Pt A), 554–567 (2020). https://doi.org/10.1016/j.ijbiomac.2020.09.182
17. Menezes, P.D.P., Andrade, T.A., Frank, L.A.: Advances of nanosystems containing cyclodextrins and their applications in pharmaceuticals. Int. J. Pharm. **559**, 312–328 (2019)
18. Plasek, B., Lakner, Z., Kasza, G., Temesi, Á.: Consumer evaluation of the role of functional food products in disease prevention and the characteristics of target groups. Nutrients **12**(1), 69, 1–19 (2019). https://doi.org/10.3390/nu12010069
19. Safety Requirements for Food Additives, Flavorings and Technological aids: TP TC 029/2012. Minsk, E'konome'nergo, p. 308 (2014)
20. Serafini, M., Peluso, I.: Functional foods for health: the interrelated antioxidant and anti-inflammatory role of fruits, vegetables, herbs, spices and cocoa in humans. Curr. Pharm. Des. **22**(44), 6701–6715 (2017). https://doi.org/10.2174/1381612823666161123094235
21. Sharma, S.K., Bansal, S., Mangal, M., Dixit, A.K., Gupta, R.K., Mangal, A.K.: Utilization of food processing by-products as dietary, functional, and novel fiber: a review. Crit. Rev. Food Sci. Nutr. **56**(10), 1647–1661 (2016). https://doi.org/10.1080/10408398.2013.794327. PMID: 25748244
22. Shikhar, G.: Solubility studies of the β-cyclodextrins inclusion complexes: a review. Int. Res. J. Pharm. **3**(10), 178–181 (2012)
23. Tarun, E.I., Kurchenko, V.P.: Comparison of antioxidant activity gallic, caffeic and chlorogenic acids. Trudy BGU **9**(1), 186–191 (2014). (In Russ.)
24. Tsygankov, V.G., Lovkis, Z.V., Stigailo, I.N., Simonenko, S.V.: Tasks and prospects for the development of functional nutrition. Trudy BGU **4**(1), 60–67 (2009). (In Russian)
25. Tutel'yan, V.A.: Food Ingredients in Modern Food Creation. DeLi plyus, Moscow (2013) (In Russian)

26. Varan, G., Varana, C., Erdo, N., Hincal, A.A., Bilensoy, E.: Amphiphilic cyclodextrin nanoparticles. Int. J. Pharmaceut. **16733**, 1–13 (2017). https://doi.org/10.1016/j.ijpharm.2017.06.010
27. Zhu, J., Huang, Q.: Nanoencapsulation of functional food ingredients. Adv. Food Nutr. Res. **88**, 129–165 (2019)

Investigation of the Acute, Subacute and Chronic Toxicity of Curcuminoids with Native and Hydroxypropylated β-cyclodextrin Complex Nanostructures

Maxim Kapustin[1(✉)], Hanna Chubarova[1], Vasili Tsygankow[2], Svetlana Lodigina[3], and Ekaterina Holodova[3]

[1] Belarussian State University, Minsk, Republic of Belarus
[2] The RUE "Scientific and Practical Center of Hygiene", Minsk, Republic of Belarus
vgz@tut.by
[3] North Caucasus Federal University, Stavropol, Russia

Abstract. Toxicological and hygienic assessment of the obtained samples of curcuminoid nanostructures with cyclodestrins showed differences in hazard classes, when studied in the *T. pyriformis* model system under the conditions of an acute and subacute experiment. In a model system with protozoa, the resulting nanocomplexes were assigned to hazard classes 4 and 2. The study on outbred Wistar rats showed no toxic effect of curcuminoid nanocomplex preparations on the body of experimental animals under conditions of a chronic experiment. Based on the results of assessing the biochemical and hematological parameters of the blood in the control and experimental groups, a significant positive effect of oral administration of curcuminoid nanostructures with native and hydroxypropylated β-cyclodextrin on rats was shown. The administration of drugs for 20 days showed the objective increase in the antioxidant status of the body, increased immunity and to the significant analgesic effect on rats.

Keywords: Cyclodextrins · Curcumin · Nanostructures · Acute toxicity · Subacute toxicity · Functional nutrition · Tetrachimena piriformis · Rats Wistar · Anesthetic

1 Introduction

Turmeric rhizomes (Curcuma longa L.) are widely used in various industries: food, cosmetic, pharmaceutical. Plant raw materials, on the one hand, are used in their native crushed form as spices, and, on the other hand, are used as a source of biologically active substances obtained by extraction for the preparation of purified pharmaceutical substances. The popularity of this plant as medicine is associated, first of all, with the pleiotropic activity of the mixture of biologically active substances synthesized by this plant – curcuminoids – hypocholysterolemic, antidiabetic, anti-inflammatory, antioxidant, hepatoprotective, wound healing and antiviral [1, 3, 5, 16, 18, 26, 28, 30].

V. Kurchenko et al. (Eds.): ICAETT 2021, LNNS 408, pp. 221–235, 2022.
https://doi.org/10.1007/978-3-030-96641-6_26

Recently, there has been a tendency to use not native medicinal herbs, but standardized medicinal substances obtained from plant sources, since they are more effective than conventional extracts. Furthermore, frequently scientists use chemical modification of natural biologically active substances in order to increase their biological activity, bioavailability and solubility. However, the safety of the obtained standardized extracts is not a consequence of the safety of their natural plant sources [8].

Curcuminoids are a mixture of three diarylheptanoids: curcumin (C) and its two derivatives, demethoxycurcumin (DMC) and bis-demethoxycurcumin (BDMC) [25]. The ratio of curcuminoids in curcumin is 52–63% for C, 19–27% for DMC and 18–28% for BDMK. Chemical analysis of the composition of turmeric rhizome extracts showed that plant materials contain carbohydrates (69.4%), water (13.1%), proteins (6.3%), fats (5.1%) and minerals (3.5%), essential oil (5.8%), and curcumin (3–6%) [2, 22].

Curcuminoids, being in a stable state in a protonated form, are lipophilic compounds. The hydrophobicity of these compounds is determined by the lipophilicity coefficient, denoted as logKo/w, the values of which are 3.29, 2.792, and 2.649 for C, DMC and BDMC, respectively [10, 17]. The difference in the hydrophobicity of diarylheptanoids was confirmed in previous study by HPLC [11, 12]. Consequently, due to their high hydrophobicity and poor water solubility, their practical use may be limited [17]. Their solubility in aqueous systems can be increased by creating inclusion complexes with cyclodextrins.

Cyclodextrins (CD) are produced by enzymatic hydrolysis of starch. Depending on the type of CD, they contain from 6 to 8 glucose residues that form a torus [29]. All OH groups in CD are located on the outer hydrophilic surface of the molecule, while the inner cavity is hydrophobic. In aqueous solutions, hydrophobic substances are able to integrate into this cavity, forming inclusion complexes – clathrates. Such molecular containers are able to retain molecules of non-polar substances in the inner cavity, and due to the hydrophilic outer surface, they give them greater solubility, stability, change their taste, color and odor. Moreover, pure CD show 5th class of toxicity and are "non-toxic compounds". They are recommended as a food additive (E459) [6, 9, 13, 20, 29].

Molecules of hydrophobic curcuminoids interacting with CD are able to integrate into their non-polar cavity, forming clathrates. As a result, the solubility of curcuminoids in the aqueous phase increases. Such nanostructures of curcuminoids will have new structural and functional properties, increased bioavailability and altered toxicological and hygienic properties [6, 9, 13, 14, 20, 29].

In recent years, intensive research has been carried out on the bioavailability, safety and efficacy of curcuminoid preparations. Due to the pronounced biological activity and low toxicity of curcuminoids, they can be used as part of functional food [4, 7, 15, 19, 23, 24, 27]. Standardized extract was evaluated for safety in an acute oral toxicity study and found to be safe up to 5 g/kg rat body weight. On repeated oral administration in either sex of the rats for 90 days did not demonstrate any significant toxic or adverse effects. The no-observed adverse effect level (NOAEL) of extract in the repeated oral toxicity study was found to be 1000 mg/kg in both male and female animals [22].

Nanostructures of cucuminoids with CD represent new forms of biologically active substances that can be promising for use in the food, pharmaceutical, and cosmetic industries.

The main purpose of this research was to carry out the toxicological characterization of curcuminoids with native beta-cyclodextrin and hydroxypropylated beta-cyclodextrin nanocomplexes in the protist and mammalian model systems in order to expand nanocoplexes' application range.

2 Materials and Methods

The following object were use in the framework of this study:

- nanocomplexes of curcuminoids with native beta-cyclodextrin and hydroxypropylated beta-cyclodextrin with molar ratio 1:2;
- sterile culture of the ciliate Tetrahymena pyriformis in the stationary growth phase, maintained in a standard nutrient medium at 25 °C;
- outbred male Wistar rats, healthy individuals with a clean and smooth coat, with normal motor activity, passed a 2-week quarantine, distributed into groups using the randomization method and identified.

To obtain complexes of inclusion of curcuminoids with cyclodextrins, a combined method of co-dissolution and lyophilization was used. For this, the required amounts of curcuminoids and cyclodextrins were calculated to achieve the required molar ratio. The curcuminoid preparation was dissolved in acetone, since it dissolves curcuminoids well (up to 40 g per 100 ml of acetone). Cyclodextrins were dissolved in distilled or deionized water, heated to 80–90 °C, and incubated for 5–10 min until complete dissolution. If it was necessary to obtain a sterile solution, the cyclodextrin solution was boiled for 15 min, then cooled to 35 °C, kept for 30 min, and boiled again for 15 min. Then the hot cyclodextrin solution was placed on a magnetic stirrer and cooled with stirring to 60 °C. Then, with vigorous stirring, solutions of curcuminoids were added dropwise to aqueous solutions of cyclodextrins. In this case, a gradual turbidity of the solution and the formation of nanostructure aggregates of inclusion complexes were observed. Stirring of the solution was continued until it cooled to room temperature, then in a refrigerator at a temperature not exceeding 5 °C. After that, the entire volume of the mixture was mixed and placed in a plastic Petri dish with small holes made in the lid. The Petri dish with the mixture was kept at a temperature of 76 °C below zero for 60–120 min, depending on the volume of the mixture. Then it was placed on a freeze dryer heated to 9 °C for 24–48 h until a constant weight and temperature of the product was achieved. The dry powder of nanostructures was ground to a homogeneous state in a mortar, weighed and stored in hermetically sealed vials at a temperature below minus 10 °C.

Determination of the parameters of acute, subacute and chronic toxicity of nanostructures of inclusion complexes of curcuminoids with beta-cyclodextrin and hydroxypropylated-β-CD: LD_{16}, LD_{50}, LD_{84}, K_{kum}, was carried out on a sterile culture of the ciliate *Tetrahymena pyriformis* in the stationary growth phase standard nutrient medium at 25 °C. The duration of the acute experiment was 5 h, during the experiment on the study of subacute toxicity – 24 h, during the study of chronic toxicity – during the entire life cycle. The study of the toxicity of the complexes was carried out in triplicate at different times with two parallel tests. The toxic effect was assessed by the alternative life-death state. Based on the results of assessing the average lethal dose and cumulative properties, the toxicity class of the investigated substances in an acute experiment was established.

The study of the toxicity of the complex of curcumin with β-cyclodextrin and hydroxypropyl-β-cyclodextrin in an experiment with laboratory animals was carried on outbred male Wistar rats. Wistar rats were administered compositions of curcumin with beta-cyclodextrin (curcumin content 68 mg/g) and curcumin with hydroxypropylated beta-cyclodextrin (curcumin content 57.5 mg/g) for 20 days. The compositions were diluted in distilled water. As a result, the rats received 500 mg/kg in terms of pure curcumin.

The experiments included healthy individuals with a clean and smooth coat, with normal motor activity, who had undergone a 2-week quarantine. The animals were housed in standard plastic rat cages in accordance with group housing guidelines. Sawdust of deciduous trees served as a bedding. Access to water and food was free, the light regime was natural. The diet included grains (oats, wheat in excess), vegetables, standard briquettes containing mineral food ingredients and vitamins.

Before the start of the study, animals meeting the criteria for inclusion in the experiment were assigned to groups using the method of randomization. Animals that did not meet the criteria were excluded from the study during quarantine.

Places of keeping animals and premises were subjected to periodic sanitization, which did not affect the results of the study. In the premises for keeping animals, it is prohibited to carry out any manipulations with animals, the use of detergents, insecticides and other chemicals that can cause stress or affect the health of laboratory animals.

To prevent environmental pollution, animal carcasses and biological waste were stored in specially equipped freezers prior to removal.

To ensure individual observation during the study, the animals were identified.

Hematological parameters were studied using a Mindray blood analyzer, BC-5300 Vet Auto Hematology Analyzer.

Determination of serum LDH, SOD, reduced glutathione, μM SH-g/mgHb and free thiol groups were determined in hemolyzed blood spectrophotometrically using an SF-26 spectrophotometer Glutathione transferase and glutathione reductase were determined in hemolyzed blood using an FP-901 M analyzer.

The study of biochemical parameters of blood serum and urine was carried out on a KFK-2-UHL 4.2 colorimeter and an Uriscan pro II analyzer.

At various stages of the experiment, the analgesic properties of curcumin nanocompositions were carried out. On the 7-th day of the experiment, the tail heat immersion test was carried out. The tail immersion test is based on the spinal flexor reflex in response to immersion of the tail in hot water. Painful irritation was simulated by immersing the tail in hot water at a temperature of 52 °C (from 45 to 54 °C), and the latent period of tail withdrawal was recorded. To avoid tissue damage, the tail was not heated for more than 30 s. The criterion for the analgesic effect was considered a significant increase in the latent period of the reaction.

On the 20-th day, the summation-threshold indicator was determined using the ISE-01 electronic stimulator.

3 Results

Parameters of acute, subacute and chronic toxicity of nanostructures of inclusion complexes of curcuminoids with beta-cyclodextrin and hydroxypropilated beta-cyclodextrin: LD_{16}, LD_{50}, LD_{84}, K_{kum} were determined on a sterile culture of the ciliate Tetrahymena pyriformis in a stationary growth phase maintained in a standard nutrient medium at 25 °C. The toxic effect was assessed by the alternative life-death state. Based on the results of assessing the average lethal dose and cumulative properties, the toxicity class of the investigated substances in an acute experiment was established.

When studying the toxicity of complexes of nanostructures with beta-cyclodextrin and hydroxypropilated beta-cyclodextrin in an experiment with laboratory animals, outbred Wistar rats were injected with curcumin nanostructure compositions with beta-cyclodextrin for 20 days. The clinical picture was described and differences were established according to the data of autopsy and macroscopic examination of the studied organs, between the animals of the experimental and control groups. According to the results of assessing the biochemical and hematological parameters of the blood of rats in the control and experimental groups, the effect of oral administration of curcuminoid nanostructures on rats was shown.

3.1 Toxicological and Hygienic Assessment of Curcuminoids Nanostructures with Beta-CD in the *Tetrahymena Pyriformis* Model System

In the study of the toxicity of the β-cyclodextrin with curcumin complex at a molar ratio of 2:1 nanostructures in acute and subacute experiments, 100,000 ciliates in the phase of slow growth were introduced into 1 ml of a suspension of the following concentrations: 0.1; 0.2; 0.5; 0.7; 1.0 mg/ml. The exposure time of this complex samples with protozoa at given concentrations in determining acute toxicity was 5 h. Single-celled organisms in the concentration of the complex of nanostructures of β-cyclodextrin with curcumin (2:1) 0.1 and 0.2 mg/ml did not differ from the control samples. At a concentration of this complex of 0.5 mg/ml, there was a decrease in the population size by 17–21% compared to the control level. In this concentration, single round ciliates with a spinning character of movement appeared. In samples containing 0.7 mg/ml of the complex, the lethality of protozoa was 41–54%, the number of ciliates with morphological and functional changes increased. At a concentration of 1.0 mg/ml, death of 85–90% of ciliates was observed.

In the study of subacute toxicity (exposure time was 24 h) of the β-cyclodextrin with curcumin complex at a molar ratio of 2:1 nanostructures, an increase in the toxic effect of LD_{16}, LD_{50}, and LD_{84} was observed in comparison with those of the acute experiment. In samples containing this complex at a concentration of 0.2 mg/ml, the lethality of protozoa was 19–25%, at a concentration of 0.5 mg/ml the lethality of protozoa was 43–48%, and the content of the complex 0.7 mg/ml caused the death of the test object at the level of 98–100%.

The probit analysis of ciliates lethality in acute and subacute experiments under the influence of the β-cyclodextrin complex with curcumin nanostructures was used to calculate the parameters of acute and subacute toxicity of this complex, shown in the Table 1.

Table 1. Toxicity parameters of the curcumin with β-cyclodextrin (1:2 M:M) complex nanostructures in acute and subacute experiments based on the results of evaluation in the *T. pyriformis* model system

Parameters of toxicity	Toxicity value	Hazard class
Acute toxicity		
LD16, mg/ml	0,47 ± 0,002	–
LD50, mg/ml	0,71 ± 0,038	2
LD84, mg/ml	0,96 ± 0,023	–
Subacute toxicity		
	0,25 ± 0,003	
LD16, mg/ml	0,44 ± 0,015	–
LD50, mg/ml	0,64 ± 0,032	–
LD84, mg/ml	0,62 ± 0,028	–
$K_{kum\ acuta}$	0,25 ± 0,003	4

3.2 Toxicological and Hygienic Assessment of Curcuminoids Nanostructures with Hydroxypropylated-Beta-CD in the Tetrahymena Pyriformis Model System

When studying the toxicity of the complex of curcumin with hydroxypropylated β-cyclodextrin at a molar ratio of 1:2 nanostructures in an acute experiment in *T. pyriformis* model system, 100,000 infusoria in the phase of slow growth were added to 1 ml of a suspension with the following concentrations: 0.2; 0.1; 0.075; 0.025; 0.01 mg/ml. The exposure time of nanostructure samples with protozoa at given concentrations in determining acute toxicity was 5 h.

Single-celled organisms at concentrations of nanostructures of the complex of 0.01 mg/ml and 0.025 mg/ml did not differ from the control samples. At a concentration of complex nanostructures 0.075 mg/ml, a decrease in the population size by 13–16% compared with the control level was observed. At this concentration, single round-shaped ciliates with a slowed-down "flinching" nature of movement appeared. In samples containing complex nanostructures at a concentration of 0.1 mg/ml, the lethality of protozoa was 51–78%, the number of ciliates with morphological and functional changes increased (Fig. 1).

0.2 mg/ml concentration of curcumin with hydroxypropylated β-cyclodextrin complex nanostructures at a molar ratio 1:2 caused the death of 98–100% of ciliates.

In the study of subacute toxicity (exposure time 24 h) of the complex of curcumin with hydroxypropylated β-cyclodextrin at a molar ratio of 1:2, an increase in the toxic effect according to LD_{16} was observed in comparison with that of the acute experiment. The LD50 and LD_{84} values remained at the level of the values established in the acute experiment. In samples containing nanostructures of the complex at a concentration of 0.05 mg/ml, the lethality of protozoa was 15–18%, at a concentration of 0.1 mg/ml

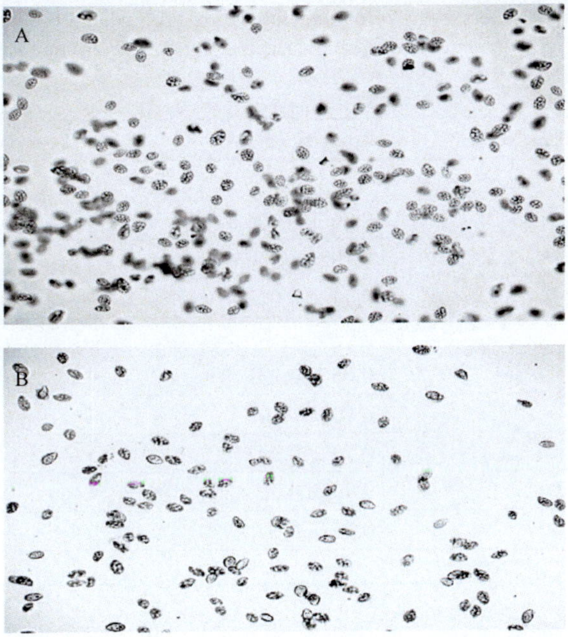

Fig. 1. Photos of the *Tetrahymena pyriformis* population (exposure time – 5 h). A. cultivation in a nutrient medium (control), photograph under a microscope at 100x magnification B. cultivation in a nutrient medium containing the curcumin:hydrohypropylated-beta-CD nanostructures with molar ratio 1:2 at a concentration of 0.1 mg/ml, photograph under a microscope at 100x magnification

the lethality of protozoa was 48–54%, and the concentration of nanostructures of the complex 0.2 mg/ml caused complete death of test object.

By the method of probit analysis of ciliates lethality in acute and subacute experiments under the influence of the complex of curcumin with hydroxypropylated β-cyclodextrin at a molar ratio of 1:2 nanostructures, the parameters of acute and subacute toxicity are calculated and shown in Table 2.

3.3 Toxicological and Hygienic Assessment of Curcuminoids Nanostructures with Beta-CD and Hydroxypropylated Beta-CD in the Wistar Rat Model System

When studying the toxicity of curcuminoids with β-CD and hydroxypropylated-β-CD complexes nanostructures in an experiment with laboratory animals, outbred Wistar male rats were orally administered nanostructure compositions for 20 days.

The clinical picture, according to the autopsy and macroscopic examination of the organs under study, did not reveal significant differences between the animals of the experimental and control groups. On visual inspection, the coat was shiny and neat. There were no foci of baldness. Discharge from natural holes was absent. The fore and hind limbs did not show any changes. No limb deformity was observed. The teeth were preserved.

Table 2. Toxicity parameters of the curcumin with hydroxypropylated-β-cyclodextrin (1:2 M:M) complex nanostructures in acute and subacute experiments based on the results of evaluation in the *T. pyriformis* model system

Parameters of toxicity	Toxicity value	Hazard class
Acute toxicity		
LD16, mg/ml	0,06 ± 0,004	–
LD50, mg/ml	0,102 ± 0,003	2
LD84, mg/ml	0,14 ± 0,003	–
Subacute toxicity		
	0,05 ± 0,001	
LD16, mg/ml	0,10 ± 0,002	–
LD50, mg/ml	0,15 ± 0,002	–
LD84, mg/ml	0,99 ± 0,05	–
$K_{kum\ acuta}$	0,05 ± 0,001	4

Visible mucous membranes were pale, shiny, smooth.

When examining the thoracic and abdominal cavities, no disturbances in the position of the internal organs were noted.

The pleura, pericardium and peritoneum were thin, shiny, smooth.

Submandibular lymph nodes and salivary glands had an oval or round shape, uniform pinkish or slightly yellowish colour. The density is moderate.

The diameter of the aorta was uniform throughout. The intima of the aorta was smooth, shiny, whitish in color. The size and shape of the heart did not show any changes. The heart muscle was moderately dense, uniformly brownish in color. The heart valves were thin, smooth, shiny. A small amount of liquid blood was contained in the cavities of the heart.

The lungs collapsed when the chest was opened. The size and shape have not changed from normal. The surface of the lungs had a uniform pale pink color.

The lung tissue was fluffy to the touch. The lumen of the trachea and large bronchi was wide.

The mucous membrane is shiny, pale pink, smooth.

The mucous membrane of the esophagus was shiny, smooth, pale in color. The size and shape of the stomach were not changed from normal. The gastric mucosa was shiny, smooth, pinkish in color.

The lumen of the stomach was filled with food contents. The duodenum lumen was not changed relative to normal, the intestinal mucosa was shiny, smooth, pale pink. The mucous membrane of the small intestine was also pale pink, shiny, smooth.

The mucous membrane of the large intestine had a slightly grayish tint, was smooth, shiny.

The shape and size of the liver were not changed from normal. The liver surface was smooth, uniform dark red color. Liver tissue was dark red on the cut. The liver capsule was thin and transparent. The consistency of the liver was of normal density.

The shape of the pancreas did not show any changes. The gland had a lobular structure, a pale pink color and a moderately dense texture.

The size and shape of the spleen were not changed from normal. The surface of the spleen had a uniform dark cherry color and was smooth. The consistency of the spleen was moderately firm. On the section of the organ, grayish small-cell follicles were distinguished. The size and shape of the kidneys were also not changed from normal. The kidney capsule was easily removable. The surface was smooth, uniformly brownish-grayish in color. On the section of the organ, the cortex and medulla are clearly distinguishable. The consistency of the kidneys was moderately firm. The adrenal glands were rounded, whitish-yellow in color, and moderately dense in texture. The section clearly showed a dark brown medulla. The bladder was filled with clear, light colored urine. The mucous membrane of the bladder was smooth, shiny, pale in color.

The study of the antioxidant system of animals did not reveal significant differences. The data on the indicators of the antioxidant systems of the blood of experimental animals are presented in Table 3.

Table 3. Indicators of the antioxidant system of blood in control and experimental groups of outbred Wistar male rats after 20 days of curcuminoids:cyclodextrins nanostructures oral administration

Studied indicators, units of measurements	Experimental groups		
	Control	curcuminoids:HP-β-CD nanostructures (1:2 M:M)	curcuminoids:β-CD nanostructures (1:2 M:M)
SH-groups, μM/mgHb	155,19 ± 3,19	149,76 ± 5,65	152,94 ± 6,73
Reduced glutathione μM SH-g/mgHb	21,86 ± 0,45	21,10 ± 0,80	21,50 ± 0,93
SOD, μg/ml	37,63 ± 1,28	40,30 ± 1,37	40,73 ± 2,08
LDH, μM HADN/mg protein	20,24 ± 0,34	23,93 ± 0,97*	22.36 ± 0,34*
Glutathione reductase, μM/g Hb min	5,44 ± 1,32	5,20 ± 1,18	4,41 ± 1,14
Glutathione transferase, μM/g Hb min	1,69 ± 0,29	2,58 ± 0,36	2,54 ± 0,61

* - statistically significant changes in relation to control at $p < 0.05$

In the experimental groups of animals that received various nanocompositions of curcumin, there was a slight but significant increase in lactate dehydrogenase, which indicates a positive effect of nanostructures on the antioxidant defense system of the rat organism.

The data of studying the biochemical parameters of the blood of rats of the control and experimental groups are shown in Table 4.

Table 4. Indicators of the biochemical parameters of blood in control and experimental groups of outbred Wistar male rats after 20 days of curcuminoids:cyclodextrins nanostructures oral administration

Studied indicators, units of measurements	Experimental groups		
	Control	curcuminoids:HP-β-CD nanostructures (1:2 M:M)	curcuminoids:β-CD nanostructures (1:2 M:M)
Glucose, mmol/l	4,79 ± 0,06	4,76 ± 0,07	4,64 ± 0,06
Protein, g/l	65,9 ± 0,40	66,36 ± 4,62	66,01 ± 0,37
Urea, mmol/l	4,80 ± 0,07	4,84 ± 0,07	4,97 ± 0,16
Lipids, g/l	4,22 ± 0,28	2,96 ± 0,1	3,02 ± 0,07

* - statistically significant changes in relation to control at $p < 0.05$

According to the data obtained, the study of blood biochemical parameters did not reveal significant differences between the control and experimental groups of animals.

The analysis of the hematological parameters of the rats blood of the control and experimental groups is shown in Table 5.

The study of data given in Table 5 showed that there were also no significant differences between the studied groups in terms of hematological blood parameters.

The study of blood count indices given in Table 6 showed that in animals receiving curcumin in the form of nanostructures, there was a significant increase in the number of neutrophils and a decrease in the number of lymphocytes.

This increase in the number of neutrophils is within the normal range of the reaction, but may indicate the cytostatic effect of curcumin, which could lead to a decrease in lymphocyte division and differentiation of cells in the myeloid direction, which include neutrophils.

The data of various indicators of urine study in control and experimental groups are shown in Table 7. The analysis also did not reveal differences between the groups, with the exception of changes in urine acidity and its shift in the alkaline range. That may be due to a decrease in the reabsorption of Na^+ in the renal tubules.

On the 7th day of the experiment, the tail heat immersion test was carried out. The "tail immersion test" showed a greater analgesic efficacy of the curcumin composition with hydroxypropyl-β-dextrin (Table 8), indicating a high bioavailability of the turmeric extract.

The analgesic effect of curcuminoids nanostructures was confirmed when studying the summation-threshold indicator 20 days after the administration of the nanocompositions.

Table 5. The analysis of the hematological parameters of blood in control and experimental groups of outbred Wistar male rats after 20 days of curcuminoids:cyclodextrins nanostructures oral administration

Studied indicators, units of measurements	Experimental groups		
	Control	curcuminoids:HP-β-CD nanostructures (1:2 M:M)	curcuminoids:β-CD nanostructures (1:2 M:M)
Erythrocytes, 10^{12}/l	7,31 ± 0,45	7,03 ± 0,57	6,36 ± 0,70
Hemoglobin, g/l	136,75 ± 9,02	134,25 ± 10,91	118,61 ± 11,62
Hematocrit, vol. fract.	0,36 ± 0,02	0,35 ±,0,03	0,32 ± 0,03
Leukocytes, 10^9/l	14,22 ± 1,38	13,06 ± 1,55	17,55 ± 4,29
Platelets, 10^9/l	693,75 ± 100,52	757,25 ± 80,19	647,75 ± 156,79
Average volume of erythrocytes, μm^3	49,60 ± 1,06	50,50 ± 0,49	50,00 ± 0,76
Average platelet volume, μm^3	5,75 ± 0,16	5,88 ± 0,13	5,85 ± 0,13
Average concentration of hemoglobin in erythrocyte, MCHC g/l	377,00 ± 18,73	380,00 ± 2,86	374,00 ± 2,48
Average hemoglobin content in erythrocyte, pg	18,73 ± 0,34	19,13 ± 0,18	18,70 ± 0,29

* - statistically significant changes in relation to control at $p < 0.05$

Table 6. The analysis of Blood formula in control and experimental groups of outbred Wistar male rats after 20 days of curcuminoids:cyclodextrins nanostructures oral administration

Studied indicators, units of measurements	Experimental groups		
	Control	curcuminoids:HP-β-CD nanostructures (1:2 M:M)	curcuminoids:β-CD nanostructures (1:2 M:M)
Neutrophils, %	11,78 ± 1,63	23,90 ± 2,54**	21,80 ± 0,72**
Lymphocytes, %	79,13 ± 2,47	62,30 ± 1,79*	67,15 ± 1,68*
Monocytes, %	3,35 ± 0,81	8,20 ± 1,59*	5,00 ± 1,11
Eosinophils, %	4,85 ± 1,11	5,20 ± 1,59	5,20 ± 1,07
Basophils, %	0,90 ± 0,14	0,40 ± 0,18	0,85 ± 0,30

* - statistically significant changes in relation to control at $p < 0.03$

** - statistically significant changes in relation to control at $p < 0.02$

Table 7. The analysis of urine indicators in control and experimental groups of outbred Wistar male rats after 20 days of curcuminoids:cyclodextrins nanostructures oral administration

Studied indicators, units of measurements	Experimental groups		
	Control	curcuminoids:HP-β-CD nanostructures (1:2 M:M)	curcuminoids:β-CD nanostructures (1:2 M:M)
Bilirubin, mg/100 ml	0,0	0,0	0,0
Urobilinogen, mg/100 ml	0,0	0,0	0,0
Ketone, mg/100 ml	0,0	0,0	0,0
Protein, mg/100 ml	0,0	0,0	0,0
Nitrite, mg/100 ml	0,0	0,0	0,0
Glucose, mg/100 m	0,0	0,0	0,0
pH	7,5	8,2	8,5
Relative density, g/ml	1,01	1,02	1,01
Ascorbic acid, mg/100 ml	0,0	0,0	0,0

* - statistically significant changes in relation to control at $p < 0.03$
** - statistically significant changes in relation to control at $p < 0.02$

Table 8. The analgetic effect in experimental groups of outbred Wistar male rats after 7 and 20 days of curcuminoids:cyclodextrins nanostructures oral administration

Studied indicators, units of measurements	Experimental groups		
	Control	curcuminoids:HP-β-CD nanostructures (1:2 M:M)	curcuminoids:β-CD nanostructures (1:2 M:M)
Tail immersion test after 7 days of composition administration, sec	414,38 ± 68,60	608,13 ± 48,28*	497,00 ± 66,51
The summation threshold indicator after 20 days of the introduction of the compositions, volts	1,09 ± 0,11	1,90 ± 0,27**	1,69 ± 0,19**

* - statistically significant changes in relation to control at $p < 0.02$
** - statistically significant changes in relation to control at $p < 0.01$

4 Discussion and Conclusion

According to the average lethal dose in an acute experiment on *T. pyriformis*, the curcumin with β-cyclodextrin complex (1:2 M:M) nanostructures belong to the 2-nd hazard class (it is a highly hazardous compound). According to the results of the assessment for *T. pyriformis* in a subacute experiment, this complex nanostructures belong to the 4-th hazard class (it is a low-hazard compound). Due to the fact that the classification of the object under study to the hazard class is carried out according to the parameter, the value of which corresponds to the highest hazard class, according to the results of the primary toxicological and hygienic assessment for *T. pyriformis*, the curcumin with β-cyclodextrin complex (2:1 M:M) nanostructures belong to 2-nd hazard class.

In the study of the toxicity of curcumin with hydroxypropylated-β-cyclodextrin complex at a molar ratio of 1:2 nanostructures in acute and subacute experiments it was shown that in terms of the average lethal dose in an acute experiment on *T. pyriformis*, the nanostructures belong to the 2-nd hazard class. According to the results of the assessment for *T. pyriformis* in a subacute experiment, this nanostructures belong to the 4-th hazard class (it is a low-hazard compound). Due to the fact that the classification of the object under study to the hazard class is carried out according to the parameter, the value of which corresponds to the highest hazard class, then according to the results of the primary toxicological and hygienic assessment for *T. pyriformis*, the nanostructures of curcumin complex with hydrohypropylated-β-cyclodextrin at a molar ratio 1:2 refered to the 2-nd hazard class (is a highly hazardous compound).

Toxicological and hygienic assessment of curcuminoids nanostructures with beta-CD and hydroxypropylated beta-CD in the Wistar rat model system haven't reveal any traits of toxical effect. The data of autopsy and macroscopic examination of the organs under study did not reveal significant differences in the clinical picture between the animals of the experimental and control groups.

The study of the antioxidant system of animals did not reveal significant differences. In the experimental groups of animals that received various nanocompositions of curcumin, there was a slight but significant increase in lactate dehydrogenase, which indicates a positive effect of nanostructures on the antioxidant defense system of the rat organism.

The data of studying the biochemical parameters of the blood of rats of the control and experimental groups did not reveal significant differences between the control and experimental groups of animals.

The study of rats' blood showed that there were also no significant differences between the studied groups in terms of hematological blood parameters.

The study of blood count indices showed that in animals receiving curcumin in the form of nanostructures, there was a significant increase in the number of neutrophils and a decrease in the number of lymphocytes. This increase in the number of neutrophils is within the normal range of the reaction, but may indicate the cytostatic effect of curcumin, which could lead to a decrease in lymphocyte division and differentiation of cells in the myeloid direction, which include neutrophils.

The data of various indicators of urine study in control and experimental groups did not reveal differences between the groups, with the exception of changes in urine acidity and its shift in the alkaline range. That may be due to a decrease in the reabsorption of Na^{+} in the renal tubules.

The "tail immersion test", that was carried out on the 7-th day, showed a greater analgesic efficiency of the curcumin nanocomposition with hydroxypropyl-β-dextrin, indicating a high bioavailability of the turmeric extract.

The analgesic effect of curcuminoids nanostructures was confirmed when studying the summation-threshold indicator 20 days after the administration of the nanocompositions.

References

1. Amalraj, A., Pius, A., Gopi, S.: Biological activities of curcuminoids, other biomolecules from turmeric and their derivatives – a review. J. Tradit. Complem. Med. **7**(2), 205–233 (2017)
2. Bagchi, A., et al.: Extraction of curcumin. IOSR J. Environ. Sci. Toxicol. Food Technol. **1**, 1–16 (2012)
3. Barchitta, M., et al.: Nutrition and wound healing: an overview focusing on the beneficial effects of curcumin. Int. J. Mol. Sci. **20**, 1119–1132 (2019)
4. Batra, H., Pawar, S., Bahl, D.: Curcumin in combination with anti-cancer drugs: a nanomedicine review. Pharmacol. Res. **139**, 91–105 (2019)
5. Fadus, M.C., Lau, C., Bikhchandani, J.: Curcumin: an age-old anti-inflammatory and anti-neoplastic agent. A review. J. Tradit. Complem. Medicine **7**(3), 339–346 (2017)
6. Fernández, M.A., et al.: Complex systems that incorporate cyclodextrins to get materials for some specific applications. Carbohydr. Res. **480**, 12–34 (2019)
7. Fu, S., et al.: Bioaccessibility of curcuminoids in buttermilk in simulated gastrointestinal digestion models. Food Chem. **179**, 52–59 (2015)
8. Huang, S., Chen, F., Cheng, H., Huang, G.: Modification and application of polysaccharide from traditional Chinese medicine such as Dendrobium officinale. Int. J. Biol. Macromol. **157**, 385–393 (2020)
9. Jansook, P., Ogawa, N., Loftsson, T.: Cyclodextrins: structure, physicochemical properties and pharmaceutical applications. Int. J. Pharm. **535**(1–2), 272–284 (2018)
10. Jia, S., et al.: Identification and characterization of curcuminoids in turmeric using ultra-high-performance liquid chromatography-quadrupole time of flight tandem mass spectrometry. J. Chromatogr. A. **1521**, 110–122 (2017)
11. Kapustin, M.A., Chubarova, H.S., Kurchenko, V.P. (eds.): Belarusian medicines: International Scientific Conference. NAS IBOC, Minsk, Belarus (2016)
12. Kapustin, M.A., Chubarova, H.S., Kurchenko, V.P.: Isolation of curcuminoids from Curcuma longa L. and its composition study by chromatographic methods. Proc. BSU **11**(2), 248–262 (2016)
13. Kapustin, M.A., Chubarova, H.S., Kurchenko, V.P. (eds.): Innovations in food technology, biotechnology and chemistry: International Scientific Conference. PC "Science", Saratov, RF (2017)
14. Kapustin, M.A., et al.: Methods of obtaining nanocomplexes of biologically active substances with cyclic oligosaccharides, analysis of their physicochemical properties and use in food production. Proc. BSU **11**(1), 73–100 (2016)
15. Kaur, I.P.: Antimutagenicity of curcumin and related compounds against genotoxic heterocyclic amines from cooked food: the structural requirement. Food Chem. **111**, 573–579 (2008)
16. Khan, H., Ullah, H., Nabavi, S.M.: Mechanistic insights of hepatoprotective effects of curcumin: therapeutic updates and future prospects. Food Chem. Toxicol. **124**, 182–191 (2019)
17. Kharat, M., McClements, D.J.: Recent advances in colloidal delivery systems for nutraceuticals: a case study – Delivery by Design of curcumin. J. Colloid Interface Sci. **557**, 506–518 (2019)

18. Kotha, R.R., Luthria, D.L.: Curcumin: biological, pharmaceutical, nutraceutical, and analytical aspects. Molecules **24**(16), 2930–2956 (2019)
19. Laokuldilok, N., Thakeow, P., Kopermsub, P.: Optimisation of microencapsulation of turmeric extract for masking flavor. Food Chem. **194**, 695–704 (2016)
20. Menezes, P.D.P., Andrade, T.A., Frank, L.A.: Advances of nanosystems containing cyclodextrins and their applications in pharmaceuticals. Int. J. Pharm. **559**, 312–328 (2019)
21. Murugan, S., et al.: Safety evaluation of standardized extract of curcuma longa (NR-INF-02): a 90-day subchronic oral toxicity study in rats. Biomed. Res. Int. **2021**, 1–14 (2021)
22. Patil, S.S., Bhasarkar, S., Rathod, V.K.: Extraction of curcuminoids from Curcuma longa: comparative study between batch extraction and novel three phase partitioning. Prep. Biochem. Biotechnol. **49**(4), 407–418 (2019)
23. Rathore, S., et al.: Curcumin: a review for health benefits. Int. J. Res. Rev. **7**, 273–290 (2020)
24. Sahne, F., Mohammadi, M., Najafpour, G.D.: Enzyme-assisted ionic liquid extraction of bioactive compound from turmeric (*Curcuma longa* L.): isolation, purification and analysis of curcumin. Ind. Crop. Prod. **95**, 686–694 (2017)
25. Siviero, A., Gallo, E., Maggini, V.: Curcumin, a golden spice with a low Bioavailability. J. Herb. Med. **5**, 57–70 (2015)
26. Smirnov, Y.A.: Antiviral potential of turmeric (*Curcuma longa* L.). Tradit. Medicine **2**(61), 12–19 (2020)
27. Souto, E.B., et al.: New nanotechnologies for the treatment and repair of skin burns infections. Int. J. Mol. Sci. **21**, 393–410 (2020)
28. Wiggers, H.J., et al.: Curcumin, a multitarget phytochemical: challenges and perspectives. Stud. Nat. Prod. Chem. **53**, 243–276 (2017)
29. Zhang, D., et al.: Cyclodextrin-based delivery systems for cancer treatment. Mater. Sci. Eng. C Mater. Biol. Appl. **96**, 872–886 (2019)
30. Zielinska, A., et al.: Properties, extraction methods, and delivery systems for curcumin as a natural source of beneficial health effects. Medicina (Kaunas) **56**(7), 336–354 (2020)

Protein-Polysaccharide Complex for the Production of an Emulsion Product

Elena Melnikova, Ekaterina Stanislavskaia(✉), Ekaterina Shabalova, and Natalia Antonova

Voronezh State University of Engineering Technologies, Voronezh, Russia
melnikova@molvest.ru

Abstract. The aim of the work is to develop a protein-polysaccharide complex imitating the flavor of milk fat for use in mayonnaise sauce technology. In the course of the research, standard methods used to assess the quality of emulsion products were used. The use of low-esterified pectin together with a microparticulate of whey proteins increases the stability of the emulsion. This confirms the feasibility of using a protein-polysaccharide complex in mayonnaise sauce technology. The use of the developed food composition allows to increase the stability of the emulsion to 100%, to obtain a homogeneous structure of the product, resistant to delamination. The increase in the stability of the emulsion with the addition of a protein-polysaccharide complex is explained by the creation of protective boundary layers around fat droplets by biopolymers. The use of pectins promotes the formation of a mesh structure, which strengthens the dispersed phase and slows down its destruction. Microparticulate particles are evenly dispersed in pectin and tightly surround fat droplets. The system formed by the protein-polysaccharide complex prevents the flocculation of fat droplets. The energy value of the sauce is 322.6 kcal/100 g, which is 50% lower than the control sample. At the same time, the sauce is characterized by standard quality indicators (the viscosity of the product is 148 MPa · s, active acidity is 6.22 pH units, titrated acidity is 0.84°) and has no significant differences in organoleptic properties from the control.

Keywords: Sauce · Fat replacement · Microparticulation · Pectin · Emulsion

1 Introduction

Modern trends in the nutrition of the population dictate the need to expand the range of food products. There are new products, in particular sauces. These are multicomponent food systems, which include consistency stabilizers, thickeners, emulsifiers, which make it possible to obtain emulsions stable to coalescence and oxidation. Among the various hydrocolloids, pectin is the most preferred to increase the stability of emulsion products. It is a heteropolysaccharide in which the elementary element is to varying degrees the methoxylated residues of D-galacturonic acid. Due to its hydrophobicity, it can form a film around the fat globule, simultaneously exhibiting emulsifying and stabilizing properties [7, 8].

V. Kurchenko et al. (Eds.): ICAETT 2021, LNNS 408, pp. 236–243, 2022.
https://doi.org/10.1007/978-3-030-96641-6_27

It is promising to use food additives not only to obtain a stable consistency of sauces, but also to increase their biological value, reduce calories and maintain consistently high quality indicators. Microparticulated whey protein concentrates [1] imitating the flavor of milk fat are relevant for use in sauce technology [3]. Such food compositions make it possible to improve the organoleptic properties of food products while reducing their caloric content and fat mass fraction [2, 6, 9–12].

2 Materials and Methods

Considering the relevance, the aim of the work was to develop a protein-polysaccharide complex imitating the flavor of milk fat for use in mayonnaise sauce technology. To achieve this goal, the following tasks were identified.

1. To obtain and investigate a protein-polysaccharide complex of fat for a low-calorie mayonnaise sauce.
2. To develop a recipe-component solution of mayonnaise sauce based on it.
3. To study the functional and technological properties, quality and safety of mayonnaise sauce during storage.

The research was carried out in the conditions of the Department of Technology of Animal Products and the Center for Collective Use of the Voronezh State University of Engineering Technologies, OOO "Mallab", the laboratory of the ANO STC "Kombikorm" (Voronezh, Russia). The objects of the study were whey protein concentrate (WPC) produced by PAO "Voronezhsky" Dairy Plant (Voronezh, Russia); samples of a complex food additive obtained on its basis, as well as using highly esterified (HE) and low-esterified (LE) pectins, mayonnaise sauces, purchased in the retail chain of Voronezh, using the obtained complexes.

To obtain a protein food composition imitating the flavor of milk fat, whey protein concentrate was dissolved in distilled water to a protein mass fraction of 8.6% and subjected to thermomechanical treatment (microparticulation). A pectin solution with a mass fraction of 5% was prepared separately: pectin was dissolved in water heated to 50 °C, dispersed on a laboratory homogenizer at a rotor rotation speed of 18000 rpm for 6–7 min. To obtain protein-polysaccharide complexes, the microparticulate of whey proteins was mixed with a solution of pectin [5].

During the study of sauce samples, standard methods were used (according to GOST 31762-2012 Mayonnaise and mayonnaise sauces. Acceptance rules and test methods) used to assess the quality of emulsion products. The effective viscosity is determined using a rotary viscometer SV-10. Its principle of operation is based on the relationship between the power expended to excite the vibration of two thin sensor plates with a frequency of 30 Hz and a constant amplitude of 1 mm, and the viscosity of the liquid. Solutions with a mass fraction of 1% protein and deodorized sunflower oil were used to determine the emulsifying ability of complex additives. A series of emulsions with a fat phase content from 10 to 80% was prepared. Emulsification was performed on a laboratory homogenizer with a rotation speed of 3000 min^{-1} at a rate of oil addition of 1 drop per second. The resulting emulsion is poured using a syringe into test tubes (5 mm in diameter and 100 mm high) and thermostated at 85 °C for 20 min. Next, the tubes were cooled with running water and centrifuged for 20 min at a rotational speed of 6000 min^{-1}. The criterion for the stability of emulsions at the initial ratio of fat and water phases was the ratio of the height of the emulsion layer to the total height (in % by volume). Based on the results obtained, a diagram of the stability of the emulsion was constructed in the axes: the initial volume fraction of the fat phase is the ratio of the volumes of the phases as a percentage. The fat peroxide number was determined by the iodometric method based on the oxidation of potassium iodide with fat peroxides and hydroperoxides of fat in a solution of acetic acid and chloroform and titration of the released iodine with a solution of sodium thiosulfate. The microstructure of the objects of study was evaluated using a binocular microscope Altami BIO 6. Mathematical processing of the experiment was carried out by methods of mathematical statistics according to the data of 5–10 experiments in a three-fold sequence.

3 Results

The emulsifying ability of food additives: protein and protein-polysaccharide complex was studied (Fig. 1).

When using low-esterified pectin together with a microparticulate of whey proteins, the stability of the emulsion increases.

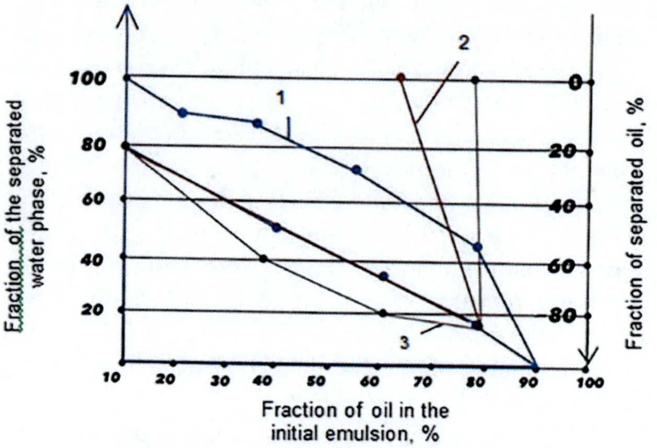

Fig. 1. Stability diagram of the emulsion: 1 – WPC, 2 – WPC+HE pectin, 3 – WPC+LE pectin

To develop a low-calorie mayonnaise sauce, "Provancal" mayonnaise, developed according to a traditional recipe, was chosen as a control. To prepare model prototypes in the standard mayonnaise recipe, vegetable oil, skimmed milk powder and starch were replaced with various protein and protein-polysaccharide compositions.

The use of acetic acid significantly reduced the pH of the product. Globular proteins in complexes with carboxyl-containing polysaccharides are precipitated at a pH below the isoelectric point (IEP) of the protein. At the same time, there is a shift in pH to the acidic region relative to the IEP. Macromolecules of protein and anionic polysaccharide are characterized by total charges opposite in sign at a pH below IEP. In this case, an insoluble complex is formed. The presence of acetic acid contributes to the separation of the sauce and the separation of the insoluble protein-polysaccharide complex. In this connection, acetic acid was excluded from the recipe of sauces (Table 1).

Table 1. Recipes of experimental samples of sauces

The component	The proportion of component in the sample №					
	1 - control	2	3	4*	5	6*
Vegetable oil	65,4	32,7	32,7	32,7	32,7	32,7
Egg powder	5	5	5	5	5	5
Skimmed milk powder	1,6	–	–	–	–	
Sugar	1,5	1,5	1,5	1,5	1,5	1,5
Salt	1,1	1,1	1,1	1,1	1,1	1,1
Sodium	0,05	0,05	0,05	0,05	0,05	0,05
Mustard powder	0,75	0,75	0,75	0,75	0,75	0,75
Acetic acid	0,75	–	–	–	–	–
Starch	0,5	–	–	–	–	–
WPC	–	58,15	48,9	48,9	48,9	48,9
Pectin solution 5% HE	–	–	10	10	–	–
Pectin solution 5% LE	–	–	–	–	10	10
Water	23,35	–	–	–	–	–

* - samples 4 and 6 are obtained by joint microparticulation of WPC and pectin.

The qualitative indicators of the obtained samples are presented in Table 2.

Table 2. Physico-chemical parameters of sauce samples

The indicator	The value of the indicator for the sample					
	1	2	3	4	5	6
Mass fraction of fat, %	57	35	35	35	35	35
Acidity, deg.	1,0	0,84	0,84	0,84	0,84	0,84
Stability of the emulsion of the undestroyed emulsion, %	98	92	95	89	100	99
Active acidity, unit pH	6,91	6,28	6,51	6,42	6,22	6,29
Viscosity, MPa·s	250	115	224	189	148	212

The stability of emulsions in the presence of protein-polysaccharide complexes after storage for a day was high. The emulsion stratification was completely absent in samples 5 and 6. The structure of such emulsions was more dense and homogeneous compared to emulsions stabilized only by a microparticle of whey proteins (sample 2). In samples 3 and 4, the emulsion stratification was noted. Despite the high viscosity value, the samples of sauces 4 and 6 were characterized by an inhomogeneous structure of the dispersed system, which is confirmed by photographs of the microstructure (Fig. 2).

The best qualitative indicators were characterized by a sample of mayonnaise sauce No. 5, developed on the basis of a mixture of microparticulate of whey proteins and low-esterified pectin. The proportion of low-esterified pectin in the finished product was selected in accordance with its effect on viscosity and organoleptic characteristics (Table 3).

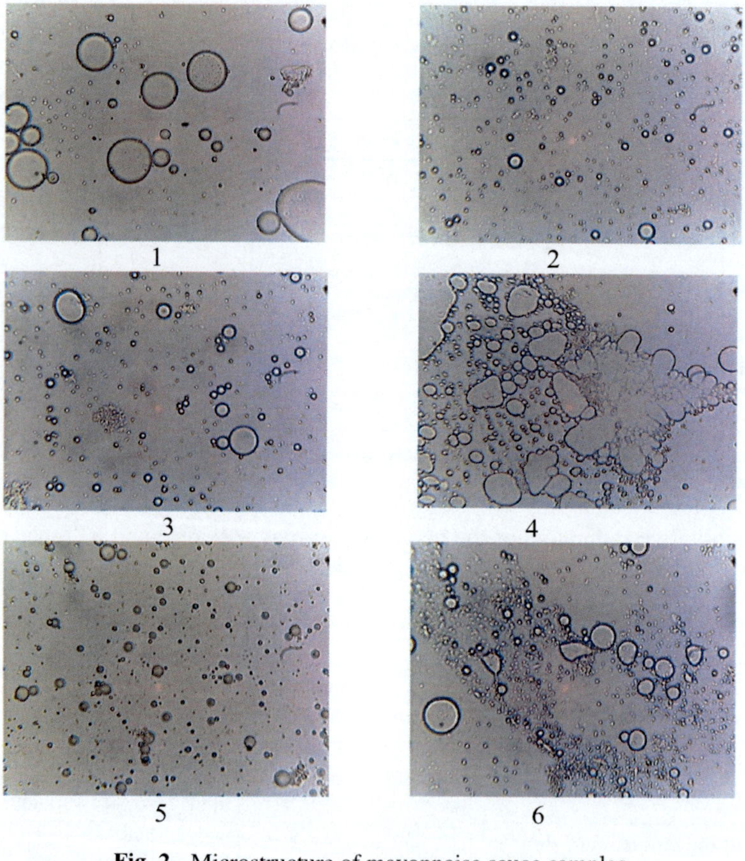

Fig. 2. Microstructure of mayonnaise sauce samples

The recommended dosage of the protein-polysaccharide complex in the mayonnaise sauce formulation was 589 kg/ton, the ratio of a solution of low-esterified pectin and microparticulates of whey proteins was 1:4.9.

Table 3. Effect of the mass fraction of pectin solution (5%) on the quality of sauce samples

The indicator	The value of the indicator for the sample with proportion of the pectin solution, %			
	5	10	15	20
Viscosity, MPa·s	112	148	169	187
Appearance, consistency	Liquid, homogeneous	Thick, homogeneous, moderately viscous	Thick, homogeneous, moderately viscous, stretching	Thick, homogeneous, moderately viscous, stretching

The qualitative indicators of the new emulsion product are presented in Table 4.

Table 4. The qualitative indicators of the sause

The indicator	The value of the indicator
Appearance, consistency	Thick, homogeneous, moderately viscous consistency
Taste and smell	Sweet-sour with an aftertaste of mustard
Color	Yellow, uniform over entire the mass
Acidity, deg	0,84
Stability of the emulsion of the undestroyed emulsion, %	100
Active acidity, unit pH	6,22
Viscosity, MPa·s	148

The energy value of the sauce was 322.6 kcal/100 g, which is 50% lower than the control sample. At the same time, the sauce was characterized by standard quality indicators and had no significant differences in organoleptic properties from the control.

A study of the storage capacity of the product performed in accordance with traditional methods (Table 5), allowed us to conclude that the shelf life of the sauce was 25 days. During this period, the product met the safety requirements for fat-and-oil products.

Table 5. Changing the quality indicators of the sauce during storage

The indicator	Acceptable level	The value of the indicator during storage, day			
		5	15	25	33
Peroxide number, meq/kg	No more than 10	0,4	1,1	3,5	5,5
Titrated acidity, deg	No more than 1	0,84	0,87	0,91	0,96
Stability of the emulsion, %	No less than 98	100	100	100	99
Yeast, CFU/g	No more than 500	–	–	No less than 10	No Less Than 10
Mold, CFU/g	No more than 50	–	–	No less than 10	11
E. coli group bacteria	Not allowed in 0.1 g	–	–	–	–

4 Discussion and Conclusion

Analysis of the stability diagram of the emulsion (Fig. 1) allows us to conclude that in the presence of pectin, the emulsifying ability of the protein component increases. The aqueous phase is almost completely bound into an emulsion, its separation in the presence of protein-polysaccharide complexes does not occur.

Pectins with a low degree of methoxylation are differed by a large number of free negatively charged carboxyl groups. In low-esterified pectins, the conformation of the macromolecule changes with the replacement of the intramolecular electrostatic bond of metal ions with a stronger intramolecular chelate bond, in the formation of which not only carboxylic, but also hydroxyl groups of galacturonic acid participate. The formation of chelate bonds strengthens the structure of the inotropic gel, which affects the increase in the aggregate stability of emulsions stabilized with pectin [4].

When preparing mayonnaise sauce, the addition of a protein-polysaccharide complex gel formation, increases viscosity and improves the moisture-retaining ability of the product. Product samples using low-esterified pectin are characterized by greater stability of the emulsion. Low-esterified pectins form gels in the presence of calcium, which binds the carboxyl groups of two pectin acid molecules, and are characterized by a higher linear charge density of the molecules, on which the strength and method of binding of calcium cations depend.

The increase in the stability of the emulsion with the addition of a protein-polysaccharide complex is explained by the creation of protective boundary layers around fat droplets by biopolymers. The use of pectins promotes the formation of a mesh structure, which strengthens the dispersed phase and slows down its destruction. Microparticulate particles are evenly dispersed in pectin and tightly surround fat droplets. The system formed by protein-polysaccharide complexes prevents the flocculation of fat droplets. With joint microparticulation (sauce samples 4, 6), pectin interacts with positively charged groups on the surface of a protein molecule at the stage of preparation of a complex food additive. In the resulting product, the number of reactive active groups is significantly reduced, the stabilization of the fat phase practically does not occur.

The performed studies made it possible to obtain a complex food additive based on a microparticulate of whey proteins and low-esterified pectin. The complex is characterized by a high emulsifying ability and can be effectively used in the technology of emulsion products. The use of the resulting additive in the mayonnaise sauce makes it possible to reduce the energy value of the product while maintaining high quality indicators.

References

1. Bannikova, A.V., Evdokimov, I.A.: The scientific and practical principles of creating products with increased protein content. Foods Raw Mater. **3**(2), 3–12 (2015)
2. Baranov, S.A., Evdokimov, I.A., Gordienko, L.A., Shramko, M.I.: The effect of whey protein microparticulate on the indicators of fermented milk drinks. Dairy Ind. **9**, 59–62 (2020)
3. Evdokimov, I.A., et al.: Composition and properties of whey protein microparticulates. Dairy Ind. **4**, 40–44 (2021)
4. Kulikova, V.V., Barybina, L.I., Oboturova, N.P., Datsko, V.A.: Complexation of milk whey protein concentrate with anionic polysaccharides. Food Ind. **3**, 62–64 (2014)
5. Melnikova, E.I., Stanislavskaya, E.B., Baranova, K.: Application of whey protein ingredients for the production of milk fat imitators. VGUIT Bull. **82**(3), 90–95 (2020)
6. Di Cagno, R., De Pasquale, I., De Angelis, M., Buchin, S., Rizzello, C.G., Gobbetti, M.: Use of microparticulated whey protein concentrate, exopolysaccharide-producing Streptococcus thermophilus, and adjunct cultures for making low-fat Italian Caciotta-type cheese. J. Dairy Sci. **97**(1), 72–84 (2014)
7. Krzeminski, A., Prell, K.A., Busch-Stockfisch, M., Weiss, J., Hinrichs, J.: Whey protein-pectin complexes as new texturising elements in fat-reduced yoghurt systems. Int. Dairy J. **36**(2), 118–127 (2014)
8. O'Chiu, E., Vardhanabhuti, B.: Utilizing whey protein isolate and polysaccharide complexes to stabilize aerated dairy gels. J. Dairy Sci. **100**(5), 3404–3412 (2017)
9. Olivares, M.L., Shahrivar, K., de Vicente, J.: Soft lubrication characteristics of microparticulated whey proteins used as fat replacers in dairy systems. J. Food Eng. **245**, 157–165 (2019)
10. Richard, I.: Microparticulated whey proteins for improving dairy product texture. Int. Dairy J. **67**, 73–79 (2017)
11. Torres, I.C., Mutaf, G., Larsen, F.H., Ipsen, R.: Effect of hydration of microparticulated whey protein ingredients on their gelling behaviour in a non-fat milk system. J. Food Eng. **184**, 31–37 (2016)
12. Torres, I.C., Amigo, J.M., Knudsen, J.C., Tolkach, A., Mikkelsen, B.Ø., Ipsen, R.: Rheology and microstructure of low-fat yoghurt produced with whey protein microparticles as fat replacer. Int. Dairy J. **81**, 62–71 (2018)

Laser Doppler Flowmetry and Tissue Oxygenation Monitoring in Assessing of the Survival Rate of an Elongated Skin Flap Under the Biomaterials Application

Zulfira Siraeva[1,2](✉), Renata Mullina[2], Olga Kravtsova[1], and Ruslan Mullin[1,3]

[1] Institute of Fundamental Medicine and Biology, Kazan (Volga Region) Federal University, Kazan, Russia
zsiraeva@yandex.ru, okravz@yandex.ru
[2] Kazan State Medical University, Kazan, Russia
mullina.rena@mail.ru
[3] Republic Clinical Hospital, Kazan, Russia
rusdan@mail.ru

Abstract. Necrosis of the distal part of the flap, especially in elongated flaps, is one of the complications of vascularized skin grafting due to insufficient blood supply. A promising approach aimed at improving the survival of skin flaps is the development and use of biomaterials. At the same time the range of methods approved for use in the clinic and allowing non-invasive assessment of the effectiveness of biomaterials is limited. The aim of the work is to substantiate the possibility of using the laser Doppler flowmetry method and monitoring tissue oxygenation with the moorVMS-PC software (Windows ™) to assess the survival rate of an elongated skin flap under the action of biomaterials. A modified model of an elongated dorsal (width to length ratio 1:4) skin flap is proposed. A decrease in the area of flap necrotization with the use of the Zn-dopped matrix by 3.3 times compared with the control group and the absence of necrotization with the use of the non-dopped matrix was found. It has been shown that microcirculation in the caudal part of the flap without the use of matrices decreases in the caudal part of the flap and increases in the cranial part; tissue oxygenation increases in the cranial direction. There were no differences in blood flow rates when using a non-dopped matrix, which indirectly indicates the creation of favorable conditions for flap vascularization. For the first time, using the laser Doppler flowmetry method and monitoring tissue oxygenation with the moorVMS-PC software for Windows ™, it was indirectly shown that the studied matrices have an effect on microcirculation, tissue oxygenation and skin graft survival in a modified model of an elongated skin flap with a width to length ratio of 1:4. The LDF method is sensitive to the action of biomaterials and can be used in preclinical studies for the initial assessment of the potential bioactivity of new biomaterials and in clinical studies to assess the effect of biomaterials on the survival of a skin flap.

Keywords: Laser Doppler flowmetry · Tissue oxygenation · Microcirculation · Skin flap · Tissue necrotization · Matrix

V. Kurchenko et al. (Eds.): ICAETT 2021, LNNS 408, pp. 244–252, 2022.
https://doi.org/10.1007/978-3-030-96641-6_28

1 Introduction

Vascularized skin grafting as a method of surgical treatment is used for skin defects treatment accompanied by exposure of the underlying structures: nerves, blood vessels, tendons, bones. The most common skin grafting with local flaps with a randomized type of blood supply [6, 12].

One of the main complications of vascularized skin grafting is necrosis of the distal part of the flap resulting from insufficient blood supply [1, 11, 12]. The length to width ratio of the flap is considered an important factor influencing the survival of this type of flap [7, 11]. It is believed that the optimal ratio of width to length for complete engraftment of the flap should be less than 1:3 [1, 10–12, 17]. This is due to the fact that when cutting out a flap, there is a decrease in vascularization (up to 80%), which significantly limits the use in the clinic of more popular flaps with a larger ratio of the linear dimensions of the sides [3, 11, 17].

One of the promising approaches aimed at accelerating the healing of skin defects is the use of 3D matrices based on natural compounds that enhance tissue vascularization [9, 12]. At the same time, the range of methods approved for use in the clinic and allowing non-invasive assessment of tissue vascularization is limited.

In recent years, the method of laser Doppler flowmetry (LDF) has attracted the attention of scientists and clinicians to assess tissue microcirculation in pathology in various fields of medicine. Thus, the method was tested in neurology for the diagnosis of diseases of the peripheral nervous system [16], in dental implantology for the diagnosis of changes in vascular blood flow in the connective tissue of the mucous membranes of the oral cavity [8]; in cardiology for the detection of skin microcirculation in adolescents with obesity in the preclinical phase of cardiovascular diseases [4], endocrinology for the study of skin perfusion in chronic skin ulcers [14] and others [8, 13, 15] areas. At the same time, no studies were found on the use of the LDF method in vascularized skin grafting using biomaterials.

In this regard, the purpose of the work is to substantiate the possibility of using the LDF method and tissue oxygenation to assess the survival rate of an elongated skin flap with a side length ratio of more than 1:3 under the action of 3D matrices. Research objectives: to develop a modified model of a skin flap with a width to length ratio of more than 1:3; conduct a visual analysis of the degree of tissue necrosis of the skin flap without and with the use of 3D matrices; to study the intensity of microcirculation and tissue oxygenation of the skin flap without and with the use of 3D matrices.

2 Materials and Methods

2.1 Vival Model

Adult male Wistar rats (18 animals, 306 ± 21 g) were received from Vivarium of Academy of Medical and Technical sciences (Russia). Experiments were carried out according to the protocol for the protection of experimental animals [2, 5].

2.2 Matrices

Experimental non-dopped matrix и Zn-dopped matrix were synthesized by cryopolymerization method [9]. Before the surgical treatment, the matrix membranes were cut into samples 1.5 × 6.0 cm in size, treated with a solution of penicillin (5 U/ml) and streptomycin (5 mg/ml) and washed with a sterile isotonic solution to remove excess of antibiotics.

2.3 Skin Flap Model

A fascial skin flap of the dorsal surface of the rat body on the cranial feeding pedicle [3] with modification of the linear dimensions of the flap (increasing the aspect ratio of the flap to an unfavorable ratio of 1:4 for the engraftment of the flap) was used as a model.

2.4 Animal's Treatment

2.4.1 Before Surgery Treatment

Rats were subjected to inhalation anesthesia with 2.5% isoflurane using an Eickemeyer evaporator equipped with a DeVilbiss oxygen concentrator. The animals were fixed and depilated in a rectangular area of 80 cm^2 (8.0 × 10.0 cm) on the dorsal upper thoracic region along the axis of the spine. Antiseptic treatment of the operating field was carried out by three times treatment with a 70% solution of ethyl alcohol.

Using a marker, a 9.0 cm^2 rectangular skin flap with sides of 1.5 × 6.0 cm was projected onto the depilated area and divided into 4 equal zones (Fig. 1A, 1B). Further, using the modules for monitoring microcirculation (Fig. 1D) and recording tissue oxygenation (Fig. 1E), the readings of tissue microcirculation (Fig. 1F) and oxygenation (Fig. 1G) were recorded in each zone, which were taken as baseline control values. Zone boundaries were counted at the target suture points (Fig. 1B).

2.4.2 Surgical Treatment

After modeling the flap and registering microcirculation and oxygenation, the skin surface of the prepared animal was treated with a 70% alcohol solution. A skin flap was cut out with a surgical scalpel and retracted from the fascia of the dorsal muscle (Fig. 2A). The wounds were desinfected by rinsing with 0.01% myramistine solution (LLC Infamed, Russia) and washed with sterile isotonic solution. In control group (group 1, n = 6), the flap was sutured without matrix (Fig. 1A, Fig. 2B); in animals from 2nd experimental group (n = 6) before suturing non-doped matrix were applied in the form of a membrane (1.5 × 6.0 cm) under the skin flap; animals of group 3 were treated with zinc-doped matrix membranes (experiment, n = 6; Zn-doped matrix). To avoid infection, the treatment of wounds was carried out with Miramistin® after suturing, then once a day during the entire experiment (6 days). After suturing, a sterile gauze napkins were applied to the wound and fixed with a bandage (Leico) (Fig. 2C).

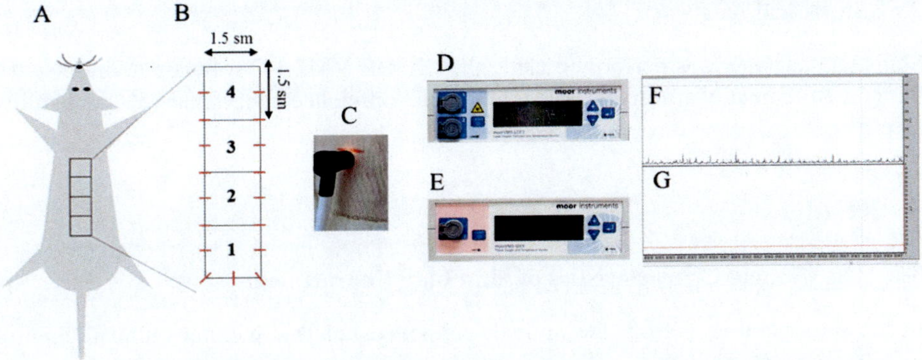

Fig. 1. Scheme of marking a skin flap on the dorsal surface of a laboratory animal's body, followed by data registration using the moorVMS system. A - cutting out a skin flap; B - dimensional characteristics of the skin flap, localization of skin sutures (highlighted in red), 1–4 - numbering of zones; C - recording probe; D - microcirculation monitoring module - moorVMS-LDF (https://www.moor.co.uk/en-us/products/monitoring/laser-doppler-monitor/); E - tissue oxygenation measurement module - moorVMS-OXY (https://us.moor.co.uk/product/moorvms-oxy-superficial-tissue-oxygenation-monitor/202) F, G - program interface and representative microcirculation curves (F) and tissue oxygenation (G) obtained using the moorVMS-LDF (D) and moorVMS-OXY (E) modules, respectively.

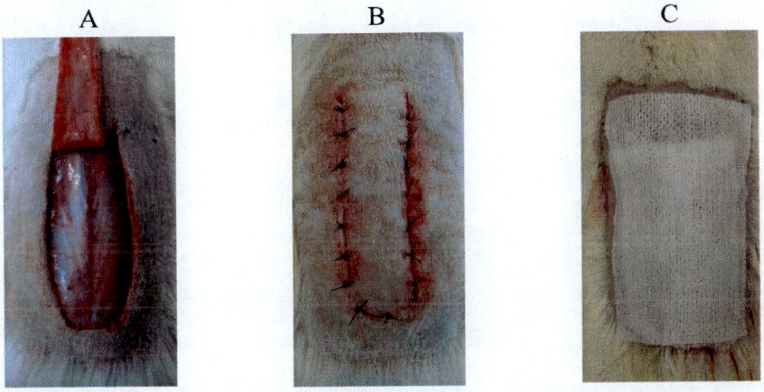

Fig. 2. An experimental animal with a cut skin flap before suturing (A), after suturing (B) and wound fixing (C).

2.5 Microcirculation and Tissue Oxygenation Registration

LDF [13] was performed before surgery and before removing the animals from the experiment (6 days after injury). Within 6 days after the operation, the degree of graft engraftment was visually assessed by the external condition of the tissues, edema, tissue necrosis and inflammation signs.

2.6 Statistical Analysis

Statistical analysis was performed using the MOOR VMS-PC software (Windows ™) with measurement of arithmetic means and their confidence intervals at 95% probability level.

3 Results

3.1 The Clinical Characteristics of Skin Flap Engraftment

In the postoperative period, the animals remained healthy, without clinical signs of infection and intoxication.

On the 3rd day after the surgery in animals from 1st group, zone 4 acquired cyanosis, the tissues were swollen, the suture edges of the wound in zone 4 turned black, which indicates the course of necrotic processes. No pronounced changes in the flap tissues in animals from 2nd and 3rd groups were registered on the 3rd day.

Figure 3 shows representative photographs of the condition of the skin flap after 6 days surgery.

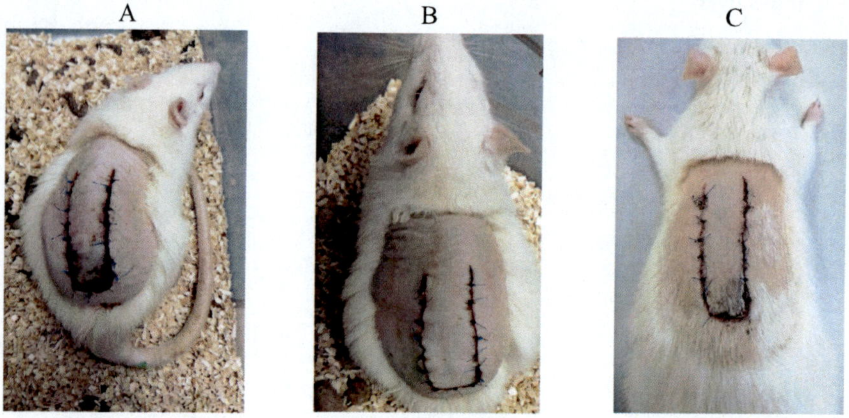

Fig. 3. Photographs of a skin flap after 6 days injury. A - group I, B - group II, B - group III.

Visual analysis revealed significant inflammation of the skin flap in animals in group I, especially in the caudal part, which manifested itself in tissue edema, blue discoloration of zones 3 and 4, and marginal necrosis (Fig. 3A). In contrast to animals of group I, the flap in group II was without signs of inflammation, necrosis, tissue edema (Fig. 3B). In group III animals, the condition of the flap tissues was generally comparable to group II, with the exception of a minor area in zone 4 corresponding to the wound edge suturing line (Fig. 3C). In the latter case, a slight darkening of the tissues may be due to their damage during suturing.

The degree of necrotic zones (mean ± confidence interval) is represented in Table 1.

Table 1. The degree of flap tissue necrosis in experimental animals after 6 days injury

Experimental group	Necrotic tissue area	
	mm^2	% from total area
I	247.5 ± 21.9	27.5
II	0*	0*
III	75.0 ± 8.8*	8.3*

*Statistical difference at $p \leq 0.05$.

Analysis of the degree of necrosis of the dorsal zones showed that the area of flap necrosis when using the Zn-doped-matrix decreased 3.3 times, and when using non-doped-matrix, tissue necrosis did not occur.

3.2 Intensity of Microcirculation and Tissue Oxygenation

To characterize the blood flow in the flap zones, a value was recorded that integrally reflects the intensity of microcirculation and expressed in relative perfusion units (PU) [13].

The mean blood microcirculation in the skin flap in group 1 decreased 1.67 and 2.17 times in zones 1 and 2, respectively, but increased 1.4 and 1.83 times, respectively, in zones 3 and 4 (Fig. 4A). At the same time, oxygenation in all zones increased in the caudal-cranial direction by 1.42–3.58 times (Fig. 4B).

In 2nd group the preoperative values of both microcirculation (Fig. 4A) and tissue oxygenation (Fig. 4B) did not differ significantly from the values recorded on the 6 days after flap modeling, which indirectly indicates the restoration of blood flow in the flap tissues.

At the same time, in 3rd group the average value of blood microcirculation in the skin flap decreased by 1.35 and 1.38 times in zones 1 and 2, respectively, in zone 3 the level of microcirculation decrease decreased to 1.54 times, and in zone 4 it did not had significant differences with the preoperative microcirculation value (Fig. 4A). Analysis of oxygenation in the skin of animals of group III revealed elements of similarity with the data of group I: postoperative values gradually increased in the direction from caudal zone 1 to cranial zone 4 by 1.46, 1.97, 2.7 and 4.17 times (Fig. 4B). This is probably due to a slight necrotization of the caudal zone, and the increase in oxygenation is due to the specific action of the biomaterial. This issue requires additional research.

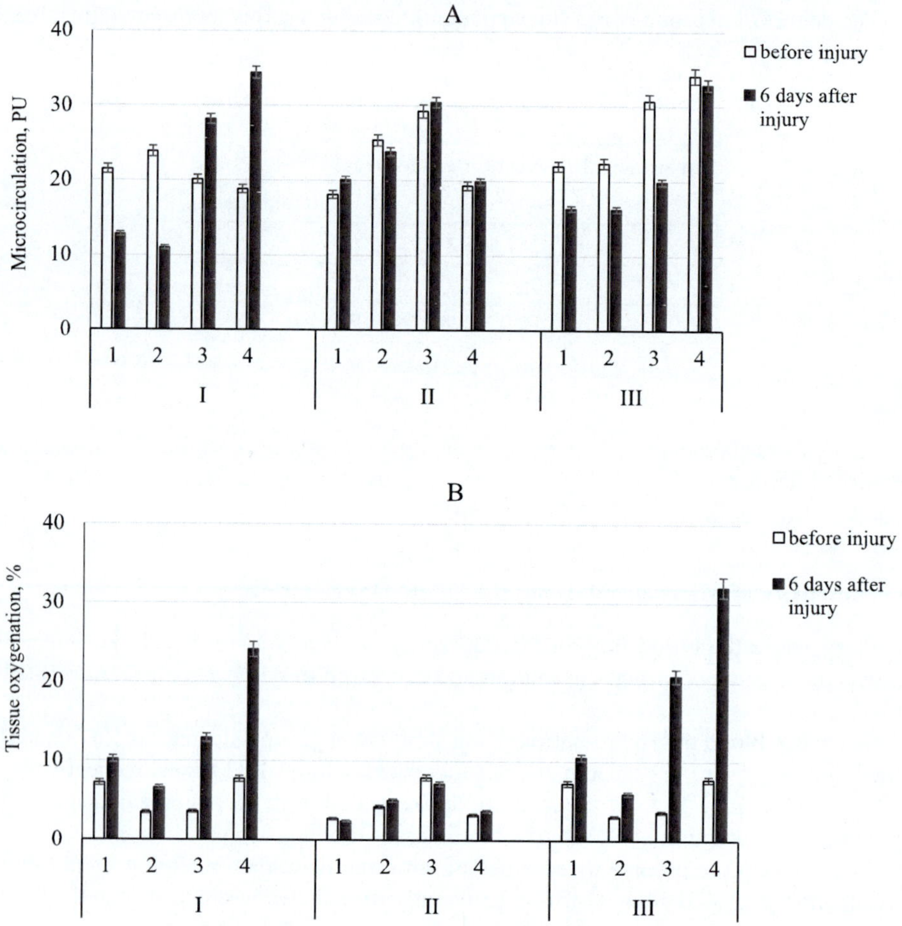

Fig. 4. Indicators of microcirculation and tissue oxygenation in the flap zones. 1–4 - flap zones; I, II, III - groups of experimental animals

4 Discussion and Conclusion

The rat's skin is similar in many aspects to human skin, although some morphological characteristics may differ depending on the region of the body. In connection with the generality of the structure for the study of the bioactivity of matrices, we modified the model of a rat skin flap [8]. The essence of the modification was to lengthen the flap to a width-to-length ratio of 1:4.

When constructing the test program, it was expedient to assess the biological effect of the developed matrices, from our point of view, was the primary study of the clinical picture of engraftment and the assessment of the degree of microcirculation of regenerating tissues in the wound.

The LDF method is sensitive to the action of biomaterials and can be used in preclinical studies for the initial assessment of the potential bioactivity of new biomaterials and in clinical studies to assess the effect of biomaterials on the survival of a skin graft. The results obtained are of great applied importance in the development and search for new applications of medical technologies and software for detecting disorders of the microvasculature.

Acknowledgments. This paper has been supported by the Kazan Federal University Strategic Academic Leadership Program (PRIORITY-2030).

References

1. Antonio, N.N., Costa, A.M.A., Marques, R.G.: Viability of randomized skin flaps – an experimental study in rats. J. Surg. Res. **207**, 92–101 (2017)
2. Directive 2010/63/EU of the European Parliament and of the Council of 22 September 2010 on the protection of animals used for scientific purposes. Text with EEA relevance. http://eur-lex.europa.eu/legal-content/EN/TXT/?uri=celex:32010L0063
3. Dorset-Martin, W.A., Wysocki, B.: Rat models of skin wound healing. In: Conn, M. (ed.) Models for Biomedical Research, pp. 631–638. Humana Press Inc. (2008)
4. Fusco, E., et al.: Preclinical vascular alterations in obese adolescents detected by Laser-Doppler Flowmetry technique. Nutr. Metab. Cardiovasc. Dis. **30**(2), 306–312 (2020)
5. Guide for the Care and Use of Laboratory Animals. The National Academies Press (2011)
6. Hallock, G., Morris, S.F.: Skin grafts and local flaps. Plast. Reconstr. Surg. **127**(1), 5e–22e (2011)
7. Karayannopoulou, M., et al.: Locally injected autologous platelet-rich plasma enhanced tissue perfusion and improved survival of long subdermal plexus skin flaps in dogs. Vet. Comp. Orthop. Traumatol. **5**, 379–386 (2014)
8. Kouadio, A.A., Jordana, F., Koffi, N.J., Le Bars, P., Soueidan, A.: The use of laser Doppler flowmetry to evaluate oral soft tissue blood flow in humans: a review. Arch. Oral Biol. **86**, 58–71 (2018)
9. Luong, D., et al.: Transition metal-doped cryogels as bioactive materials for wound healing applications. Mater. Sci. Eng. C-Mater. Biol. Appl. **103**, 109759, 1–13 (2019)
10. Mullin, R.I., Bogov, A.A., Topyrkin, V.G., Masgutov, I.G., Khannanova, I.G.: Vascularized skin flap plastic surgery with medial-dorsal flap in reverse flow in treatment of patients with combined injuries of fingers. Pract. Med. **2**(4), 92–95 (2014). (in Russian)
11. Mullin, R.I., Masgutov, R.F., Ibragimova, L.Y., Topyrkin, V.G., Filimonova, A.A.: Vascularized dermepenthesis flaps with axial blood supply in patients with combined injuries of the hand. Pract. Med. **7**(55), 83–88 (2011). (in Russian)
12. Pei, Y., et al.: Biomaterial scaffolds for improving vascularization during skin flap regeneration. Chin. J. Plast. Reconstr. Surg. **2**(2), 109–119 (2020)
13. Rajan, V., Varghese, B., van Leeuwen, T.G., Steenbergen, W.: Review of methodological developments in laser Doppler flowmetry. Lasers Med. Sci. **24**(2), 269–283 (2009). https://doi.org/10.1007/s10103-007-0524-0
14. Raposio, E., Bertozzi, N., Moretti, R., Grignaffini, E., Gricco, M.P.: Laser Doppler flowmetry and transcutaneous oximetry in chronic skin ulcers: a comparative evaluation. Wound **29**(7), 190–195 (2017)
15. Siraeva, Z., Đặng, T., Yergeshov, A., Salikhova, T., Lopukhov, L., Abdullin, T.: Evaluation of skin irritation in rats using simultaneous laser Doppler flowmetry and oxygenation monitoring. BioNanoScience **6**(4), 384–387 (2016). https://doi.org/10.1007/s12668-016-0242-9

16. Stoyneva, Z.: Clinical application of laser Doppler flowmetry in neurology. In: Bartels, E., Bartels, S., Poppert, H. (eds.) New Trends in Neurosonology and Cerebral Hemodynamics (2012). Perspect. Med. **1**, 89–93. Elsevier GmbH
17. Zhou, K., Zhang, Y., Lin, D., Tao, X., Xu, H.: Effects of calcitriol on random skin flap survival in rats. Sci. Rep. **6**, 18945 (2016)

Development of Raw Semi-dry Sausages Enriched with Colloidal Chelate Complexes of Essential Nutrients

Aleksey Golik, Natalya Oboturova, Andrey Blinov(✉), Tamara Bacholdina, and Uktamjon Rajabov

Federal State Autonomous Educational Institution of Higher Education "North Caucasus Federal University", Stavropol, Russian Federation

Abstract. Microelement analysis of the hair of Russia population indicates a clear insufficient intake of essential nutrients into the human body. The development of functional food products based on a popular food product is one of the most economically and practically feasible solutions. The purpose of the work was to develop a functional product that can cover the daily human need for essential microelements. The paper presents a method for obtaining a colloidal chelate complex of essential nutrients containing riboflavin, lysine and microelements zinc, manganese, iron, copper and cobalt. As a result of spectrophotometric studies, absorption and fluorescence spectra were obtained. Their analysis showed characteristic absorption and fluorescence bands that are unique for this compound. The microstructure of the complex was studied by scanning electron microscopy. It was found that the compound crystallizes in the form of acicular crystallites with a width of up to 300 nm and a length of several microns. The theoretical concept of the number of microelements in the complex is confirmed experimentally by energy dispersive X-ray spectroscopy. A recipe for the production of a meat product (dry-cured semi-dry sausages) enriched with a complex of microelements has been developed. Organoleptic characteristics of the dry-cured semi-dry sausages have been investigated.

Keywords: Functional nutrition · Microelements · Sausage · Complex · Vitamins · Amino acids

1 Introduction

Vitamins, amino acids and microelements are an important part of the diet of each of us. All these nutrients enter the human body and are absorbed by them mainly with food [2]. One way or another, numerous studies of the hair elemental analysis of the population to identify the microelement status of the population indicate a systematic deficiency of certain nutrients in the body of the Russian Federation population and the whole world [5, 8, 13, 15].

One of the solutions to the problem of nutrient deficiency is the fortification of popular foods, such as milk, bread, sausage, drinking water, etc. complexes of micro- and macronutrients [10, 14, 17].

V. Kurchenko et al. (Eds.): ICAETT 2021, LNNS 408, pp. 253–259, 2022.
https://doi.org/10.1007/978-3-030-96641-6_29

As is known, microelements enter our body in various forms – more often they are organic and inorganic, less often they are colloidal, which is considered the most assimilable form. The discoverer of such forms is A. Albert, he proved the ability to bind riboflavin and folic acid with metals [1]. B. Oberle, Y. Fridman, N. Kebets and V. Serov were the first to establish the formation of mixed ligand compounds of metals with amino acids and vitamins, simulating the active centers of metalloenzymes [7, 8, 16, 18, 20]. At the moment, researchers are developing various forms of vitamins and minerals with increased functional properties [6, 14, 19].

2 Materials and Methods

We have developed a new colloidal chelate form of a complex of essential microelements, vitamins and amino acids. The complex contains microelements zinc, iron, copper, manganese and cobalt, the amino acid L-lysine and vitamin B2. As a product to be enriched, a meat product, dry-cured sausages, was chosen and its production recipe was developed.

2.1 Method for the Synthesis of a Colloidal Chelate Complex of Essential Nutrients

At the first stage, 56.45 g of riboflavin, then 21.14 g of lysine hydrochloride are dissolved in a 0.3 N sodium hydroxide solution. At the second stage, zinc (1.780 g), manganese (6.504 g), cobalt (0.003 g), iron (1.599 g) and copper (0.054) sulfates are added to the vitamin and amino acid solution, then the solution is mechanically stirred with continuous nitrogen bubbling for 12 h at the boiling point of a water bath until the bright yellow color of the suspension turns orange, sometimes dark orange. The finished product is filtered on a nutsch filter, washed with cold water and air-dried in a dark place.

2.2 Research Methods

Spectrophotometry was chosen to study the optical characteristics of the colloidal chelate complex of essential nutrients. This method is based on the measurement and study of absorption spectra in the UV and visible regions of the electromagnetic radiation spectrum of the sample [9].

Scanning electron microscopy was used to study the morphology of the colloidal chelate complex of essential nutrients. The studies were carried out on a scanning electron microscope MIRA-LMH with a system for determining the elemental composition AZtecEnergy Standart/X-max 20 (standard) from Tescan [3].

Energy dispersive spectroscopy was used to confirm the composition of the complex. The EDX method is based on the interaction of a certain X-ray excitation source and a sample [4].

2.3 Development of a Recipe for Dry-Cured Sausages

The recipe for dry-cured sausages was based on the classic recipes of similar products. Thus, the preparation of dry-cured sausages, as well as other meat products, begins with such technological processes as acceptance of raw materials, stripping, cutting into cuts, deboning and trimming. Fresh meat of all types of slaughter farm animals and poultry meat are used as raw materials. Next, the meat is injected with brine, and then rubbed with a dry mixture, significantly speeding up the salting process. Minced meat is prepared in vacuum cutters. The prepared minced meat is filled into the food casing using vacuum syringes, and then the resulting sausages are sent to ripening in climatic chambers for 8–10 days at a temperature from 0 to +4 °C. After ripening, the process of drying sausages follows at a temperature of 12 °C, a relative humidity of about 75–78% for 15–20 days.

3 Results

As a result of the colloidal chelate complex of essential nutrients studies, its optical characteristics were obtained. The absorption and fluorescence spectra of a 0.0001% aqueous solution of the complex are shown in Figs. 1 and 2.

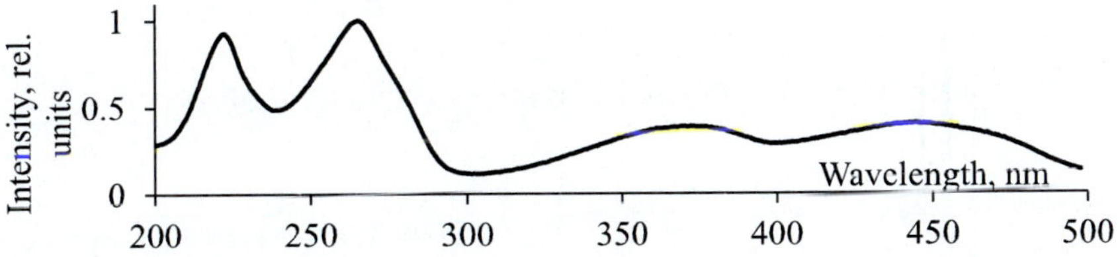

Fig. 1. Absorption spectrum of a colloidal chelate complex of essential nutrients.

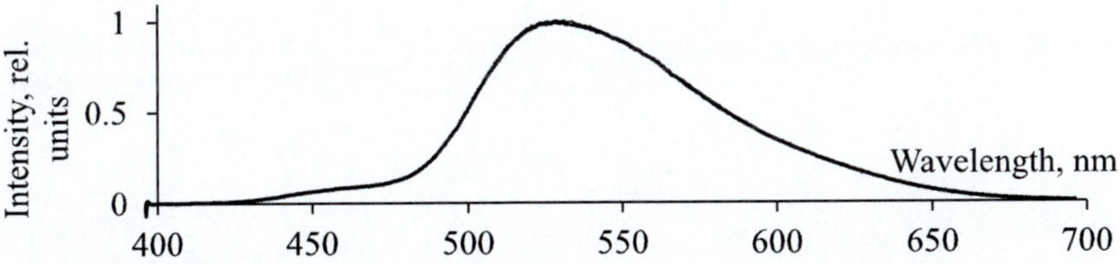

Fig. 2. Fluorescence spectrum of a colloidal chelate complex of essential nutrients.

As a result of the study by scanning electron spectroscopy, SEM images of the colloidal chelate complex of essential nutrients were obtained, which shown in Fig. 3.

As a result of the study of the composition, the theoretical content of microelements in the sample of the colloidal chelate complex of essential nutrients was calculated and the EDX spectrum was obtained by energy dispersive X-ray spectroscopy, which demonstrates the real composition. The EDX spectrum and the theoretical compositional calculations and the actual composition are presented in Figs. 4 and 5.

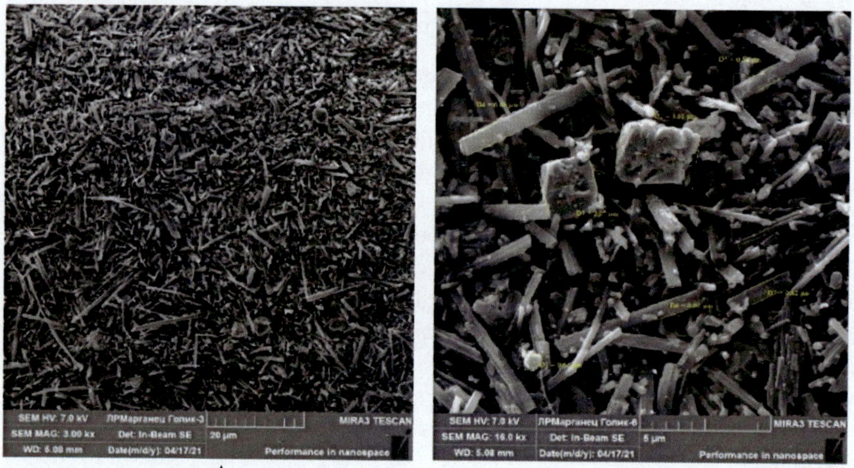

A B

Fig. 3. SEM-images of a sample of the colloidal chelate complex of essential nutrients: A) 3,000 times magnification, B) 16,000 times magnification.

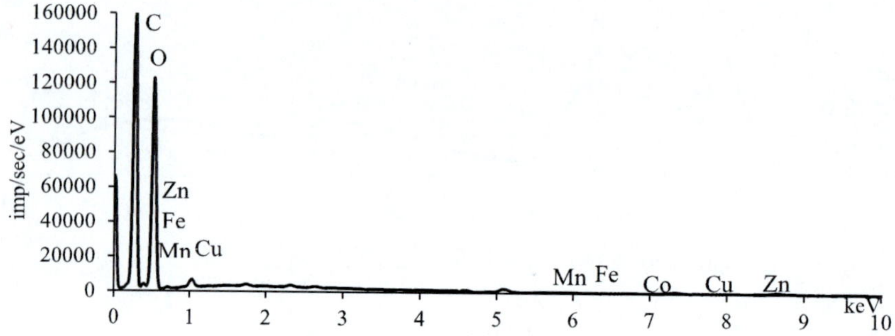

Fig. 4. EDX spectrum of a sample of a colloidal chelate complex of essential nutrients.

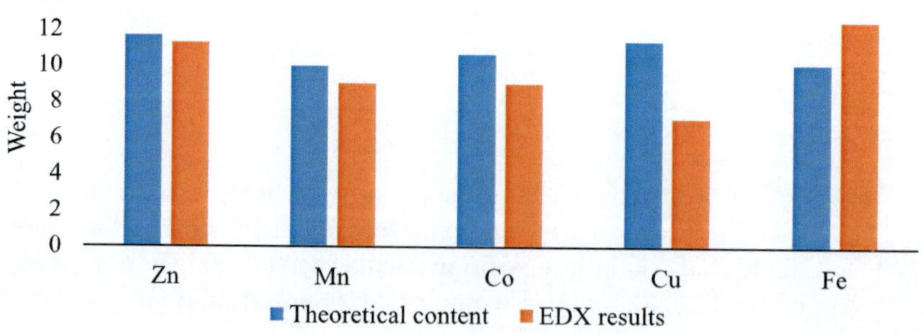

Fig. 5. Mass content of microelements in the colloidal chelate complex of essential nutrients.

Based on modern trends, the recipe for dry-cured semi-dry sausages was compiled, presented in Table 1.

Table 1. Recipe for control and experimental samples of dry-cured semi-dry sausages

Name of raw materials and spices	Production rate for different samples, kg/100 kg	
	Control	*Experienced*
Beef trimmed 1 grade	30	30
Rabbit meat	26	26
Broiler chicken thigh fillet	36	36
Chicken fat	8	8
Auxiliary raw materials, kg/100 kg		
Nitrite salt, 0.6%	1.1	1.1
Salt	1.5	1.5
White pepper	0.13	0.13
Dried mint	0.025	0.025
Glucono delta lactone	0.8	0.8
Colloidal chelated essential nutrient complex	–	0.42
Complex food additive for the production of all types of semi-finished products from meat and poultry meat	0.2	0.2

According to the proposed recipe, a batch of 2 types of dry-cured semi-dry sausages was made. One type of sausages contained a colloidal chelate complex of essential nutrients in the recipe, the second acted as a control. To evaluate the organoleptic characteristics of a batch of sausages, a group of 25 people was assembled. The results of the organoleptic evaluation of the products are presented in Fig. 6.

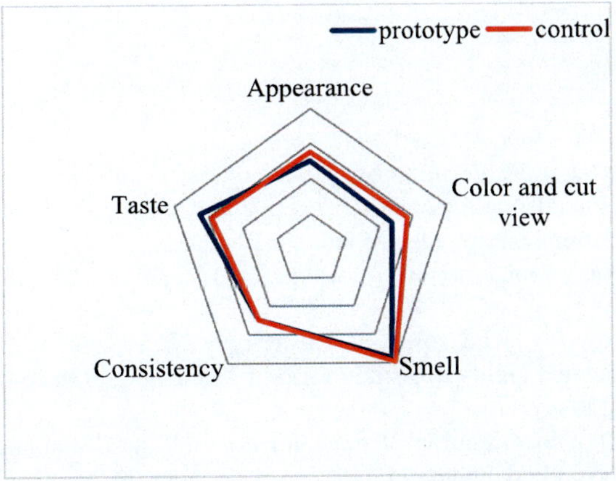

Fig. 6. The results of the organoleptic evaluation of the samples.

4 Discussion and Conclusion

As a result of the colloidal chelate complex of essential nutrients study, the following data were obtained.

Analysis of the absorption spectra showed the presence of four characteristic absorption bands at 445, 375, 265, and 225 nm. These bands corresponds to the transitions of the molecule from the ground state S0 to the excited states S1, S2, S3, and S4. These bands correspond to the transitions of electrons in the colloidal chelate complex of essential nutrients with connecting the π orbital to the π^* orbital. In the fluorescence spectrum of the colloidal chelate complex of essential nutrients, there is only one band with a maximum at 530 nm, which corresponds to bright green fluorescence. Its appearance is responsible for the successive transition in riboflavin from the highest excited states to the ground excited state. According to the obtained optical characteristics, it is possible to detect this compound in any product in which this complex is included, because these are the unique optical characteristics of this compound.

Analysis of SEM-images showed that a sample of a colloidal chelate complex of essential nutrients has an acicular structure, the width of individual crystals ranges from 100 to 300 nm, and the length reaches several microns.

From the EDX spectra of the colloidal chelate complex of essential nutrients, it can be judged that the obtained data on the composition coincide with the theoretical calculations of the composition of this compound.

As a result of the development of the recipe for an element-balanced functional food product, a recipe for dry-cured semi-dry sausages was compiled and a test batch of the fortified product was made.

According to the results of organoleptic indicators study, it can be concluded that the complex will not significantly affect the taste and color of the product, the indicators of the fortified product, which makes it possible to enrich food products with a colloidal chelate complex of essential nutrients.

Acknowledgments. The work was carried out using the Center for Collective Use of the North Caucasus Federal University (NCFU) with the financial support of the Ministry of Education and Science of the Russian Federation (project No. RF-2296.61321X0029, agreement No. 075-15-2021-687).

References

1. Albert, A., Serjeant, E.P.: Quantitative studies of the avidity of naturally occurring substances for trace metals. 4. A potentiometric method for highly insoluble complexes, and its application to complexes of adenine anion*. Biochem. J. **76**(3), 621–624 (1960)
2. Barybina, L.I., et al.: Multicomponent meat products for sports nutrition. J. Hyg. Eng. Des. **28**, 81–84 (2019)
3. Blinov, A.V., Serov, A.V., Kravtsov, A.A., Kostenko, K.V., Snezhkova, J.J., Gvozdenko, A.A.: New colloidal chelated highly digestible form of zinc essential trace element. J. Hyg. Eng. Des. **27**, 95–101 (2019)
4. Blinov, A.V., et al.: Investigation of the influence of Zinc-containing compounds on the components of the colloidal phase of milk. Arab. J. Chem. **14**(7), 103229 (2021)

5. Cai, J., et al.: Comparations of major and trace elements in soil, water and residents' hair between longevity and non-longevity areas in Bama, China. Int. J. Environ. Health Res. **31**(5), 581–594 (2021)
6. Dominiczak, M.H.: Vitamins and minerals. Med. Biochem. E-Book **3**, 75 (2018)
7. Fridman, Y.D., Kebets, N.M., Nanaeva, M.T., Zurdinov, A.Z., Sabirova, T.S., Atarskaya, L.I.: Mutual influence of ascorbic and γ-aminobutyric acids on their biological properties during complexation. Pharm. Chem. J. **23**(11), 879–883 (1989)
8. Fridman, Y.D., Moldogazieva, A.M., Sarbaev, D.S., Smashevskii, N.D., Kopelevich, V.M., Gunar, V.I.: Synthesis, properties, and vitamin activity of mixed-ligand compounds of metals with pantothenic acid and amino acids. Pharm. Chem. J. **22**(9), 666–669 (1988)
9. Gutnova, T.S., Kompantsev, D.V., Gvozdenko, A.A., Kramarenko, V.N., Blinov, A.V.: Vitamin D nanocapsulation. Izvestiya Vysshikh Uchebnykh Zavedenii Khimiya Khimicheskaya Tekhnologiya **64**(5), 98–105 (2021)
10. Ivanov, V., et al.: Trends and expected benefits of the breaking edge food technologies in 2021–2030. Ukrainian Food J. **10**(1), 7–36 (2021)
11. Izydorczyk, G., Mironiuk, M., Baśladyńska, S., Mikulewicz, M., Chojnacka, K.: Hair mineral analysis in the population of students living in the Lower Silesia region (Poland) in 2019: comparison with biomonitoring study in 2009 and literature data. Environ. Res. **196**, 110441 (2020)
12. Jaccoulet, E., Daniel, T., Dammak, D., Prognon, P., Caudron, E.: Interest of flow injection spectrophotometry as an orthogonal method for analyzing biomolecule aggregates: application to stressed monoclonal antibody study. Spectrochimica Acta A Mol. Biomol. Spectrosc. **99**(6), 2620–2627 (2021)
13. Lisetskaya, L.G.: Elemental hair profile of children in rural areas of the Irkutsk region. Hum. Ecol. **2021**(2), 13–19 (2021)
14. Oh, C., Keats, E.C., Bhutta, Z.A.: Vitamin and mineral supplementation during pregnancy on maternal, birth, child health and development outcomes in low-and middle-income countries: a systematic review and meta-analysis. Nutrients **12**(2), 491 (2020)
15. Sadovoy, V.V., Selimov, M.A., Shchedrina, T.V., Nagdalian, A.A.: Usage of biological active supplements in technology of prophilactic meat products. Res. J. Pharm. Biol. Chem. Sci. **7**(5), 1861–1865 (2016)
16. Salnikova, E.V., Detkov, V.Y., Skalny, A.V.: Accumulation of essential and conditionally essential trace elements in the hair of residents of Russia. Trace Elements Med. (Moscow) **17**(2), 24–31 (2016)
17. Satyanarayana, S., Reddy, K.V.: Stability constants of the binary complexes of Co(II), Ni(II), Cu(II) and Zn(II) with thiamine and ternary complexes with thiamine as primary ligand and thymine or uracil as secondary ligand. Indian J. Chem. **28**(A), 169–171 (1989)
18. Skalny, A.V., et al.: Comparative hair trace element profile in the population of Sakhalin and Taiwan Pacific Islands. Biol. Trace Elem. Res. **184**(2), 308–316 (2017). https://doi.org/10.1007/s12011-017-1204-y
19. Tolmachev, V.A., Pogidaeva, E.A., Golubeva, L.V., Blinov, A.V.: Effect of the colloidal chelated form of zinc on the characteristics of the enriched milkcontaining product with milk fat substitute. Αγαη **8**(5), 55 (2019)
20. Zhukov, R.B., Eremenko, O.N., Ospichuk, G.V.: Red cattle breed's feeding rations with selenium- enriched components from yeast and chlorella. Int. Trans. J. Eng. Manage. Appl. Sci. Technol. **12**(11r), 7 (2021)

Analysis of the Biochemical Composition and Antioxidant Capacity of *Vaccinium Corymbosum* L. Leaves

Alla Deeva[1](✉), Polina Shabunya[2], Sviatlana Fatykhava[2], Alena Spirydovich[1], and Vladimir Reshetnikov[1]

[1] Central Botanical Garden of the National Academy of Sciences of Belarus, Minsk, Belarus
V.Reshetnikov@cbg.org.by

[2] Institute of Bioorganic Chemistry of the National Academy of Sciences of Belarus, Minsk, Belarus
fsa1981@tut.by

Abstract. A comparative study of the composition of biologically active substances in the ethanol extracts of the leaves of nine *Vaccinium corymbosum* L. varieties collected in the southern agroclimatic zone of Belarus was performed. The analysis of the total phenolic content demonstrated that, in the process of vegetation, the total phenolic content in blueberry leaves increased in all varieties under study with various degrees of intensity. Based on the results of the HPLC-MS-UV analysis of the anthocyanin spectrum of the leaves, glycosides of two anthocyanidins were detected. The measurement of antioxidant capacity in a model system with $ABTS^{+\cdot}$ radical cation in the extracts of *V. corymbosum* L. leaves collected in autumn showed an increase in this indicator by 21–43% compared to the flowering phase, which is consistent with an increase in the synthesis of phenolic compounds and, in particular, anthocyanins during this vegetative phase of plant development.

Keywords: Anthocyanins · Ascorbic acid · Phenolic compound · HPLS-MS-UV · Antioxidant capacity · *Vaccinium corymbosum* L.

1 Introduction

At present, synthetic antioxidants are broadly used in various sectors of the food and pharmaceutical industries, however, their long-term use may cause toxic and other side effects for the human body [9]. Therefore, there is a constant search for and study of new sources of safe natural antioxidants [12].

One of the sources of antioxidants is the fruits of tall blueberry (*Vaccinium corymbosum* L.). This deciduous shrub of the Ericaceae family is grown as a berry crop in warm, temperate, subtropical and tropical regions [20]. According to the 2017 data, the area of tall blueberry plantations was 341 thousand hectares with crop yield of 7.4–7.6 t/ha. Apart from their high nutritional value, the fruits of *Vaccinium corymbosum* L. are characterized by the presence of a broad range of antioxidant compounds, both low molecular

V. Kurchenko et al. (Eds.): ICAETT 2021, LNNS 408, pp. 260–269, 2022.
https://doi.org/10.1007/978-3-030-96641-6_30

weight (ascorbic acid, carotenoids, anthocyanins and other phenolic compounds), and high molecular weight (ferments - catalase, superoxide dismutase, various peroxidases, etc.) [2, 3, 8, 18].

Phenolic compounds in the fruits are represented by quercetin, isorhamnetin, syringetin and their glycosides, catechin, epicatechin, chlorogenic acid, laricitrin [1, 14]. The lipophilic fraction contains triterpenes (oleanolic acid, ursolic acid, lupeol and lanosterol), sterols (β-sitosterol), and fatty acids [25].

It should be noted that the phenolic profile of blueberry fruits depends on the place where they are growth, climatic conditions, and on the genetic factors of the variety and the stage of maturity [15, 21, 23, 26, 27].

Biologically active compounds are found not only in the fruits of *V. corymbosum* L., but also in other parts of the plant. At tall blueberry plantations, the bushes are pruned annually, and a large number of leaves are discarded. Thus, the leaves are a by-product in the industrial cultivation of blueberries and can be a promising source for the production of food additives whose biologically active properties are directly related to the composition of secondary metabolites of raw materials.

Therefore, the purpose of our work was to study the content of ascorbic acid and the total phenolic content, the qualitative composition of the anthocyanin complex in autumn period, and to assess the antioxidant capacity during the growing season in the leaves of nine tall blueberry varieties (Bluegold, Bluecrop, Brigitta Blue, Denise Blue, Duke, Elizabeth, Hardyblue, Nelson, Northland).

2 Materials and Methods

The experimental material was collected at the collection plantations of the Applied Research Laboratory for the Introduction and Technology of Non-Traditional Berry Plants of the Central Botanical Garden of the National Academy of Sciences of Belarus (Head of Laboratory, Ph.D. (biology) N.B. Pavlovsky, located in the Hancavičy District of Brest Oblast (N 52°44', E 26°22'). The study subject was nine varieties of tall blueberries of different maturity classes.

The biochemical analysis of the plant samples was performed at the Biotechnology Department of the Central Botanical Garden (city of Minsk). The pigment content was determined in freshly harvested leaves. Plant material for the analysis was collected from 10–15 plants, from the middle part of the shoots by a random sample. The samples were collected three times in the growing season during the main phenological phases: flowering (phase 1 - May), fruiting (phase 2 - August) and after the leaves changed colour in autumn (phase 3 - October). To avoid the destruction of anthocyanin pigments, the compounds were extracted using ultrasound (Bandelin Sonoplus HD 2070).

All measurements were done in four replicates. The reliability of the obtained experimental data was evaluated using the biological statistics method [10]. Microsoft Excel Analysis ToolPak was used for the statistical processing of the results. The hypotheses about the equality of two means were tested using the Student's t-test [10]. The assessment of the reliability of the obtained results was carried out at significance level of $p < 0.05$.

The total phenolic content in the dry matter of the extracts was determined using the modified Folin–Ciocalteu assay for the quantification of phenolics [6] by measuring the optical density using an Agilent 8453 spectrophotometer. To prepare the calibration curve, gallic acid solution was used with concentration range 0.15–1.0 g/l.

The quantification of ascorbic acid in blueberry leaves was carried out according to the method described in [4, 11].

The antioxidant properties of the blueberry leaf extracts were assessed in the system with $ABTS^{+\cdot}$ radical cations. The stock solution of the $ABTS^{+\cdot}$ radical cation (AppliChem, Germany) was prepared according to the method [16]. Potassium persulfate was used as an oxidizing agent. The antioxidant capacity of the extracts was assessed as compared to that of the standard substance - Trolox ((±)6-hydroxy-2,5,7,8-tetramethylchroman-2-carboxylic acid) (Sigma-Aldrich, Germany).

For the identification and quantitative analysis of anthocyanins in the leaves of *V. corymbosum* L. collected in autumn the HPLC-mass spectrometry and UV detector were used. Ethanol extracts of blueberry leaves of 3 mL each were evaporated to dryness in nitrogen flow, the precipitate was dissolved in 0.5 mL methanol.

The solutions of delphinidin-3-galactoside (95% HPLC, Extrasynthese) and cyanidin-3-O-glucoside chloride (96% HPLC, Extrasynthese) standards were prepared in the range of concentrations from 5 to 100 μg/mL. Separation was performed on the column Agilent Zorbax Eclipse Plus C18 (3 × 100 mm; 1,8 μm) at +25 °C using Agilent 1200 liquid chromatograph. The mobile phase consisted of two solvents: **A** –5% formic acid (v/v) and **B** –100% acetonitrile. The composition of the mobile phase changed from 6% to 23.5% of phase B over 20 min with eluent flow rate at 0.25 mL/min. The detection was conducted at 510 nm wavelength. The injection volume was 5 μl. The temperature in the autosampler was +10 °C. The ionization interface was electrospray in the positive ion mode. For the Agilent 6410 Triple Quad tandem mass spectrometer, the target mass detection mode was used. Detector operation parameters: drying gas temperature +350 °C; drying gas flow rate 9 l/min; spray pressure 30 psi; capillary voltage 3000 V; fragmentor voltage and collision cell energy were 135 and 25 V, respectively. The analysis of chromatograms and mass spectra was performed using Agilent MassHunter Workstation Software version B.01.03 and Agilent ChemStation (Agilent Technologies Inc., USA).

3 Results and Discussion

The maximum total phenolic content was determined in the samples of blueberry leaves of the early maturing variety - Duke - for all phenological phases under study (from 62.7 ± 1.5 mg/g DW to 119.5 ± 5.8 mg/g DW) (Fig. 1).

For Bluegold, Bluecrop, Denise Blue, Duke and Hardyblue varieties, the total phenolic content was quite high, and it was over 49.5 ± 2.1 mg/g DW during the flowering period, 71.4 ± 3.9 mg/g DW during the fruiting period and 87.5 ± 4.3 mg/g DW during the phase of complete change in the seasonal colour of the leaves for these taxa. For phases 1 and 2 of the growing season, the lowest phenolic content in the leaves was observed in the late-maturing Elizabeth variety, and for phase 3 the lowest total phenolic content was observed in the leaves of the late-maturing Brigitta Blue variety. It should be noted that in the leaves of such late-maturing varieties as Brigitta Blue, Elizabeth, Nelson and the medium-maturing Northland variety, the content of soluble

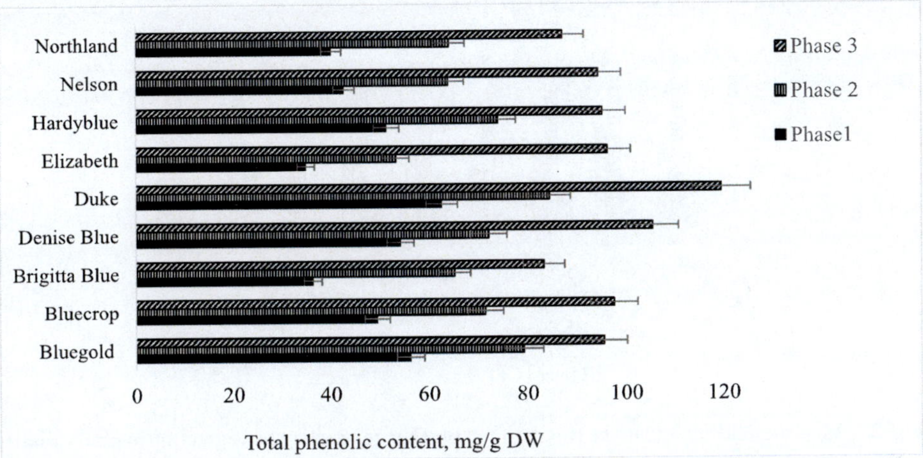

Fig. 1. Total phenolic content in the ethanol extracts of the leaves of tall blueberry varieties under study

phenolic compounds increased 1.5-fold or more during the fruiting phase compared to the flowering phase. The subsequent observation during phase 3 revealed a 1.5-fold or a bigger increase in the phenolic content level for the late-maturing Denise Blue, Elizabeth and Nelson varieties. The analysis of the data obtained showed that, during the growing season, the total phenolic content in the leaves of tall blueberry increased in all studied varieties with varying degrees of intensity. In early and medium-maturing varieties, the accumulation process is slower; in late-maturing varieties, it is more rapid, especially in the autumn period. It appears that enhanced synthesis of phenolic compounds is a genetically determined and necessary factor of survival in the changing environment for the plants of the *Vaccinium L.* genus.

Ascorbic acid is a powerful antioxidant, a cofactor of some enzymes and is involved in the regulation of plant cell division [19]. In the human body and in animals, it regulates blood clotting ability, normalizes capillary permeability, and is essential for hematopoiesis (Fig. 2).

Our studies have shown that most late-maturing blueberry cultivars (except for Denise Blue variety) are distinguished by high levels of ascorbic acid during all vegetative phases (up to 1.7 ± 0.07 mg/g DW during the flowering period, up to 5.4 ± 0.2 mg/g DW during fruiting and up to 2.8 ± 0.1 mg/g DW in autumn (Fig. 3). During the flowering period, the maximum content of ascorbate was found in the leaves of early maturing Duke variety (1.6 ± 0.04 mg/g DW), medium maturing Bluegold (1.3 ± 0.03 mg/g DW) and Bluecrop (1.3 ± 0.05 mg/g DW), and late-maturing Nelson variety (1.7 ± 0.06 mg/g DW).

Maximum ascorbic acid during the fruiting phase was accumulated in the leaves of the late season Nelson variety (5.5 ± 0.1 mg/g DW). At the same time, the fruiting period is characterized by a 3-fold increase in the ascorbic acid content on average compared to the flowering phase. In autumn, the level of ascorbic acid decreases and ranges from 1.8 ± 0.03 mg/g DW (Denise Blue) to 2.9 ± 0.08 mg/g DW (Bluecrop).

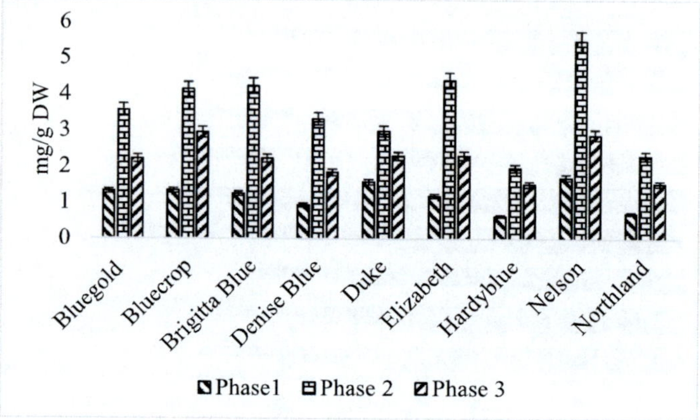

Fig. 2. Ascorbic acid content in the leaves of tall blueberries during the vegetation of nine varieties of different maturity classes

The study revealed that the content of soluble phenolics and ascorbic acid depends on the growing period - during the fruiting period their levels increase as compared to the flowering period. Similar observations are described in [17] for the end of the fruiting period in large-fruited cranberries (*V. macrocarpon* Ait.) when there is a sharp increase in the concentration of polyphenols in the plant. This may be due to the creation of a reserve that ensures the functioning of the plant, as well as the protection of reproductive and emerging generative organs through the accumulation of phenols and ascorbate in plant berries [19, 24].

In autumn, a change in the colour range of the leaves of *V. corymbosum* L. is observed, which is explained by the photoprotective function of anthocyanins. Anthocyanins protect the destroyed chlorophyll from exposure to light rays, limiting the formation of oxygen radicals [5].

To assess the qualitative and quantitative content of individual anthocyanin pigments in blueberry leaves collected in autumn, we used HPLC-mass spectrometry and a UV detector. Based on the HPLC analysis of the native aqueous-alcoholic extract from the leaves of *V. corymbosum* Bluegold variety, seven derivatives of four anthocyanidins were detected: delphinidin galactoside, delphinidin glucoside, cyanidin galactoside, cyanidin glucoside, cyanidin arabinoside, peonidin glucoside, malvidin hexoside (Fig. 3). The results obtained illustrate that there was methylation of cyanidin and delphinidin derivatives in the leaves of the Bluegold taxon, as according to the anthocyanidin biosynthesis mechanism [7, 22], when subjected to methyltransferase, cyanidin glycoside is converted into peonidin (3'-O-methyl cyanidin) glycoside; and in the same way petunidin (3'-O-methyl delphinidin) and malvidin (3', 5'-O-dimethyl delphinidin) glycosides are synthesized from delphinidin glucoside.

The leaves of the remaining studied varieties contained only cyanidin derivatives. Comprehensive data on the content of individual anthocyanins in the leaves of various blueberry varieties is presented in Table 1.

The leaf extracts of medium maturing blueberry varieties - Bluegold and Bluecrop - were characterized by the highest anthocyanin content of 307.8 and 698.3 μg/g DW.

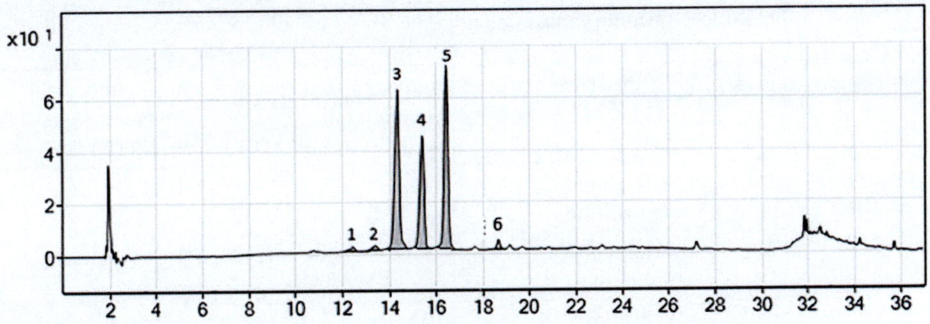

Fig. 3. HPLC-UV chromatogram of *Vaccinium corymbosum* L. leaf extract, Bluegold variety ($\lambda = 510$ nm): 1 - delphinidin-galactoside (retention time about 12.5 min); 2 - delphinidin-glucoside (retention time about 13.4 min); 3 - cyanidin-galactoside (retention time about 14.4 min); 4 - cyanidin glucoside (retention time about 15.4 min); 5 - cyanidin arabinoside (retention time about 16.4 min); 6 -peonidin glucoside (99.2% peak) + malvidin hexoside (0.8%) (retention time about 18.7 min).

Late season varieties, such as Denise Blue (45.1 μg/g DW), Elizabeth (81.9 μg/g DW), Nelson (23.4 μg/g DW) had a lower anthocyanin content in comparison to early and medium maturing varieties.

Thus, in early and medium maturing varieties, the concentration of anthocyanins in leaves is on average 4–17 times higher than in late-maturing varieties, which fact, considering an increased content of ascorbic acid in Brigitta Blue, Elizabeth, and Nelson varieties may be an indication that in the plants of late-maturing taxa in a stressful situation the extent of anthocyanins contribution into the plant's antioxidant system is the highest among other low molecular weight antioxidants. The presence of cyanidin arabinoside and cyanidin galactoside in late- and early maturing varieties may indicate a higher ability of these anthocyanins to protect photolabile compounds. This assumption is consistent with the data from the work by Page J. and Towers N., where the cells containing cyanidin 3-O-glucoside and cyanidin 3-O-(6′-O-malonylglucoside) protected the tissues of *A. chamissonis* by absorbing excess light quanta and increasing the protective functions of the plant [13].

It is known that the key properties of antioxidants is their ability to be easily oxidized and to participate in radical and redox reactions. Therefore, we studied the antioxidant capacity of the leaf extracts of nine blueberry varieties in a model system with $ABTS^{+\bullet}$ cation radicals (Fig. 4).

As a rule, the easier the process of a substance oxidation, the higher its antioxidant activity as an inhibitor of a radical process. As we can see from Fig. 4, the measured antioxidant capacity (AOC) (μmol TE/g DW) of the extracts of all studied blueberry varieties are approximately at the level of 93.1 ± 2.1 (Elizabeth) – 123.9 ± 3.5 (Duke) during flowering phase. During fruiting phase, a slight reduction or increase not more than by 15% of the AOC level was observed, except for the Brigitta Blue variety. The measurement in autumn showed an increase in AOC by 21–43% compared to the flowering phase, which is consistent with an increase in the synthesis of phenolic compounds and, in particular, anthocyanins during this vegetative phase of plant development. This

Table 1. Anthocyanin content in the leaves of tall blueberry

Variety	Peak no.	Substance	Individual anthocyanins content in the leaves, µg/g dry weight (DW)	Total anthocyanin content, µg/g dry weight (DW)
Bluegold	1	Delphinidin galactoside	12.0	307.8
	2	Delphinidin glucoside	12.4	
	3	Cyanidin galactoside	103.3	
	4	Cyanidin glucoside	70.1	
	5	Cyanidin arabinoside	97.3	
	6	Peonidin glucoside (99.2% peak) + Malvidin hexoside (0.8%)	12.7	
Bluecrop	1	Cyanidin galactoside	348.2	698.3
	2	Cyanidin glucoside	38.8	
	3	Cyanidin arabinoside	311.3	
Denise blue	1	Cyanidin galactoside	18.5	45.1
	2	Cyanidin glucoside	10.4	
	3	Cyanidin arabinoside	16.1	
Duke	1	Cyanidin galactoside	93.8	162.9
	2	Cyanidin arabinoside	69.1	
Elizabeth	1	Cyanidin galactoside	49.4	81.9
	2	Cyanidin arabinoside	32.5	
Nelson	1	Cyanidin galactoside	11.9	23.4
	2	Cyanidin arabinoside	11.6	

* The error in mean did not exceed 5%.

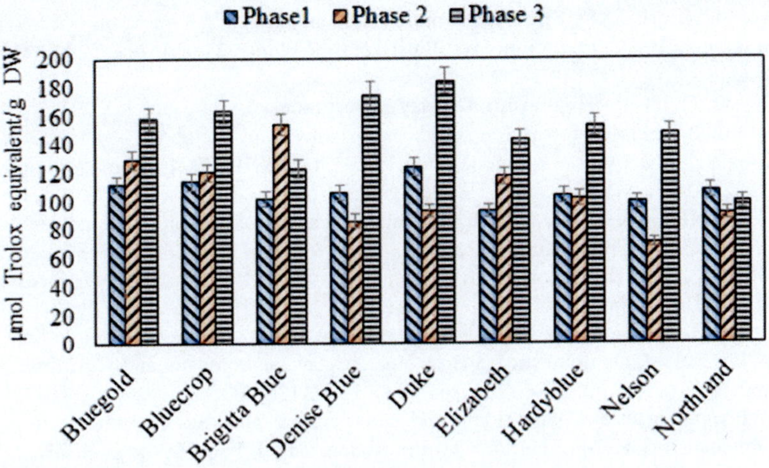

Fig. 4. Antioxidant capacity of the leaves of tall blueberries in the vegetation process of nine varieties of different maturing classes

assumption is supported by the correlation between the AOC level and the total phenolic content in the extracts of the leaves of nine *V. corymbosum* L. varieties collected in autumn. The correlation coefficient was 0.85 and the significance of the correlation relationship is based on the fact that the calculated Student's t-test value was greater than the corresponding table value.

The results obtained can be used to assess the quality of raw blueberry leaves during the development of herbal medicinal products with an increased content of biologically active substances.

References

1. Aaby, K., Hvattum, E., Skrede, G.: Analysis of flavonoids and other phenolic compounds using high-performance liquid chromatography with coulometric array detection: relationship to antioxidant activity. J. Agric. Food Chem. **52**, 4595–4603 (2004). https://doi.org/10.1021/jf0352879
2. Berezina, E.V., Brilkina, A.A., Veselov, A.P.: Content of phenolic compounds, ascorbic acid, and photosynthetic pigments in Vaccinium macrocarpon Ait. dependent on seasonal plant development stages and age (the example of introduction in Russia). Sci. Hortic. **218**, 139–146 (2017). https://doi.org/10.1016/j.scienta.2017.01.020
3. Bujor, O.C., Ginies, C., Popa, V.I., Dufour, C.: Phenolic compounds and antioxidant activity of lingonberry (Vaccinium vitis-idaea L.) leaf, stem and fruit at different harvest periods. Food Chem. **252**, 356–365 (2018). https://doi.org/10.1016/j.foodchem.2018.01.05
4. Ermakov, A.I., Yarosh, N.P., Arasimovich, V.V.: Methods of biochemical research of plants. Agropromizdat, pp. 85–122. Leningrad Department, Russia. (1987). (in Russian)
5. Field, T.S., Lee, D.W., Halbrook, N.M.: Why leaves turn red in autumn? The role of anthocyanins in senescing leaves of red-osier dogwood. Plant Physiol. **127**(2), 566–574 (2001)
6. Ikawa, M., Schaper, T.D., Dollard, C.A., Sasner, J.J.: Utilization of folin-ciocalteu phenol reagent for the detection of certain nitrogen compounds. J. Agric. Food Chem. **51**, 1811–1813 (2003). https://doi.org/10.1021/jf021099r

7. Jaakola, L., Kaisu, M.R., Sirpa, K., Hohtola, A.: Activation of flavonoid biosynthesis by solar radiation in bilberry (Vaccinium myrtillus L) leaves. Planta. **218**(5), 721–728 (2004). https://doi.org/10.1007/s00425-003-1161-x
8. Koshovyi, O., et al.: Highbush blueberry (Vaccinium corymbosum L.) leaves extract and its modified arginine preparation for the management of metabolic syndrome—chemical analysis and bioactivity in rat model. Nutrients **13**(8), 2870 (2021). https://doi.org/10.3390/nu13082870
9. Liu, Z., Li, G., Long, C., Xu, J., Cen, J., Yang, X.: The antioxidant activity and genotoxicity of isogarcinol. Food Chem. **253**, 5–12 (2018). https://doi.org/10.1016/j.foodchem.2018.01.074
10. McDonald, J.H.: Handbook of Biological Statistics, 3rd edn., p. 317. Sparky House Publishing, Baltimore (2014)
11. Nweze, C.C., Abdulganiyu, M.G., Erhabor, O.G.: Comparative analysis of vitamin C in fresh fruits juice of Malus domestica, Citrus sinensi, Ananas comosus and Citrullus lanatus by iodometric titration. Int. J. Sci. Environ. **4**(1), 17–22 (2015)
12. Ouerghemmi, I., et al.: Antioxidant and antimicrobial phenolic compounds from extracts of cultivated and wild-grown tunisian ruta chalepensis. J. Food Drug Anal. **25**(2), 350–359 (2017). https://doi.org/10.1016/j.jfda.2016.04.001
13. Page, J., Towers, N.: Anthocyanins protect light-sensitive thiarubrine phototoxins. Planta **215**(3), 478–484 (2002). https://doi.org/10.1007/s00425-002-0769-6
14. Pertuzatti, P.B., Barcia, M.T., Gómez-Alonso, S., Godoy, H.T., Hermosin-Gutierrez, I.: Phenolics profiling by HPLC-DAD-ESI-MSn aided by principal component analysis to classify rabbiteye and highbush blueberries. Food Chem. **340**, 127958 (2021). https://doi.org/10.1016/j.foodchem.2020.127958
15. Prior, R.L., et al.: Antioxidant capacity as influenced by total phenolic and anthocyanin content, maturity, and variety of Vaccinium species. J. Agric. Food Chem. **46**(7), 2686–2693 (1998). https://doi.org/10.1021/jf980145d
16. Re, R., Pellegrini, N., Proteggente, A., Pannala, A., Yang, M., Rice-Evans, C.: Antioxidant activity applying an improved ABTS radical cation decolorization assay. Free Radical Biol. Med. **26**(9–10), 1231–1237 (1999). https://doi.org/10.1016/s0891-5849(98)00315-3
17. Rupasova, Zh.A., Ignatenko, V.A., Rusalenko, V.G., Rudakovskaya, R.N.: Development and metabolism of large-fruited cranberries in the Belarusian woodland. Science and Technology, Belarus, Minsk (1989). (in Russian)
18. Siddiq, M., Dolan, K.D.: Characterization of polyphenol oxidase from blueberry (Vaccinium corymbosum L.). Food Chem. **218**, 216–220 (2017). https://doi.org/10.1016/j.foodchem.2016.09.061
19. Smirnoff, N.: The function and metabolism of ascorbic acid in plants. Ann. Bot. **78**, 661–669 (1996)
20. Strik, B.C., Finn, C.E., Moore, P.P.: Blueberry cultivars for the pacific northwest. Oregon State Univ. Exten. Serv. **656**, 1–13 (2014)
21. Sun, Y., et al.: Comparative phytochemical profiles and antioxidant enzyme activity analyses of the southern highbush blueberry (Vaccinium corymbosum) at different developmental stages. Molecules **23**, 2209 (2018). https://doi.org/10.3390/molecules23092209
22. Tanaka, Y.: Flower colour and cytochromes P450. Phytochem. Rev. **5**(2/3), 283–291 (2006)
23. Uleberg, E., et al.: Effects of temperature and photoperiod on yield and chemical composition of northern and southern clones of bilberry (Vaccinium myrtillus L.). J. Agric. Food Chem. **60**(42), 10406–10414 (2012). https://doi.org/10.1021/jf302924m
24. Vermerris, W., Nicholson, R.: Phenolic Compound Biochemistry. Springer, Heidelberg (2008). https://doi.org/10.1007/978-1-4020-5164-7
25. Vrancheva, R., Ivanov, I., Dincheva, I., Badjakov, I., Pavlov, A.: Triterpenoids and other non-polar compounds in leaves of wild and cultivated vaccinium species. Plants **10**(1), 94 (2021). https://doi.org/10.3390/plants10010094

26. Zoratti, L., Jaakola, L., Häggman, H., Giongo, L.: Anthocyanin profile in berries of wild and cultivated Vaccinium spp. along altitudinal gradients in the alps. J. Agric. Food Chem. **63**(39), 8641–8650 (2015). https://doi.org/10.1021/acs.jafc.5b02833
27. Zorenc, Z., Veberic, R., Stampar, F., Koron, D., Mikulic-Petkovsek, M.: Changes in berry quality of northern highbush blueberry (Vaccinium corymbosum L.) during the harvest season. Turk. J. Agric. For. **40**, 855–864 (2016). https://doi.org/10.3906/tar-1607-57

Check for updates

Cryostability of Bifidobacteria in Milk and Vegetable Mixtures

Marina Kozhuhova, Elena Terkun, Tatiana Barkhatova, Larisa Rylskaya, and Roman Drozdov(✉)

Kuban State Technological University, Krasnodar, Russia

Abstract. The effect of freezing and low-temperature storage on the survival of bifidobacteria in media based on milk and whey with the addition of Jerusalem artichoke, carrot and beet vegetable purees was studied. The samples were fermented using probiotic starter culture ABY-3, then frozen to −18 °C and stored for 6 months. It was found that in combined media containing plant components, bifidobacteria exhibit higher cryostability than in milk and whey without additives. The cryoprotective effect is enhanced with an increase in the concentration of vegetable purees from 10 to 50% and is the most apparent in the samples with addition of Jerusalem artichoke puree. The results obtained provide additional information on the cryostability of bifidobacteria and make it possible to create new probiotic foods with a long shelf life.

Keywords: Probiotics · Bifidobacteria · Milk · Whey · Herbal ingredients · Freezing · Cryostability

1 Introduction

Probiotic microorganisms which are normally present in the intestinal microflora have diverse positive effects on human health: they improve the digestive system functions, increase immunity, normalize carbohydrate and lipid metabolism, promote the absorption of calcium, iron, magnesium, and have anti-inflammatory and anti-allergenic effects [1–4]. The classic probiotics include lacto- and bifidobacteria. Under the influence of external and internal unfavorable factors the number of beneficial microorganisms in the intestine is reduced which can cause various diseases, including cancer. One of the ways to restore the normal balance of intestinal microflora is the regular use of functional foods containing probiotic microorganisms [3]. It has been proved that the greatest positive effect on human health and well-being is provided by synbiotic products containing both pre- and probiotics, which act in synergy [4, 5]. Prebiotics are substances that can stimulate the growth and activity of probiotic microorganisms, improve their adhesion to the intestinal walls. Such properties are possessed by plant oligo-and polysaccharides which are not hydrolyzed by digestive enzymes, for example, fructooligosaccharides, xylooligosaccharides, pectin, inulin and resistant starch [2, 3, 5].

The study of probiotic microorganisms, as well as factors affecting their activity and viability, is the subject of a significant number of works by Russian and foreign

V. Kurchenko et al. (Eds.): ICAETT 2021, LNNS 408, pp. 270–276, 2022.
https://doi.org/10.1007/978-3-030-96641-6_31

scientists [6–10]. Recently, the attention of researchers has been drawn to the study of the stability of probiotics to cold temperatures. It has been found out that the survival of lacto- and bifidobacteria during freezing and storage depends on the specificity and adaptive capabilities of a particular species and strain of the microorganism, the stage of fermentation, the composition of medium, the freezing regime, the conditions and duration of storage, the presence and type of cryoprotectant [11–17]. Various substances of plant origin can act as cryoprotectants: mono- and oligosaccharides, inulin, pectin, amino acids, antioxidants [15, 16]. It was also found that plant fibers not only contribute to the accelerated growth and reproduction of probiotic microorganisms, but also increase their resistance in adverse conditions. The effect of immobilization of microbial cells on the surface of fibers is observed which is an additional protective factor [11, 17]. In this regard, the study of plant components as sources of prebiotic substances and natural cryoprotectants is of scientific and practical interest which will make it possible to purposefully create new food products that have probiotic activity and are able to maintain this activity at low temperatures during long-term storage.

2 Materials and Methods

The purpose of this work is to determine the survival rate of bifidobacteria during freezing and low-temperature storage in media based on milk and whey by adding vegetable purees from carrots, beets and Jerusalem artichoke. Samples for study were prepared on the basis of milk with a fat mass fraction of 2.5% and milk (curd) whey with adding vegetable puree from carrots, beets and Jerusalem artichoke in an amount of 10 to 50% of the mixture weight. The samples were thoroughly mixed and sterilized for 15 min at (96 $\pm$ 1) °C, then cooled to (37 $\pm$ 2) °C and inoculated with pre-activated ABY-3 starter culture "CHR-Hansen" (Denmark). The species composition of the starter culture is a consortium of lactic acid microorganisms and bifidobacteria: Lactobacillus delbrueckii subsp. bulgaricus, Lactobacillus acidophilus La-5, Streptococcus thermophilus, Bifidobacterium animalis subsp. lactis BB-12. The choice of the starter culture is due to its ability to ferment both dairy and plant media as well as the content of B. lactis BB-12 bifidobacteria, which are characterized by high probiotic activity, resistance to oxygen, high acidity, other unfavorable factors and good adhesion to the intestinal walls [13]. Ripening was carried out for 24 h until the acidity reached 85-95° T, then the samples were frozen in a thin layer for 30 min to a temperature of -18 °C, then they were stored at the same temperature for 6 months. As control samples were used whey and milk without vegetable additives. On the first day after freezing and every 3 months of storage the number of viable cells of bifidobacteria in the experimental and control samples was determined by the standard method of inoculating the samples on agar selective nutrient medium, culturing the crops and counting typical colonies. The inoculation was carried out as follows: 1 cm^3 of the last three dilutions of the product were inoculated into two tubes with a high column of the regenerated culture medium GMK-1. The tubes were thermostated at 37 $\pm$ 1 °C for 72 $\pm$ 3 h under aerobic conditions. After the end of the incubation, the number of colonies in test tubes was counted, where from 5 to 150 colonies had grown, and the number of bifidibacteria in the samples was determined taking into account the dilutions [18]. To assess the survival rate of bifidobacteria during storage, the coefficient of the average monthly decrease in cell concentration (Cam) and the indicator of viability (V) were calculated by using the following formulas:

$$V = \frac{l}{C_{am}}, \quad (1)$$

$$C_{am} = (lgCFU_0 - lgCFU_n)/\tau, \quad (2)$$

where lg CFU_0 and lg CFU_n are the logarithms of the cellular concentration of microorganisms at the initial and final stages of storage; τ - storage duration, months.

3 Results

The results of the study of the survival of bifidobacteria during freezing and storage in milk-plant media are shown in Fig. 1. As you can see, against the background of the general trend towards a decrease, the number of viable cells of bifidobacteria in the test samples is higher than in the control samples during the entire period of storage at sub-zero temperatures. After 6 months, the number of viable cells of bifidobacteria in the control sample decreased by 2.2 lg, and in the test sample by 1–2 lg, depending on the type of the plant component and its concentration in the mixture. The slightest difference between the initial and final probiotic content was observed in milk-plant mixtures containing 40–50% vegetable puree, i.e. an increase in the proportion of vegetable puree in the mixture contributed to the better survival of bifidobacteria. The analysis of experimental data reflecting the influence of the nature of the plant component on the viability of the studied microorganisms shows that the cryoprotective properties are the most apparent in Jerusalem artichoke. Thus, the survival rate of bifidobacteria V, determined by the formula (1), in samples with the addition of 50% vegetable puree from Jerusalem artichoke was 6.6, from carrots – 4.6 and from beets – 3.6. The results of the study of the survival of bifidobacteria during freezing and storage in media based on whey with vegetable purees are given in Fig. 2. The comparison of the graphs in Figs. 1 and 2 shows that the main regularities for milk-plant media are nearly the same in the case of using whey. The difference is that the initial and final concentrations of bifidobacteria in whey-based media are lower, especially for the samples with added carrot and beet puree. In the process of freezing and low-temperature storage of whey-herbal mixtures, the concentration of probiotics decreased, and after 6 months the survival rate in the samples containing 50% Jerusalem artichoke, carrot and beet puree was 6.0; 4.6 and 3.5, respectively. These values are inferior to corresponding indicators determined for milk-plant media, but not significantly.

4 Discussion

The increase in the survival rate of bifidobacteria established in the course of studies when plant components are added to the milk or whey base is consistent with the data obtained by other sresearchers [11, 17, 19] and suggests the presence of cryoprotective properties in the studied vegetable purees. The functions of cryoprotectants can be performed

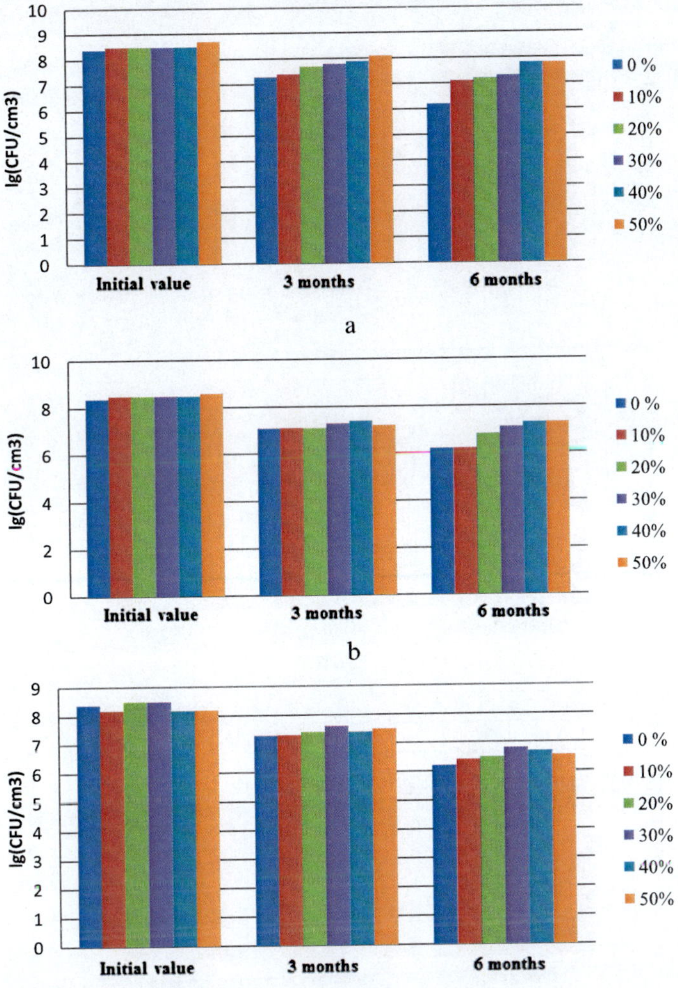

Fig. 1. Change in the concentration of bifidobacteria in milk-based samples containing 0–50% vegetable puree from Jerusalem artichoke (a), carrots (b), beets (c), during storage at −18 °C.

by mono- and oligosaccharides, antioxidants, including carotene from carrots and beet betanin, as well as insoluble plant fibers, on the surface of which probiotic cells are able to adsorb and become more resistant to cold, like the immobilized cultures [13].

The comparison of different types of vegetable purees showed that the cryoprotective properties are the most evident in Jerusalem artichoke puree. This can be explained by the high content of fructo-oligosaccharides and inulin in Jerusalem artichoke, the effectiveness and mechanism of protective action of which are noted in works [20–23]. Fructose and its oligomers not only prevent deep cell dehydration during ice formation due to osmotic forces, but also interact in a certain way with biological membranes preserving their integrity during freezing and thawing [23].

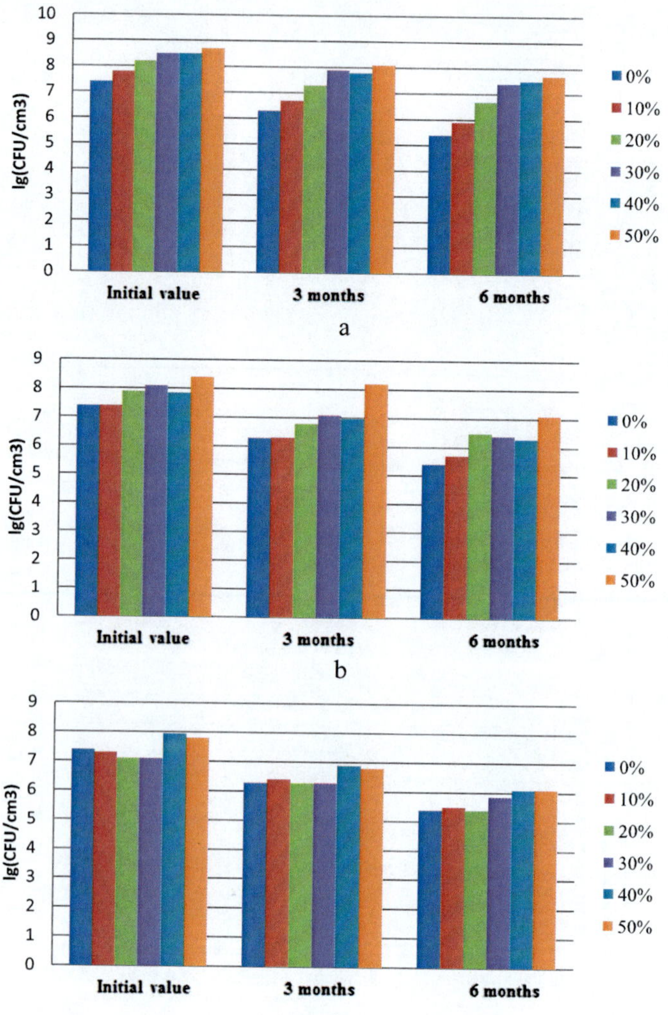

Fig. 2. Change in the concentration of bifidobacteria in whey-based samples containing 0–50% vegetable puree from Jerusalem artichoke (a), carrots (b), beets (c), during storage at −18 °C.

Whey is of great interest as a basis for functional food products, since it contains a significant amount of minerals, water-soluble vitamins, whey proteins, biologically active peptides, while being a low-calorie, affordable and technological raw material. Due to the high content of organic acids, curd whey has good organoleptic characteristics. However, high acidity makes it difficult for whey to get fermented by probiotic cultures, since probiotics slowly grow and multiply in the acidic medium. We have previously found that the combination of whey with vegetable purees can significantly intensify the fermentation process by raising the pH, increasing the concentration of fermentable sugars and enriching the medium with growth factors: easily assimilated nitrogen, vitamins and minerals [24]. In this regard, there is a direct relationship between the mass fraction of the vegetable component in the mixture and the initial amount of bifidobacteria for the samples based on whey. This dependence also is observed during subsequent freezing and low-temperature storage.

The positive effect of herbal supplements on the survival of bifidobacteria creates the prerequisites for the development of new types of frozen probiotic products based on milk and whey. The number of bifidobacteria in such products is regulated and should be at least 106 CFU/cm^3 at the end of the shelf life. All the milk-based mixtures studied meet this requirement, as well as whey-based mixtures containing carrot and Jerusalem artichoke puree in an amount of 20–50% and beet puree in an amount of 40–50%.

5 Conclusion

As a result of the study, it was established that bifidobacteria B. Lactis BB-12 in the composition of the multi-species probiotic starter culture ABY-3 exhibit higher cryostability in combined milk-based or whey-based media with added vegetable purees in contrast to milk and whey bases without additives. The cryoprotective effect increases with an increase in the concentration of plant components in the mixture in the following order: beets, carrots, Jerusalem artichoke. The results obtained provide additional information about the effect of the medium composition on cryostability of probiotic cultures and create the preconditions for creating new combined milk-based or whey-based food products of high probiotic activity which retain their functional properties during freezing and storage.

References

1. Youssef, M., et al.: Probiotic supplements: their strategies in the therapeutic and prophylactic of human life-threatening diseases. Int. J. Mol. Sci. **22**(20) (2021). Article №: 11290
2. Li, H.-Y., et al.: Effects and mechanisms of probiotics, prebiotics, synbiotics, and postbiotics on metabolic diseases targeting gut microbiota: a narrative review. Nutrients **13**(9) (2021). Article №: 3211
3. Bengmark, S.: Pre-, pro-, synbiotics and human health. Food Technol. Biotechnol. **48**(4), 464–475 (2010)
4. Sanders, M.E., Marco, M.L.: Food formats for effective delivery of probiotics. Ann. Rev. Food Sci. Technol. **1**, 65–85 (2010)
5. Lim, S.-M., Jeong, K.-S., Lee, N.-G., Park, S.-M., Ahn, D.-H.: Synergy effects by combination with lactic acid bacteria and frutooligosaccharides on the cell growth and antimicrobial activity. Food Sci. Biotechnol. **20**(5), 1389–1397 (2011). https://doi.org/10.1007/s10068-011-0191-6
6. Ryabtseva, S.A., Bratsikhina, M.A., Ganina, V.I.: Preservation of the viability of the starter microflora. Dairy Ind. **1**, 22–23 (2010)
7. Pasko, O.V.: Enhancing the growth and protection of crops with probiotic properties. Dairy Ind. **7**, 31–32 (2010)
8. Desai, A.R., Powell, I.B., Shah, N.P.: Survival and activity of probiotic lactobacilli in skim milk containing prebiotics. J. Food Sci. **69**, 57–60 (2004)

9. Rokka, S., Rantamäki, P.: Protecting probiotic bacteria by microencapsulation: challenges for industrial applications. Eur. Food Res. Technol. **231**(1), 1–12 (2010). https://doi.org/10.1007/s00217-010-1246-2
10. Nualkaekul, S., Charalampopoulos, D.: Survival of lactobacillus plantarum in model solutions and fruit juices. Int. J. Food Microbiol. **146**(2), 111–117 (2011)
11. Parussolo, G., Busatto, R.T., Schmitt, J., Pauletto, R., Schons, P.F., Ries, E.F.: Synbiotic ice cream containing yacon flour and Lactobacillus acidophylus NCFM. LWT Food Sci. Technol. **82**, 92–198 (2017)
12. Homayouni, A., Azizi, A., Ehsani, M.R., Yarmand, M.S., Razavi, S.H.: Effect of microencapsulation and resistant starch on the probiotic survival and sensory properties of synbiotic ice cream. Food Chem. **111**, 50–55 (2008)
13. Homayouni, A., Ehsani, M.R., Azizi, A., Razavi, S.H., Yarmand, M.S.: Growth and survival of some probiotic strains in simulated ice cream conditions. J. Appl. Sci. **8**(2), 379–382 (2008)
14. Mohammadi, R., Mortazavian, A.M., Khosrokhavar, R., da Cruz, A.G.: Probiotic ice cream: viability of probiotic bacteria and sensory properties. Ann. Microbiol. **61**(3), 411–424 (2011). https://doi.org/10.1007/s13213-010-0188-z
15. Shamekhi, F., Shuhaimi, M., Ariff, A.B., Yazid, A.M.: Optimization of a cryoprotective medium for infant formula probiotic applications using response surface methodology. Ann. Microbiol. **62**(3), 911–921 (2012)
16. Miao, S., Mills, S., Stanton, C., Fitzgerald, G.F., Roos, Y., Roos, R.P.: Effect of disaccharides on survival during storage of freeze dried probiotics. Dairy Sci. Technol. **88**(1), 19–30 (2008)
17. Espirito Santo, A.P., et al.: Fibers from fruit by-products enhance probiotic viability and fatty acid profile and increase CLA content in yoghurts. Int. J. Food Microbiol. **154**, 135–144 (2012)
18. GOST 33924-2016. Milk and milk products. Methods for determination of the bifidobacterium. Publishing House of Standards, Moscow
19. Cavalcanti, M.H., et al.: Production of frozen probiotic fermented milk enriched with green banana biomass: the effects of freezing, acid stress conditions and bile salts on Lactobacillus paracasei subsp paracasei LBC 81 viability. J. Food Process. Preserv. **44**(2) (2020). Article №: e14318
20. Falah, F., Zareie, Z., Vasiee, A., Yazdi, F.T., Mortazavi, S.A., Behbahani, B.A.: Production of synbiotic ice-creams with Lactobacillus brevis PML1 and inulin: functional characteristics, probiotic viability, and sensory properties. J. Food Measur. Charact. **15**(6), 5537–5546 (2021). https://doi.org/10.1007/s11694-021-01119-x
21. Akin, M.B., Akin, M.S., Kirmaci, Z.: Effects of inulin and sugar levels on the viability of yogurt and probiotic bacteria and the physical and sensory characteristics in probiotic ice-cream. Food Chem. **104**(1), 93–99 (2007)
22. Paseephol, T., Sherkat, F.: Probiotic stability of yoghurts containing Jerusalem artichoke inulins during refrigerated storage. J. Funct. Foods **1**, 311–318 (2009)
23. Santivarangkna, C., Higl, B., Foerst, P.: Protection mechanisms of sugars during different stages of preparation process of dried lactic acid starter cultures. Food Microbiol. **25**, 429–441 (2008)
24. Merkulova, E.P., Kozhukhova, M.A.: Lacto-fermented drinks based on milk whey. Izvestiya vuzov. Food Technol. **4**, 40–42 (2009)

Application of Cluster Analysis of the Genus *Pinus* Various Species Essential Oils Terpene Compounds for Their Taxonomic Identification

Hanna Chubarova[1(✉)], Maxim Kapustin[1,2], Alena Spiridovich[3], Svetlana Lodigina[4], and Alexandr Sobol[1]

[1] Belarusian State University, Minsk, Republic of Belarus
[2] North Caucasus Federal University, Pyatigorsk, Russia
[3] The Central Botanical Garden, Minsk, Republic of Belarus
[4] North Caucasus Federal University, Stavropol, Russia

Abstract. By means of GC-MS, the composition of terpene and terpenoid compounds of 6 pine species belonging to the Strobus subgenus was studied: *P. sibirica, P. pumila, P. koraiensis, P. peuce, P. griffithii, P. strobus, P x schwerinii* and 5 species, included in the subgenus Pinus: *P. mugo, P. ponderosa, P. contorta, P. sylvestris, P. rigida*. It was shown that monoterpenes dominate in the studied essential oils, its' relative content ranges in different species from 46 to 88%. The sesquiterpenes in the studied essential oils range from 2.3 to 31%. The relative content of terpenoids: alcohols, ethers, ketones range from 0.7 to 11%. On the basis of cluster analysis of terpene and terpenoid compounds in the essential oils of the studied pine species, statistical processing was carried out to identify closely related species. As a result of the analysis of monoterpene compounds, the most closely related species included in the subgenus Strobus were established, having a distance of less than 5 units from other samples. It was found that, in terms of the composition of terpenoids, the mountain pine is the most distant from other samples (distance ~4.5 units). Two groups, the first of which includes *P. strobus, P. wallichiana, P. sibirica*, and in the second – *P. sylvestris, P. peuce, P. koraiensis* and a hybrid *P.peuce Griseb.* x *P. stobus L.*, had the smallest differences among themselves and were located at a distance of 1.7 units.

Keywords: GC-MS · Pine trees · Essential oils · Cluster analysis · Monoterpenes · Sesquiterpenes · Terpenoids

1 Introduction

There are more than 100 species of pine trees, most of which are of great economic importance in forestry and ornamental gardening [1].

In Belarus, one species grows in nature – *P. sylvestris*, and in culture there are more than 20. The area of natural distribution of pine trees is practically the entire Northern Hemisphere, with the exception of the Arctic, desert zones and some tropical regions (the Indian subcontinent, Central Africa). The distribution of individual species obeys a

V. Kurchenko et al. (Eds.): ICAETT 2021, LNNS 408, pp. 277–290, 2022.
https://doi.org/10.1007/978-3-030-96641-6_32

clear pattern: in the taiga zone of Eurasia and North America there are only a few species with extensive transcontinental ranges, in mountainous tropical and subtropical regions, on the contrary, there are many species with small ranges [2, 5, 9, 10, 13, 15].

Of particular interest among researchers are pine terpenoids, which, in terms of quantitative content and qualitative composition, are outperform many other types of compounds found in medicinal plants. Substances of the terpenoid group obtained from plant materials of the genus Pinus, due to their unique pharmacological properties, are increasingly used in the treatment of various pathological conditions [2, 4–7, 10, 13–15].

These compounds successfully used both in the form of individual terpenoids and as part of complex preparations; as the main active substances and related compounds that enhance the pharmacological activity of other components.

E.g., «Biopin» ointment, containing pine resin in its composition, has an immunomodulatory, wound healing, antimicrobial, anti-inflammatory effect. It is used in the treatment of burns, wounds and purulent-inflammatory processes of the skin and subcutaneous fat. «Pinabin», a 50% solution in peach oil of the heavy fraction of essential oils obtained from pine needles, has an antispasmodic effect on the muscles of the urinary tract and bacteriostatic against gram-positive bacteria; it is used for urolithiasis and renal colic. The drug «Pinovit» contains Pinus mugo oil and has a pronounced decongestant, anti-inflammatory and antimicrobial effect, which makes it possible to use it for the symptomatic treatment of rhinitis.

The introduction of various species of pine trees in Belarus has been carried out for many years. The genus Pinus is usually divided into two subgenus: Pinus and Strobus [9]. The name of the subgenus Pinus coincides with the name of the entire genus, which, apparently, is intended to designate the species belonging to the subgenus as "real" pines or pines proper. The subgenus Strobus owes its name to one of the most widespread and important species, the American Weymouth pine (*P. strobus*). A collection of native and introduced species of the subgenera Pinus and Strobus has been created in the National Academy of Sciences of Belarus Central Botanical Garden. The practical use of introduced species is associated with the possibility of essential oils obtaining, which may differ in the terpene compounds composition from native ones. The essential oils composition of introduced species of pine trees, cultivated under local conditions, and its' biological activity have not been sufficiently studied. An important task is to determine the composition of terpene compounds of essential oils of the genus Pinus cultivated species [2, 9–11, 13, 14]. In addition, it is important, using chemosystematics, based on cluster analysis of terpene compounds of essential oils of various species of the genus Pinus, to identify their species [10].

The purpose of the investigation was to carry out the cluster analysis of the 13 pine trees species essential oils composition, of which 12 species are not indigenous and introduced in the Central Botanical Garden of the National Academy of Sciences of Belarus, to clarify their taxonomic affiliation.

2 Materials and Methods

The objects of the study were essential oils isolated from the stubby ends of branches 30–40 cm long, collected from various species of pines (Table 1) of the National Academy of Sciences of Belarus Central Botanical Garden.

Table 1. Pine species cultivated in the Central Botanical Garden of the National Academy of Sciences of Belarus and used to obtain essential oils for gas-chromatography analysis

Name	Subgenus	Distribution
P. sibirica Du Tour (Siberian cedar)	Strobus	Siberia
P. pumila (Pall.) Regel (Siberian stone pine)	Strobus	Siberia and the Far East
P. koraiensis Sieb. Et Zucc. (Korean cedar)	Strobus	Far East
P. peuce Griseb. (Rumelian pine)	Strobus	Europe (Albania, Bulgaria, Montenegro)
P. griffithii Hoff ex Thomson (Griffith pine)	Strobus	
P. strobus L. (Veimutova pine)	Strobus	North America
P.peuce Griseb. x P.stobus L.		
P.x schwerinii Fitschen. (Schwerin pine)	P. griffithii x P. strobus	
P. mugo Turra (Mountain pine)	Pinus	Western Europe
P. ponderosa Douglas (Hard pine)	Pinus	America
P. contorta Douglas ex Louden (Twisted pine)	Pinus	North America
P. sylvestris L. (Common pine)	Pinus	Europe
P. rigida Mill. (Rigid pine)	Pinus	North America

Essential oils were obtained by steam distillation [11]. Studies of the essential oils composition were carried out on an Agilent 6850 gas chromatograph equipped with an Agilent 5975B mass detector. A DB-5MS capillary column (copolymer of 5%-biphenyl–95%-dimethylsiloxane) with a length of 60 m with an inner diameter of 0.25 mm and a film thickness of a stationary phase of 0.25 μm was used. The carrier gas is helium. The thermostat temperature programming was used: initial temperature – 35 °C; lifting speed 5 °C/min up to 180 °C, lifting speed 20 °C/min up to 280 °C; 280 °C for 10 min. Sample preparation: preparation of a 1% solution of essential oil in hexane. Sample volume: 1 μL.

The percentages of essential oils were calculated from the peak areas without the use of correction factors. The qualitative analysis is based on comparing the mass spectra of the essential oil components with the corresponding data from the NIST0.5a mass spectral library.

The dendrograms presented in the article were obtained as a result of cluster analysis in the «Statistika 7» program. The initial data were pre-standardized. Dendrogram branches were connected according to the single linkage rule. The distance between objects on the dendrogram was calculated using the Euclidean distances method.

3 Results and Discussion

Essential oils obtained from the 13 pine trees species were practically colorless, with a characteristic balsamic smell of needles.

By the yield of essential oil, the most productive were Siberian cedar (0.86 ml/100 g of dry weight (DW)), Siberian stone pine (0.61 ml/100 g DW), Korean cedar (0.56 ml/100 g DW), representatives of the Siberia and the Far East flora belonging to the subgenus Strobus. Among the representatives of the subgenus Pinus, the highest yield of essential oil was typical for Mountain pine (*P. mugo*) and Hard pine (*P. ponderosa Douglas*). Comparative analysis of the composition of terpene and terpenoid compounds of essential oils of the studied pine trees species are presented in Table 2.

Table 2. Terpene compounds composition of pine essential oils

№	Chemical substance	Relative content (%) of components pine essential oils												
		Pinus strobus L.	*P.peuce Griseb. x P. stobus* L.	*Pinus mugo* Turra	*Pinus griffithii.* Hoff ex Thomson	*Pinus rigida* P. Mill.	*Pinus koraiensis* Sieb. et Zucc.	*Pinus pumila* (Pall.) Regel	*Pinus sylvestris* L.	*Pinus peuce* Griseb	*Pinus. sibirica* Du Tour	*Pinus. contorta* Douglas ex Louden	*Pinus ponderosa* Douglas	*Pinus schwerinii hybrid* Fitschen
1	2	3	4	5	6	7	8	9	10	11	12	13	14	15
	TERPENES													
	MONOTERPENES													
	Linear (acyclic) monoterpenes													
1	β-myrcene CAS#:123-35-3	1,70	1,94	1,95	5,04	0,34	3,36	2,19	1,91	1,53	0,64	1,90	1,35	3,03
2	*cis*-β-Ocimene CAS#:3338-55-4		0,97	0,22				0,46	1,06					

(*continued*)

Table 2. (*continued*)

	Total linear monoterpenes:	**1,70**	**2,91**	**2,17**	**5,04**	**0,34**	**3,36**	**2,65**	**2,97**	**1,53**	**0,64**	**1,90**	**1,35**	**3,03**
	Carbocyclic monoterpenes													
	Monocyclic monoterpenes													
3	α-Phellandrene CAS:99-83-2	0,16	0,09	0,22	0,67	0,04	0,15	0,21	0,08	0,24	0,07	0,11	0,03	0,14
4	α-Terpinene CAS:99-86-5		0,06	0,15			0,32	0,30	0,05	0,05	0,03	0,08	0,07	0,21
5	p-Cymene CAS:99-87-6	0,43	0,23	0,25	0,12	0,17	0,15	0,21	0,29	0,06	0,04	1,04	0,19	0,06
6	Limonene CAS:138-86-3	4,65	2,04	3,51	2,01	1,22	14,63	6,79	2,63	2,58	5,87	2,82	1,69	21,67
7	β-Phellandrene CAS:555-10-2	2,38	2,42	11,03	3,20	7,42	1,23	7,82	2,39	2,30	4,05	17,51	1,38	1,00
8	γ-Terpinene CAS:99-85-4		0,14	0,31	0,07		0,07	0,50	0,10	0,06	0,06	0,16	0,14	0,16
9	α-Terpinolene CAS:586-62-9	0,45	1,14	3,10	0,64		7,24	8,08	0,97	0,60	0,53	2,21	1,10	13,02
	Total monocyclic monoterpenes:	**8,07**	**6,12**	**18,57**	**6,71**	**8,85**	**23,79**	**23,91**	**6,52**	**5,88**	**10,65**	**23,91**	**4,59**	**36,27**
	Bicyclic monoterpenes													
10	α-Thujone CAS:2867-05-2							0,94				0,23	0,07	
11	α-Pinene CAS:80-56-8	25,40	32,62	11,61	30,72	17,79	30,95	16,27	40,04	47,72	54,47	7,86	9,33	14,06
12	Camphene CAS:79-92-5	5,84	5,78	1,02	3,94	0,41	7,34	5,69	7,22	12,68	0,77	0,67	0,20	4,38

(*continued*)

Table 2. (*continued*)

13	Sabinene CAS:3387-41-5	0,33	0,43	0,95	0,23	0,11	0,15	1,21	0,33	0,08	0,10	0,80	0,25	0,22
14	β-Pinene CAS:127-91-3	24,96	4,96	2,67	15,63	18,92	4,02	2,42	5,25	8,84	2,11	16,44	56,43	5,32
15	δ-3-Carene CAS:13466-78-9	15,94	12,50	24,21	2,39		0,56	34,74	13,43	1,31	1,59	35,32	15,60	9,44
	Total bicyclic monoterpenes:	**72,47**	**56,30**	**40,46**	**52,91**	**37,23**	**43,01**	**61,27**	**66,28**	**70,63**	**59,04**	**61,32**	**81,88**	**33,43**
Tricyclic monoterpenes														
16	Tricyclene CAS:508-32-7	0,34	0,62	0,36	0,21	0,16	0,70	0,46	0,71	0,60	0,21			0,39
	Total tricyclic monoterpenes:	**0,34**	**0,62**	**0,36**	**0,21**	**0,16**	**0,70**	**0,46**	**0,71**	**0,60**	**0,21**			**0,39**
	Total carbocyclic monoterpenes:	**80,87**	**63,04**	**59,39**	**59,83**	**46,23**	**67,50**	**85,64**	**73,51**	**77,12**	**69,90**	**85,23**	**86,47**	**70,09**
	Total MONOTERPENES:	**82,57**	**65,95**	**61,57**	**64,87**	**46,58**	**70,86**	**88,29**	**76,48**	**78,65**	**70,53**	**87,13**	**87,82**	**73,11**
SESQUITERPENES														
Carbocyclic sesquiterpenes														
Monocyclic sesquiterpenes														
17	β-Elemen CAS:515-13-9	0,47	0,44	0,63	0,22	1,04	0,05	0,18	0,24	0,06	0,23		0,03	0,08
18	α-Caryophyllene CAS:6753-98-6	0,42	0,56	1,10	0,20	0,61	0,28	0,14	0,36	0,20	0,15			0,85
19	Hemacrene D CAS:23986-74-5	5,64	3,31	8,72	5,70	6,28	8,78	1,10	3,05	4,27	7,32	0,67	0,14	6,07

(*continued*)

Table 2. (*continued*)

20	β-Bisabolene CAS:495-61-4							0,27						
21	α-Bisabolene CAS:25532-79-0			0,05	0,10			1,25			0,15			0,06
Total monocyclic sesquiterpenes:		**6,52**	**4,30**	**10,50**	**6,22**	**7,93**	**9,12**	**2,94**	**3,66**	**4,53**	**7,86**	**0,67**	**0,17**	**7,06**
Bicyclic sesquiterpenes														
22	β-Caryophyllene CAS:87-44-5	2,95	3,71	8,03	2,15	4,35	4,06	0,99	2,73	2,21	0,85	0,54	0,11	6,20
23	trans-α-Bergamotene CAS:17699-05-7												0,38	
24	3,7-Guaiadiene CAS:6754-04-7											4,45		
25	α-Muurolene CAS:31983-22-9	0,28	0,82	0,72	1,52	2,34	0,38	0,34	0,35	0,16	1,10	0,22	0,15	0,43
26	Bicyclo-germacrene CAS:67650-90-2	0,53	1,42	2,72	0,97	4,82	0,62	1,36	0,44	0,32	1,90		0,12	1,31
27	γ-Cadinene CAS:39029-41-9	0,35	1,22	0,42	1,27	3,07	0,64	0,55	0,56	0,32	2,45	0,34	0,32	0,92
28	δ- Cadinene CAS:483-76-1	1,58	3,40	2,32	8,75	7,16	2,44	2,04	1,57	1,12	6,17	0,89	0,86	2,67
29	Calamenene CAS:483-77-2		0,05		0,12	0,48	0,08		0,07		0,04		0,08	
Total bicyclic sesquiterpenes:		**5,69**	**10,62**	**14,22**	**14,78**	**22,23**	**8,21**	**5,30**	**5,72**	**4,13**	**12,51**	**6,44**	**2,02**	**11,52**

(*continued*)

Table 2. (*continued*)

	Tricyclic sesquiterpenes													
30	α-Cubebene CAS:17699-14-8		0,07			0,08	0,11		0,05		0,26			0,08
31	α-Longipinene CAS:5989-08-2		0,05		0,04		0,79		0,02	0,04	0,42	0,23	0,05	0,35
32	α-Copaene CAS:3856-25-5		0,13	0,06	0,04	0,22	0,18	0,08	0,10	0,04	0,40		0,04	0,19
33	Longifolene CAS:475-20-7		0,07		0,12		0,24		0,05	0,05	0,15			0,10
34	β-Cubebene CAS:13744-15-5		0,10	0,10		0,12	0,14		0,04	0,05	0,20			
35	Aromadendrene CAS:109119-91-7		0,19			0,45	0,11			0,04	0,41		0,04	0,13
	Total tricyclic sesquiterpenes:		**0,61**	**0,16**	**0,21**	**0,87**	**1,55**	**0,08**	**0,26**	**0,22**	**1,83**	**0,23**	**0,12**	**0,85**
	Total SESQUITERPENES:	**12,22**	**15,53**	**24,89**	**21,21**	**31,02**	**18,89**	**8,31**	**9,63**	**8,88**	**22,20**	**7,33**	**2,32**	**19,44**
	Total TERPENES:	**94,79**	**81,48**	**86,45**	**86,08**	**77,60**	**89,75**	**96,60**	**86,11**	**87,53**	**92,74**	**94,47**	**90,14**	**92,55**
	TERPENOIDS													
	ALCOHOLS													
36	Linalool CAS#:78-70-6							0,09					0,04	
37	Borneol CAS#:507-70-0	0,08	0,21	0,34		0,07	0,02	0,04	0,20	0,12	0,03		0,03	0,07

(*continued*)

Table 2. (*continued*)

38	Terpinen-4-ol CAS#:20126-76-5	0,10	0,11	0,37	0,05	0,06		0,26			0,05	0,33	0,19	0,06
	Total ALCOHOLS:	**0,18**	**0,31**	**0,71**	**0,05**	**0,13**	**0,02**	**0,39**	**0,20**	**0,12**	**0,09**	**0,33**	**0,26**	**0,13**
	ETHERS													
39	Eucalyptol CAS#:470-82-6			0,10										0,09
40	Estragole CAS#:140-67-0	0,45											7,64	
41	Thymyl methyl ether CAS#:1076-56-8		0,06	0,43				0,11			0,07			0,06
42	Bornyl acetate CAS#:76-49-3	1,48	10,07	7,16	0,65	2,83	8,07	0,63	10,96	11,14	1,14	1,95		4,54
	Total ETHERS:	**1,93**	**10,13**	**7,69**	**0,65**	**2,83**	**8,07**	**0,74**	**10,96**	**11,14**	**1,21**	**1,95**	**7,64**	**4,69**
	KETONES													
43	Criptone CAS#:500-02-7	0,11				0,68						0,27		
	Total KETONES:	**0,11**				**0,68**						**0,27**		
	Total TERPENOIDS:	**2,23**	**10,44**	**8,39**	**0,70**	**3,64**	**8,09**	**1,13**	**11,15**	**11,26**	**1,29**	**2,56**	**7,90**	**4,82**
	Identified components:	**97,01**	**91,92**	**94,85**	**86,77**	**81,24**	**97,84**	**97,73**	**97,26**	**98,79**	**94,03**	**97,03**	**98,04**	**97,37**

A distinctive feature of pine essential oils is a high content of pinenes. Among studied samples, it is possible to distinguish two groups of species. The first group with a high content of these bicyclic monoterpenes, such as Hard pine (*P. ponderosa Douglas*), where their amount exceeded 65%, Rumelian pine, Siberian pine and Veimutova pine, where the content of pinenes was more than half of the total amount of identified essential oil components. At the same time, Mountain pine, Siberian stone pine and Schwerin pine were largely depleted in pinene: the content of the sum of these compounds did not exceed 20% in these species. In the remaining samples, the amount of pinenes was 24.3–46.3%, which is comparable to the content of these monoterpenes in essential oils of the previously studied fir, another genus of the Pinaceae family [12].

At the same time, according to previously obtained data, conifers of the genus Juniperus (family Cupressaceae) are characterized by a significantly lower content of pinenes, not exceeding 15%. Moreover, in the majority of junipers, the prevalence of α-pinenes was characteristic, while for representatives of the Pinaceae family, the quantitative ratio of α- and β-pinenes was individual, depending on the species of coniferous species. So, if for Veimutova pine and Rigid pine the α/β-pinene ratio was approximately equal to 1, then for Mountain pine, Griffith pine, Korean pine, Common pine, Siberian pine, the predominance of α-pinene was shown, and for solid pine, on the contrary, an overwhelming amount was observed β-pinene (β/α-pinene ratio for this species was 6). The content of limonene in the studied pine samples was relatively low and comparable to its presence in essential oils of fir [12]. The content of this monocyclic monoterpene was about 22%.

A number of pine species have been found to have a significant amount of delta-3-carene. Moreover, in Mountain pine, Siberian stone pine, Twisted pine, the content of this compound was more than 20%. At the same time, delta-3-carene was not found in the essential oil of hard pine, while it was less than 2% in Korean pine, Rumelian pine and Siberian pine. Also, during earlier investigations, species specificity in the quantitative accumulation of this compound was established in firs [12]. A number of pine species were distinguished by a relatively high content of bornyl acetate (Common pine, Rumelian pine); however, among the studied samples, the amount of this compound did not exceed 12%, while for a number of junipers, the presence of more than 40% bornyl acetate from the total amount of identified compounds was established. Also, the studied fir species contained bornyl acetate in an amount exceeding 15% [12].

In general, the terpenoids of pines did not differ in diversity, so only 3 representatives of the class of alcohols, 4 – ethers and 1 ketone – krypton, found only in Veimutova pine, Rigid pine and Twisted pine, were found in composition. The essential oils of fir and juniper contained a much more varied composition of terpenoids [12].

3.1 Intrageneric Characteristics of the Composition of Pine Essential Oils

In the 13 samples of pine essential oil studied, only two acyclic monoterpenes were found: β-myrcene and cis-β-ocymene (Table 1), and the latter compound was present only in four representatives of the genus Pinus. The Griffith pine, belonging to the subgenus Strobus, was distinguished by an increased biosynthesis of β-myrcene (5%), while the Rigid pine, belonging to the subgenus Pinus, was most depleted in this component. The greatest quantitative contribution to the group of carbocyclic monoterpenes was made by bicyclic monoterpenes, including pinene and delta-3-carene, which accumulated in 9 species of the studied pines in an amount of more than 50%. Hard pine (82%), which was dominated by β-pinene, was especially distinguished by the content of this class of monoterpenes.

In the composition of essential oils in analytically significant amounts, 7 monocyclic monoterpenes were found, among which the main ones were limonene, β-pellandrene and α-terpinolene, while two types of pine, Schwerin pine and Korean pine, were distinguished by increased biosynthesis of both limonene and α- terpinolene. Siberian stone pine was distinguished by a relatively even distribution of these three monoterpenes, while β-pellandrene predominated in the essential oils of mountain pine and twisted pine (subgenus Pinus).

In the class of sesquiterpenes, bicyclic sesquiterpenes predominated quantitatively, their maximum content was found in the essential oil of Rigid pine (22.2%). At the same time, the quantitative distribution of individual sesquiterpene components of essential oils was specific for each pine species. In addition, a number of compounds were established only in certain representatives of the genus Pinus. Thus, β-bisabolic was found only in the essential oil of Siberian stone pine, trans-α-bergamoten in the essential oil of Hard pine, and 3, 7-guaadiene - in the essential oil of Twisted pine. In general, it should be noted that the amount of terpenes in all studied essential oils exceeded 77%, reaching 96.6% in the case of Siberian stone pine.

3.2 Characteristics of Pine Essential Oils When Introduced in Belarus

The oil obtained by hydrodistillation from the needles of *P. koraiensis* [3] contained limonene (27.9%), α-pinene (23.4%), and β-pinene (12.9%) as dominant components. Limonene and α-pinene also prevailed in the sample of essential oil from the Central Botanical Garden of the National Academy of Sciences of Belarus, however, the amount of β-pinene was significantly lower (4.02%).

A large amount of α-pinene (36.5%) and germacrene D (11.4%) has been found in the essential oil of Rumelian pine from Macedonia [8]. In the sample of the Rumelia pine essential oil from the Central Botanical Garden of the National Academy of Sciences of Belarus, the content of α-pinene was significantly higher, and the content of germacrene D, on the contrary, was lower (4.3%). Also, in the sample of essential oil from Central Botanical Garden, the content of bornyl acetate was higher in comparison with the sample of Macedonian origin (6.8%). In a sample of essential oil from Rumelian pine of Greek origin [8], the presence of a significant amount of citronellol (13.4%) was established, while the presence of this component was not observed in the essential oil from Central Botanical Garden. Also, significant differences between the samples of Greek and Belarussian origin were in the content of a-pinene: in the Greek plant it was 23.1%. A sample of Common pine essential oil from the Central Botanical Garden of the National Academy of Sciences of Belarus was distinguished by a significant content of α-pinene (40%) in comparison with the sample analysed in the article [1]. Even more significant differences were found in the accumulation of cadinenes: their content in the sample from the Central Botanical Garden did not exceed 2.2%.

3.3 Cluster Analysis of the Pine Essential Oils Composition

On the basis of cluster analysis of terpene and terpenoid compounds included in the essential oils of the studied pine species, statistical processing was carried out, which made it possible to identify closely related species.

As a result of its use, it was found that in terms of the composition of monoterpenes, which reach from 81 to 96% of essential oils, the essential oils of the Common pine *P. sylvestris L.* and the hybrid *P. peuce Griseb. x P. stobus L.* are the closest to each other (distance ~1 unit). The essential oils of *P griffithii Hoff ex Thomson* and *P. pumila* (Pall.) *Regel* had a pronounced individuality, the latter being at a distance of more than 5 units from other samples (Fig. 1).

It was found that according to the composition of terpenoids, which are contained in essential oils of pine trees from 0.7 to 11%, Mountain pine is the farthest from other samples (distance ~4.5 units) (Fig. 2). Two groups, the first of which includes pine Veimutova, pine Griffith, pine Siberian, and in the second – Common pine, p. Rumelian pine, Korean pine and a hybrid *P.peuce Griseb. x P. stobus L.*, had the smallest differences and were located at a distance of 1.7 units.

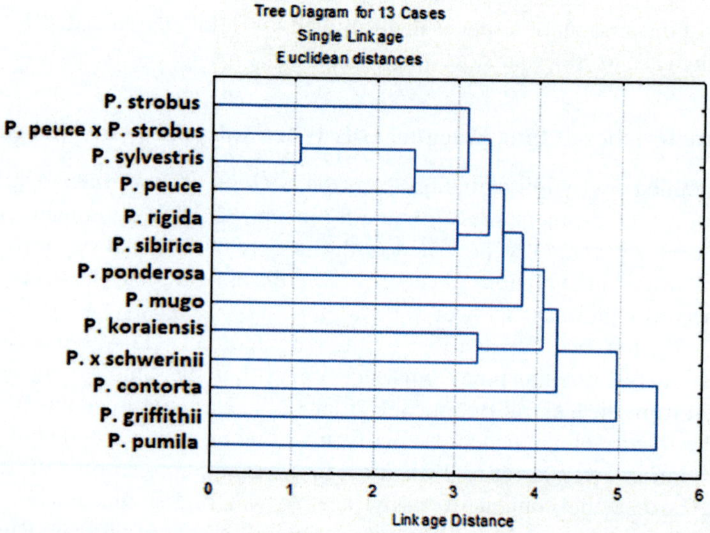

Fig. 1. Dendrogram based on the component composition of essential oils' monoterpenes of the species from genus Pinaceae

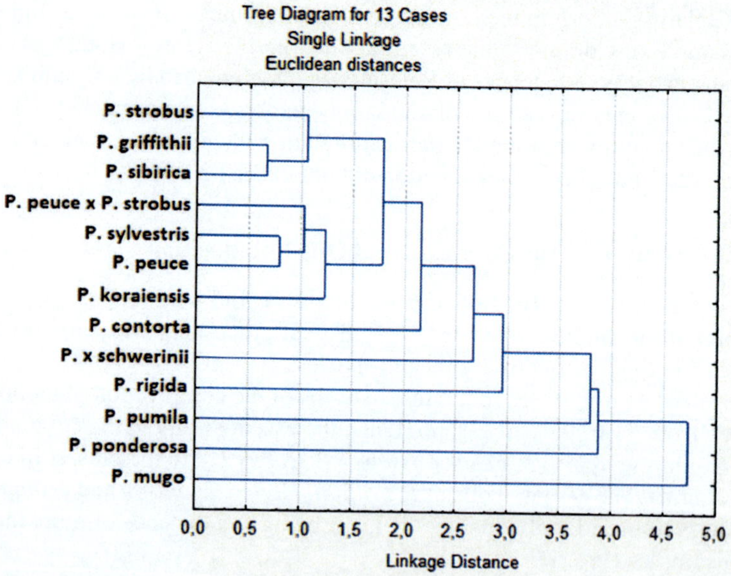

Fig. 2. Dendrogram based on the component composition of essential oils' terpenoids of the species from genus Pinaceae

4 Conclusion

Essential oils samples analysis of various pine trees species introduced into the Central Botanical Garden of the National Academy of Sciences of Belarus showed that representatives of this genus have a number of common features in the accumulation of terpene compounds in essential oils. The major terpenes of pine trees are pinenes; a number of species have been found to contain significant amounts of Δ-3-carene. Terpenoids of the studied species did not differ in diversity. A number of individual characteristics of the essential oils components composition in some species of the genus Pinus have been established in introduced species pine trees in Belarus. Cluster analysis of the essential oils components composition showed the greatest individuality for monoterpenes of the essential oil of Siberian stone pine and Mountain pine terpenoids.

References

1. Grassmann, J.: Antioxidative properties of the essential oil from Pinus mugo. J. Agric. Food Chem. **51**, 7576–7582 (2003)
2. Basholli-Salihu, M., et al.: Phytochemical composition, antiinflammatory activity and cytotoxic effects of essential oils from three Pinus spp. Pharm. Biol. **55**(1), 1553–1560 (2017)
3. Lee, J.-H.: Chemical composition and antimicrobial activity of essential oil from cones of Pinus koraiensis. J. Microbiol. Biotechnol. **18**, 497–502 (2008)
4. Djerrad, Z., Djouahri, A., Kadik, L.: Variability of Pinus halepensis Mill. Essential oils and their antioxidant activities depending on the stage of growth during vegetative cycle. Chem. Biodivers. **14**(4), e1600340, 1–27 (2017)
5. Koutsaviti, A., et al.: Antioxidant potential of pine needles: a systematic study on the essential oils and extracts of 46 species of the genus Pinus. Foods **10**(1), 142, 1–19 (2021)
6. Krauze-Baranowska, M., Mardarowicz, M., Wiwart, M., Dynowska, M.: Antifungal activity of the essential oils from some species of the genus Pinus. Z. Naturforsch. **57**(5–6), 478–482 (2002)
7. Mahdhi, A., Ghazghazi, H., El Aloui, M., Ben Salem, R., Rigane, G.: Identification and quantification of phenolic and fatty acid profiles in Pinus halepensis Mill. Seeds by LC-ESI-MS and GC: effect of drying methods on chemical composition. Food Sci. Nutr. **9**(4), 1907–1916 (2021)
8. Nikolić, B.: Variability of the needle essential oils of Pinus peuce from different populations in Montenegro and Serbia. Chem. Biodivers. **5**, 1377–1388 (2008)
9. Price, R.A., Liston, A., Strauss, S.H.: Ecology and Biogeography of Pinus. Cambridge University Press, Cambridge (1998)
10. Rodrigues, A.M., et al.: Pinus halepensis, Pinus pinaster, Pinus pinea and Pinus sylvestris essential oils chemotypes and monoterpene hydrocarbon enantiomers, before and after inoculation with the pinewood nematode Bursaphelenchus xylophilus. Chem. Biodivers. **14**(1), e1600153, 1–16 (2016)
11. Shutova, A.G.: Materials of the republican scientific. In: Molecular Medicine and Biochemical Pharmacology, Grodno (2007)
12. Shutova, A.G.: Component composition of essential oils of species of the genus Abies (Pinaceae) under conditions of introduction in Belarus. Vegetable Resour. **4**, 60–67 (2009)
13. Silori, G.K., Kushwaha, N., Kumar, V.: Essential Oils from pines: chemistry and applications. In: Malik, S. (ed.) Essential Oil Research, pp. 275–297. Springer, Cham (2019). https://doi.org/10.1007/978-3-030-16546-8_10

14. Ulukanli, Z., et al.: Chemical composition, antimicrobial, insecticidal, phytotoxic and antioxidant activities of Mediterranean Pinus brutia and Pinus pinea resin essential oils. Chin. J. Nat. Med. **12**(12), 901–910 (2014)
15. Visan, D.-C., et al.: Original contributions to the chemical composition, microbicidal, virulence-arresting and antibiotic-enhancing activity of essential oils from four coniferous species. Pharmaceuticals **14**, 1159, 1–16 (2021)

Author Index

A
Akhmadiev, Pavel, 176
Aksenov, Andrey, 114
Alexandrovich, Kravtsov Alexander, 144
Alexeevich, Gvozdenko Alexey, 144, 151
Andrey, Bratsikhin, 1, 10
Angelina, Mamay, 1, 10
Antonova, Natalia, 236
Apryatina, Kristina, 28, 106

B
Bacholdina, Tamara, 253
Bannikova, Anna, 65
Barkhatova, Tatiana, 270
Belousova, Evgeniya, 18
Blinov, Andrey, 253
Borisenko, Aleksandr, 18
Borisenko, Aleksey, 18
Borisova, Veronika, 191

C
Cherepanova, Ekaterina A., 99
Chubarova, Hanna, 221, 277
Chukhchin, Dmitry, 75

D
Deeva, Alla, 260
Dmitriy, Mamay, 1, 10
Drozdov, Roman, 270

E
Evdokimov, Ivan, 167
Evteev, Alexandr, 65

F
Falev, Danil, 114
Farkhutdinov, Rashit, 176
Fatykhava, Sviatlana, 260
Fedorova, Albina, 176
Frolov, Vadim, 28
Fukina, Diana, 136

G
Georgievich, Khramtsov Andrey, 151
Golik, Aleksey, 253
Grigorievich, Kaishev Vladimir, 151
Guramievich, Maglakelidze David, 151

H
Halavach, Tatsiana, 35, 167, 207
Holodova, Ekaterina, 221

K
Kachalova, Ekaterina, 106
Kapustin, Maxim, 221, 277
Khalimova, Lilija Kh., 99
Khodyreva, Zoja, 159
Khokh, Anna, 92
Kiseleva, I., 124
Kiseleva, Sofia V., 99
Korokin, Vitaliy, 136
Koroleva, Ekaterina, 176
Kozhuhova, Marina, 270
Krasyuk, Ekaterina, 184
Kravtsova, Olga, 56, 244
Kupriyanov, A., 81
Kurchenko, V., 81, 124
Kurchenko, Vladimir, 167

V. Kurchenko et al. (Eds.): ICAETT 2021, LNNS 408, pp. 291–292, 2022.
https://doi.org/10.1007/978-3-030-96641-6

L
Laptev, Georgy, 28
Larionova, Olga, 65
Lednev, Ivan, 136
Lodigina, Svetlana, 221, 277
Lodygin, Aleksei, 167
Lodygin, Aleksey, 207
Lodygina, Svetlana, 207
LudmilaAlieva,, 207

M
Maiorava, K., 81, 124
Maksimov, Igor V., 99
Melnikova, Elena, 236
Mullin, Ruslan, 244
Mullina, Renata, 244
Musina, Olga, 159
Mustafina, Alsu, 56

N
Nasibullina, Elnara R., 99
Nikityuk, Dmitriy, 200
Novokshanova, Alla, 200

O
Oboturova, Natalya, 253

P
Pupykina, Kira, 176, 184

R
Rajabov, Uktamjon, 253
Razinkova, Victoria, 18
Reshetnikov, Vladimir, 260
Rylskaya, Larisa, 270

S
Salomatina, Evgeniia, 46
Savchenko, Alexey, 18
Schetinin, Michael, 159
Sergeevna, Slyadneva Kristina, 144
Sergey, Babenyshev, 1, 10
Shabalova, Ekaterina, 236
Shabunya, Polina, 260
Shamuradov, David, 56
Shaydullina, Galiya, 184
Shevchenko, Aleksandr, 114
Shramko, Maria, 167
Shutskiy, Nikita, 75
Siraeva, Zulfira, 56, 244
Smirnova, Larisa, 28, 46, 106, 136
Sobol, Alexandr, 277
Spiridovich, Alena, 277
Spirydovich, Alena, 260
Stanislavskaia, Ekaterina, 236
Stefanova, Isabella, 191
Sushinskaya, N., 81, 124
Sushiskaya, Natalia, 207

T
Tarun, E., 81, 124
Terkun, Elena, 270
Toptunov, Evgeniy, 114
Tsygankow, Vasili, 221
Tyshkunova, Irina, 114

U
Ugli, Rajabov Uktamjon Ikromjon, 144

V
Vadim, Ivanets, 10
Valentin, Shipulin, 1
Vashukova, Ksenia, 75
Vaytanis, Marina, 159
Vladimirovich, Blinov Andrey, 144, 151

Z
Zaitsev, Sergey, 28, 46
Zuev, Dmitriy, 46